B+B.

Advances in Inorganic and Bioinorganic Mechanisms

Volume 3

edited by
A. G. Sykes

Department of Inorganic Chemistry
School of Chemistry
The University, Newcastle upon Tyne, England

1984

Academic Press

(*Harcourt Brace Jovanovich, Publishers*)

London   Orlando   San Diego   New York
Toronto   Montreal   Sydney   Tokyo

ACADEMIC PRESS, INC. (LONDON) LTD.
24-28 Oval Road,
London NW1 7DX

United States Edition published by
ACADEMIC PRESS, INC.
Orlando, Florida 32887

ISBN 0-12-023803-9

PRINTED IN THE UNITED STATES OF AMERICA

84 85 86 87   9 8 7 6 5 4 3 2 1

Contents

Contributors

G. V. BUXTON
Cookridge Radiation Research Centre
University of Leeds
Cookridge Hospital
Leeds LS16 6QB
England

CRISPIN G. S. ELEY
The Inorganic Chemistry Laboratory
University of Oxford
Oxford OX1 3QR
England

S. FALLAB
Institut für Anorganische Chemie der Universität Basel
CH-4056 Basel
Switzerland

JANE E. FREW
Radiation and Biophysical Chemistry Laboratory
University of Newcastle upon Tyne
Newcastle upon Tyne NE1 7RU
England

PETER JONES
Radiation and Biophysical Chemistry Laboratory
University of Newcastle upon Tyne
Newcastle upon Tyne NE1 7RU
England

P. R. MITCHELL
Institut für Anorganische Chemie der Universität Basel
CH-4056 Basel
Switzerland

GEOFFREY R. MOORE
The Inorganic Chemistry Laboratory
University of Oxford
Oxford OX1 3QR
England

MICHAEL G. SEGAL

Central Electricity Generating Board
Technology Planning and Research Division
Berkeley Nuclear Laboratories
Berkeley, Gloucestershire GL13 9PB
England

ROBIN M. SELLERS

Central Electricity Generating Board
Technology Planning and Research Division
Berkeley Nuclear Laboratories
Berkeley, Gloucestershire GL13 9PB
England

RUDI VAN ELDIK

Institute for Physical and Theoretical Chemistry
University of Frankfurt
6000 Frankfurt am Main 1
Federal Republic of Germany

KARL WIEGHARDT

Anorganische Chemie I
Ruhr-Universität Bochum
Bochum
Federal Republic of Germany

GLYN WILLIAMS

The Inorganic Chemistry Laboratory
University of Oxford
Oxford OX1 3QR
England

Preface

The aim of this series is to continue to provide authoritative reviews both in the area of inorganic mechanisms and in the rapidly expanding field of bioinorganic mechanisms.

This third volume includes an article by Fallab and Mitchell on superoxo- and peroxodicobalt(III) complexes. These complexes merit attention in their own right and also for comparative purposes, in view of current interest in the binuclear O_2 carriers hemocyanin and hemerythrin. Electron transfer at a solid–liquid interface is addressed by Segal and Sellers. This is a developing area to which Taube contributed landmark papers. The coordination and redox chemistry of hydroxylamine, which is covered by Wieghardt, has not been comprehensively reviewed previously. The article by van Eldik on reactions involving coordinated ligands complements very nicely earlier articles on oxyanions by Saito and Sasaki (Volume 1) and Gamsjäger and Murmann (Volume 2).

Pulse radiolysis of metalloproteins (Buxton) is an area of study that has produced some inconsistencies and is thus appropriate for review at this time. The class I cytochromes c are addressed by Moore and colleagues, who feature important NMR contributions. There have also been advances in understanding the structure and functional properties of peroxidases and catalases (Frew and Jones), which provide examples of formal oxidation states Fe(IV) and Fe(V).

It is of interest that four of the reviews in this volume refer to superoxide chemistry. This is an indication of the current status of this ion. Terminology in this area has not yet been standardized. The symbol $O_2{}^-$ for superoxide, indicating radical properties, is sometimes used. In the case of cytochrome c, the lower oxidation state is still widely referred to as ferrocytochrome c at a time when the term ferrous is being rapidly superseded by iron(II). The alternative designation of cytochrome c(II) is now gaining some acceptance.

Units are as in previous volumes, with necessary conversions to note: $M = mol\,dm^{-3}$ (or mol liter^{-1}), 1 calorie $= 4.2\,J$, and $10\,Å = 1\,nm = 10^3\,pm$.

Once again I would like to acknowledge the enthusiastic cooperation of all the authors in the preparation of this volume.

A. G. Sykes

Newcastle
February, 1984

Advances in Inorganic and Bioinorganic Mechanisms

Volume 3

Electron Transfer Reactions of Class I Cytochromes *c*

Geoffrey R. Moore, Crispin G. S. Eley, and Glyn Williams

The Inorganic Chemistry Laboratory
University of Oxford
Oxford, England

I. INTRODUCTION

Since class I cytochromes c constitute the most thoroughly characterised family of electron transfer proteins, they are particularly suitable for the study of electron transfer between proteins and small molecules. Their suitability for the purpose is attested to by the appearance of more than 100 papers in the past 10 years alone, all dealing with various aspects of just this topic. The purpose of the present article is to consider what the different techniques have told us about the mechanism of electron transfer between cytochrome c and small molecules.

When considering the electron transfer reactions of proteins it is useful to bear in mind that the simplest reaction scheme is as follows:

Association:

$$A^o + B^r \rightarrow (A^o \cdots B^r) \tag{1}$$

Equalisation of energy levels:

$$(A^o \cdots B^r) \rightarrow (A^o \cdots B^r)^* \tag{2}$$

Electron transfer:

$$(A^o \cdots B^r)^* \rightarrow (A^r \cdots B^o)^* \tag{3}$$

Relaxation:

$$(A^r \cdots B^o)^* \rightarrow (A^r \cdots B^o) \tag{4}$$

Dissociation:

$$(A^r \cdots B^o) \rightarrow A^r + B^o \tag{5}$$

where A^o and B^o are the oxidised counterparts of the reduced species A^r and B^r. Steps 1, 2, 4, and 5 may be very rapid and thus not appear in the rate law for the reaction, and in such cases the rate of reaction describes

$$A^o + B^r \rightarrow A^r + B^o \tag{6}$$

Nevertheless, electron transfer does occur through the agency of a reactive complex, although it is a transient one. The rate of electron transfer depends upon a number of factors, of which the most important are the thermodynamic driving force ΔG°_{12}, the individual reactivity characteristics of the reactants, which may be both structural and electronic in origin, the distance between the redox centres, and the work involved in forming reactive complexes.

The nature of the reactive complexes has been one of the major areas of research in recent years, and a number of experimental approaches have been adopted to study them. A second area of current attention is whether the electron transfer can be described by the Marcus theory. In this article we consider these two topics after first reviewing the structures and properties of the reactant and product species and briefly describing the techniques that have been used to study the reactions. We have not attempted a comprehensive review of the field, but instead have selected for discussion a number of well-characterised outer-sphere reactions of mitochondrial cytochromes *c*. Our aim has been to describe general features of the reactions that help in defining how cytochrome *c* carries out its biochemical function, but we expect that the principles discussed are also applicable to other types of electron transfer protein, such as other cytochromes, nonheme iron proteins, and copper proteins. More comprehensive reviews dealing with a wide range of inner- and outer-sphere reactions of a variety of different types of protein are those by Bennett[1] and Wherland and Gray,[2] and Sutin[3] has produced a good account of the energetics of inner- and outer-sphere reactions of mitochondrial cytochrome *c*.

II. CYTOCHROMES *c*

A. Classification of cytochromes

A determining characteristic of a cytochrome is its possession of one or more hemes, and this leads to a fairly simple classification of the cytochrome type. The three most commonly encountered types are cytochromes *a*, *b*, and *c*, and their associated hemes (Fig. 1) are sufficiently different to produce distinctive optical absorption spectra that serve as the basis for classification.[4,5] Note that it is not the nature of the axial ligands that is the determinant of cytochrome type, a point that occasionally causes confusion, since different axial ligation schemes cross the heme classification boundaries. Thus histidine–methionine ligation occurs with both cytochromes *b* and *c*[6,7] and bis(histidine) ligation occurs with all three of the major types of cytochrome.[7–9] Note also that the classification into type is not on the basis of the tertiary structure of the protein, a point that also occasionally leads to confusion. Thus the class II cytochromes *c* and *Escherichia coli* cytochrome b_{562} have similar tertiary structures,[10] but because they have different hemes they are different types of cytochrome.

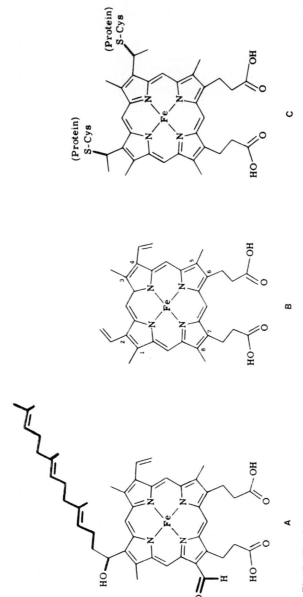

Fig. 1. The three most commonly encountered types of heme. Heme A is present in a-type oxidases and hemes B and C are present in b-type and c-type cytochromes. Heme C is covalently bound to the protein via thioether linkages resulting from the condensation of cysteine residues with the heme vinyl substituents at positions 2 and 4.

TABLE 1

The major classes of cytochromes c

Class	Comments
I	The heme is located near the N-terminus; 80–130 amino acids per heme; histidine–methionine axial ligation. This class includes mitochondrial cytochrome c
II	The heme is located near the C-terminus; 100–130 amino acids per heme; both five-coordination (histidine only) and six-coordination (histidine–methionine) occur
III	Multiheme; 30–40 amino acids per heme; bis(histidine) coordination

Further classification of type c cytochromes causes additional confusion, as names such as cytochrome c_2, cytochrome c_3, cytochrome c_{551}, and cytochrome c' are commonly encountered. These names are the result of a variety of different nomenclature schemes, all of which are in current use; the designations cytochrome c_2 and c_3 merely record the historical order of discovery. In this scheme mitochondrial cytochrome c is followed by mitochondrial cytochrome c_1 (which is part of cytochrome c reductase). Cytochrome c_{551} is the recommended International Union of Biochemistry nomenclature, which records the type of heme and the wavelength of the α band in the optical spectrum of the reduced form of the protein.[5] Cytochrome c' is the general name for high-spin cytochromes c. On the basis of amino acid sequences, Ambler[11] has divided cytochromes c into a number of groups, of which representatives of three have been characterised by X-ray crystallography: classes I, II and III. The structures and properties of these classes of cytochrome are described in the following sections (see Tables 1 and 2).

B. Structures of cytochromes *c*

1. Class I cytochromes *c*

By definition, class I cytochromes c contain histidine and methionine ligands (Fig. 2); in addition, most are monoheme. A group of diheme proteins have been isolated, and these, the cytochromes c_4, are currently being characterised from both a structural[12] and a functional (G. W. Pettigrew, personal communication) perspective.

Amino acid sequences of more than 80 mitochondrial and 50 bacterial class I cytochromes c (see compilation in Ref. [13]) and X-ray structures of 4

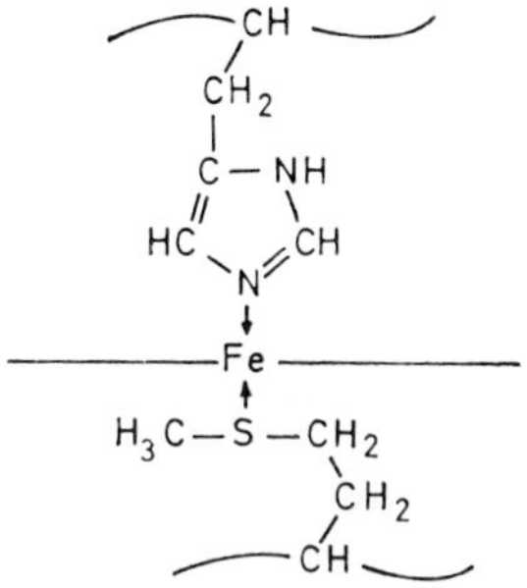

Fig. 2. Axial ligation of class I cytochromes *c*. The axial ligands are histidine and methionine.

mitochondrial and 4 bacterial class I cytochromes c[14–22] have been determined. Ribbon diagrams of five of these X-ray structures are shown in
Fig. 3. It is clear from this figure that despite the variation in size (from
82 to 129 amino acids) the overall structures are very similar, with the extra
amino acids all located in loops on the surface of a basic cytochrome fold.
Although the figure does not show it, the heme is almost entirely enclosed by
the polypeptide chain. Only one edge of the heme of mitochondrial cytochrome *c* is exposed, and that edge is not the edge bearing the heme propionates but the edge bearing heme methyl 5 (see Fig. 1). Stellwagen[23]
calculates that an area of the heme of 32 Å^2 is exposed, which is 4% of the
total heme surface and 0.06% of the total protein surface. A greater amount
of the heme of *Pseudomonas aeruginosa* cytochrome c_{551} is exposed, including part of the edge bearing the heme propionates, but it still represents only
a small fraction of the protein surface. The exposed heme edge is particularly
important because, as first noted by Wüthrich,[24] it represents a possible
site for electron transfer, a role that it fulfills for both physiological and
nonphysiological reactants, as we shall see later.

Another relevant aspect of the structures of Fig. 3 is that some of them
change with the redox state of the protein. The best-characterised protein
with regard to this conformation change is mitochondrial cytochrome *c*.
Although its conformation change is small, it extends from the heme to the
surface at the base of the molecule (as defined by the orientation of Fig. 3),
where the largest changes take place.[16,25] The occurrence of a structural
change means that in considering the energetics of redox reactions of cytochrome *c* an account must be taken of the energy required for the conformation change.

2.   Class II cytochromes *c*

Seventeen class II cytochromes have been sequenced (see compilation in
Ref. [13]), of which 13 are cytochromes *c'* and 4 are low spin with histidine

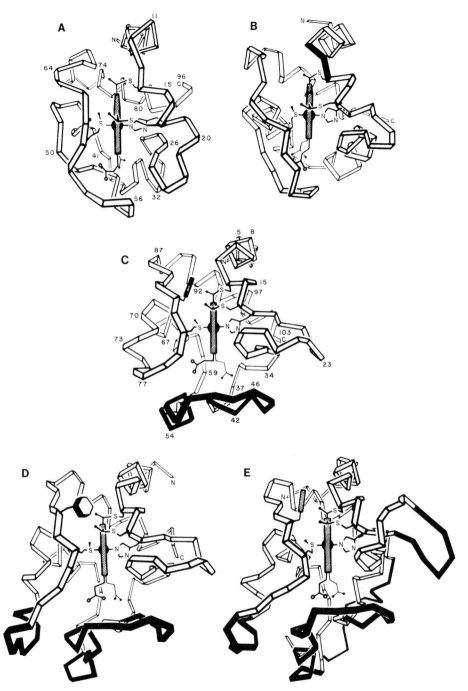

Fig. 3. Ribbon drawings of the three-dimensional structures of class I cytochromes *c*. (A) *Chlorobium thiosulfatophilum* cytochrome c_{555}; (B) *Pseudomonas aeruginosa* cytochrome c_{551}; (C) tuna cytochrome *c*; (D) *Rhodospirillum rubrum* cytochrome c_2; (E) *Paracoccus denitrificans* cytochrome c_{550}. The shaded portions represent additions to the simplest structure. All the additions are located at the molecular surfaces. (From Meyer and Kamen.[71])

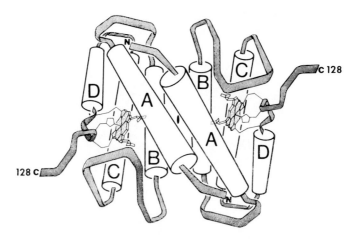

Fig. 4. Schematic representation of the *Rhodospirillum molischianum* cytochrome *c'* dimer. The cytochrome *c'* subunits are principally composed of four roughly parallel α helices, shown here as the cylinders A, B, C, and D. The left-twisted, 4-α-helical arrangement of the subunits is repeated at the subunit interface. Regions of essentially extended polypeptide chain connect and terminate each α helix. (From Weber *et al.*[26])

and methionine ligands. The sequences are homologous, although it is not clear how they should be optimally aligned, suggesting that they share a common tertiary structure. All but one of the cytochromes *c'*, that from *Rhodopseudomonas palustris*, are dimers of identical subunits. The *R. palustris* cytochrome *c'* and the low-spin proteins are monomers.

The structure of one class II cytochrome has been determined by X-ray crystallography, that of *Rhodospirillum molischianum* cytochrome *c'*.[26] The iron of this protein is five-coordinate, with an imidazole ligand at pH 7[26−28] that converts to an imidazolate ligand with a pK_a of 8.7.[29,30] The tertiary structure, which does not appear to be greatly affected by the change in axial ligand at alkaline pH, is constructed of four α helices per subunit (Fig. 4) that pack together to form both the core of each subunit and the subunit interface. The heme is located toward one end of each subunit at the protein surface, and both the axial ligand and much of one face of the heme are exposed at the surface. The heme propionates are also exposed, but the vacant axial coordination site is buried within the protein. The Fe–Fe distance within the dimer is 24 Å.

3. Class III cytochromes *c*

Various sizes of class III cytochrome *c* are known: the triheme protein from *Desulfuromonas acetoxidans*, tetraheme proteins from a variety of

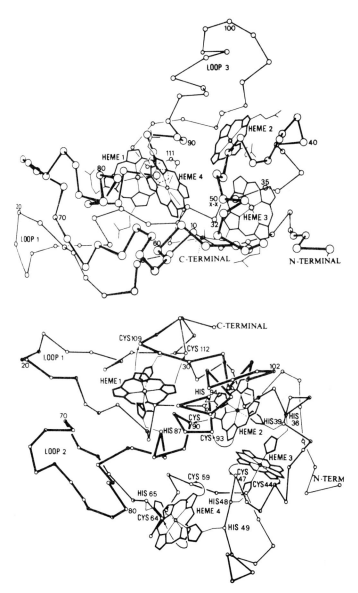

Fig. 5. The structure of *Desulfovibrio desulfuricans* (Norway 4) ferricytochrome c_3. Only the α-carbon positions and the four heme groups are illustrated, with the size of circle indicating their relative depth. The two views were selected to show the arrangement of the four hemes and the position of the three large surface loops. Iron–iron distances: Fe_1–Fe_2, 12.7 Å; Fe_1–Fe_3, 17.3 Å; Fe_2–Fe_3, 10.9 Å; Fe_1–Fe_4, 16.3 Å; Fe_2–Fe_4, 16.8 Å; Fe_3–Fe_4, 12.8 Å. Heme plane angles: hemes 1 and 2, 90°; hemes 2 and 3, 85°; hemes 1 and 4, 60°; hemes 3 and 4, 80°. (From Haser *et al.*[32])

Desulfovibrio spp., and an octaheme protein from *Desulfovibrio gigas* (see Ref. [13], and references therein). The X-ray structures of two tetraheme cytochromes c_3 have been determined, one from *Desulfovibrio vulgaris* (Miyazaki)[31] and one from *Desulfovibrio desulfuricans* (Norway).[32,33] The structures of these two proteins are very similar, and differences in size are accommodated in surface loops in much the same way as for the class I cytochromes (see Fig. 3). The structure of the Norway protein is shown in Fig. 5.

The Norway structure consists of a core formed from the four hemes and their interconnecting peptide, around which the remainder of the chain is wrapped. The environments of the four hemes are different, and although all are partially exposed at the molecular surface, their degrees of exposure are different. An interesting feature of the structure, which presumably is important for intramolecular electron transfer, is the smallness of the inter-heme distances; the shortest distance between heme pyrrole rings is about 5 Å (between hemes 1 and 2, and between hemes 2 and 3).

C.   Properties of cytochromes *c*

The properties of cytochrome *c* that are of the greatest relevance to the subject of this article are the oxidation–reduction potentials and electron self-exchange rates. In addition, knowledge of the physiological functions of the various cytochromes is needed in order to relate them to the chemical studies described here. However, this is a vast subject that has been reviewed a number of times[7,13,34,35]; thus, with the exception of mitochondrial cytochromes *c* (Section VII), physiological functions are not discussed in greater detail. Table 2 summarizes the relevant properties.

1.   Reduction potentials

Reduction potentials of electron transfer proteins are usually measured as midpoint potentials at pH 7 and are given the symbol $E_{m,7}$. Values for cytochromes *c* vary from -400 to 500 mV (see the reviews in Refs. [13, 14, and 34]; see also Table 2 for a selection of the better characterised proteins), with the class I cytochromes coming into the higher potential end of the range. With few exceptions, class I cytochromes *c* have potentials of 150–500 mV.

In general, reduction potentials are exceedingly sensitive to the protein environment, as Figs. 6–8 illustrate for a variety of class I cytochromes. The pH dependence of E_m for mitochondrial cytochromes *c* and some related

TABLE 2

Structures and properties of cytochromes c

Protein	Size (kDa)	Class	Number of hemes	$E_{m,7}$ (mV)	pH dependent	$k_{11}{}^b$ ($M^{-1}s^{-1}$)	Ref.
Horse c	11.7	I	1	260	N	10^3–10^4	[35–37]
Candida krusei c	11.9	I	1	265	Y	10^2–10^4	[38–40]
Rhodospirillum rubrum c_2	12.9	I	1	320	Y	$<10^4$	[41–43]
Pseudomonas aeruginosa c_{551}	9	I	1	260	Y	1.2×10^7	[18, 44, 45]
Pseudomonas stutzeri c_{551}	9	I	1	240	Y	6×10^6	[7, 46]
Euglena gracilis c_{552}	9.3	I	1	360	N	5×10^6	[7, 47–50]
Rhodospirillum rubrum c'	26	II	2	10	Y	$<10^4$	[7, 27, 51]
Desulfovibrio vulgaris c_3	11.7	III	4	c	c	c	[7, 52, 53]

^{a} $E_{m,7}$ is the midpoint reduction potential at pH 7; Y and N in the column indicate that the reduction potential is either pH dependent or pH independent, respectively, over the pH range 5.5–8.5. The E_m of *Candida* cytochrome c has not been shown to be pH dependent but, in view of the similarity of *Candida* cytochrome and other yeast cytochromes, it is likely to be.[39]

^{b} k_{11} is the rate of electron self-exchange. These values were all measured by NMR under a variety of conditions in ^2H_2O: horse and *Candida* cytochromes at 25°C, pH* 7, and $I = 0.1$–1.0 M (NaCl); *R. rubrum* c_2 at pH* 6 in 0.1 M NaCl; *P. aeruginosa* c_{551} at pH* 7, 0.05 M phosphate, and 42°C; *P. stutzeri* c_{551} at pH* 5.5, unbuffered, and 27°C; *E. gracilis* c_{552} at pH* 7, 0.05 M phosphate, and 29°C; *R. rubrum* c' at pH* 7, unbuffered, and 30°C.

^{c} The presence of four dissimilar hemes complicates greatly the task of accurately measuring redox properties. At pH 8 in 50 mM Tris, EPR measurements at 77 K give E_m values of -284, -310, -319, and -324 mV. It is not known if they are pH dependent. Also, aside from intermolecular electron exchange, intramolecular exchange can take place. At present, only an upper limit of 5×10^5 M^{-1} s^{-1} at 25°C and pH* 6.8 can be given for the intermolecular exchange. Under these conditions, the intramolecular exchange rate between at least two of the hemes is $\sim 5 \times 10^4$.

bacterial cytochromes c_2 are shown in Fig. 6. The pattern of pH dependence at high pH has been shown by Brandt *et al.*[54] and Davis *et al.*[55] to arise from an apparent pK_a in the oxidised form (pK_{O2}) whose value results from an ionisation of $pK_a \sim 11$ coupled to a conformation change with a $pK \sim 2$. pK_{O2} is shown by most class I cytochromes and has no physiological significance. The pH dependence at lower pH has been thoroughly analysed by Pettigrew *et al.*,[42,44,46,56] who have shown that the pattern is due to an ionisation on the ferricytochrome (pK_{O1}) and an ionisation on the ferro-cytochrome (pK_R). The shift in pK_a and consequent change in potential result from electrostatic interactions between the ionising group and the charge on the heme. NMR investigations[39,44,46] have identified the responsible ionising groups as either surface histidines or heme propionates. Not all cytochromes exhibit such a pH dependence, and most mitochondrial cytochromes are included in the group with a pH-independent E_m over the

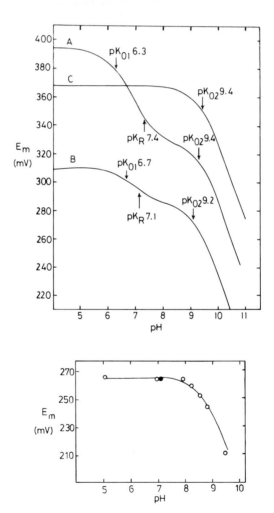

Fig. 6. Dependence on pH of the reduction potentials of bacterial cytochromes c_2 (top) and horse cytochrome c (bottom). The cytochromes c_2 are from (A) *Rhodomicrobium vannielii*; (B) *Rhodopseudomonas viridis*; (C) *Rhodopseudomonas capsulata*. (From Moore et al.[39])

range 5.5–8.0 (bottom, Fig. 6).[34,36,39] However, some yeast cytochromes have a slight pH dependence similar to curve B in Fig. 6.[39]

Figure 7 shows that the E_m of *Euglena gracilis* cytochrome c_{552} is particularly sensitive to ionic strength, while the E_m of horse cytochrome c is less sensitive.[49] This sensitivity could arise from a general ionic strength effect or from a specific ion effect (see Section V). Specific ion effects on the E_m of horse cytochrome c are illustrated in Fig. 8.[57] This figure also shows the temperature dependence of the E_m of horse cytochrome c and the $^1H/^2H$ isotope effect on it. Kreishman *et al.*[57] deduced from these data that the water structure around cytochrome c is important for maintaining the level of E_m. This may be relevant to the reactions we are interested in because to attain the minimum distance between cytochrome c and a small-molecule electron donor or acceptor, the small molecule will have to penetrate the solvation shell of the protein. We return to this point in Section VI,F and G.

The reduction potentials of class II and class III cytochromes have not been characterised to the same extent as those of class I cytochromes, partially because of the practical problems in working with low-potential, autoxidisable proteins and partially because of the difficulties in determining

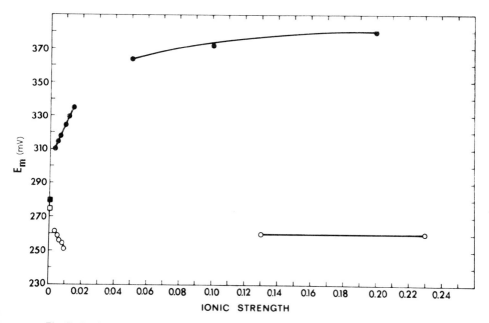

Fig. 7. Ionic strength dependence of the reduction potentials of *Euglena gracilis* cytochrome c_{552} (●) and horse cytochrome c (○) at pH 7.0 and 25°C. The ionic strength ≥0.10 was provided by NaCl; (■) and (□), extrapolated values. (From Goldkorn and Schejter.[49])

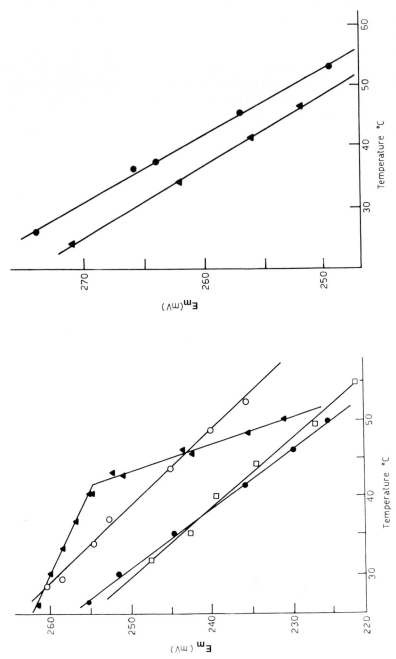

Fig. 8. Temperature dependence of the reduction potential of horse cytochrome c. The solutions contained 0.10 M sodium halide and 0.10 M sodium phosphate buffer at pH 7.0. $\bigcirc$, F$^-$; $\blacktriangle$, Cl$^-$; $\bullet$, Br$^-$; $\square$, I$^-$; (left) in H$_2$O, (right) in D$_2$O. (From Kreishman et al.[57])

the potentials of multiheme proteins. The hemes of class II cytochromes have identical environments (Fig. 4), and they are sufficiently far apart for heme—heme interactions not to be important. Therefore, their reduction potentials are relatively simple to define. However, the hemes of class III cytochromes have distinctly different environments (Fig. 5), and they are close enough to each other for heme heme interactions to affect the electron distribution. The application of electrochemical methods[58–60] should help to characterise fully these multiheme proteins.

2. Electron self-exchange rates

As with the reduction potentials, the electron self-exchange rates of class II and III cytochromes have not been adequately characterised. Indeed, with few exceptions, nor have the self-exchange rates of class I cytochromes *c* been thoroughly examined. Table 2 lists those cytochromes for which self-exchange rates (given the symbol k_{11}) have been experimentally determined. All of the measurements were obtained by NMR.

The most thoroughly studied cytochrome, that of the horse, was investigated by Gupta *et al.*[37,40] The pH dependence and ionic strength dependence of its k_{11} are shown in Fig. 9. The rate is strongly ionic strength dependent at pH 7, and it is pH independent until high pH. Above pH 8, the rate increases with increasing pH until close to the isoelectric point (at pH 10), when it begins to fall. At the isoelectric point, k_{11} is ionic strength independent. Gupta *et al.*[37] concluded from these data that at pH 7 and low ionic strength, electrostatic interactions were limiting the electron exchange. Gupta[40] also studied the self-exchange of *Candida* cytochrome *c* and found that its k_{11} was an order of magnitude lower than that of horse cytochrome *c* (Table 2), thus demonstrating that factors other than electrostatic interactions are important in determining k_{11}.

One such factor is the size of the electron transfer site on cytochrome *c*. From calculations of the diffusion-limited k_{11} of horse cytochrome *c*, Gupta *et al.*[37] calculated that the electron transfer site occupied approximately 7 $Å^2$ of the surface of cytochrome *c*. All the evidence points to the exposed edge of the heme being the electron transfer site (Sections VI and VII) and, given the assumptions in the calculations, the value calculated by Gupta *et al.*[37] agrees reasonably well with the value calculated from the X-ray structure for the area on the surface of the protein taken up by the exposed edge of the heme.[23] Although this factor does not account for the difference between horse and *Candida* cytochrome *c*, it might account for part of the difference between the slowly exchanging mitochondrial cytochrome *c*: cytochrome c_2 group and the rapidly exchanging cytochrome c_{551} group

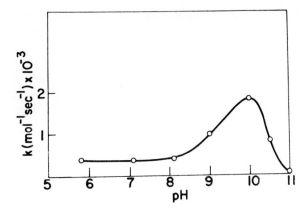

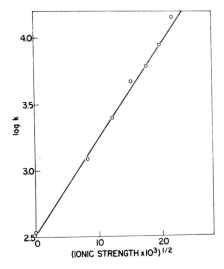

Fig. 9. The electron self-exchange rate of horse cytochrome *c*. (Top) pH dependence at 25°C and ~50 m*M* HEPES buffer; (bottom) ionic strength dependence at 30°C and pH 7.0 (the salt was not specified). (From Gupta *et al.*[37])

(Table 2). The latter group of proteins are smaller than the former group (Fig. 3) and have a greater area of their heme exposed. Also, the surface around the exposed hemes of these smaller proteins is generally less highly charged than that of mitochondrial cytochrome *c*, thus reducing unfavourable electrostatic interactions. A third factor that may be important is the oxidation-state-linked conformation change, which is considerably larger for the mitochondrial cytochromes c than for the cytochromes c_{551}.

D. Summary and mitochondrial cytochrome c

The structures of many different kinds of cytochrome c have been determined to a sufficient level of accuracy to place a systematic survey of cytochrome reactivity on a firm structural foundation. Thus the investigations of reactivity that are currently underway in many laboratories should allow general principles of hemoprotein electron transfer to be enumerated. In this article, we have selected for further discussion only one group of cytochromes—the mitochondrial class I cytochromes c—but the many investigations of this group clearly point toward a description of the electron transfer process.

The selection of the mitochondrial class I cytochromes c was not an arbitrary decision. Although the different types of cytochrome described previously are all being investigated, none is as well defined as the mitochondrial proteins; therefore, rather than present a number of incomplete accounts for different proteins, we have chosen to describe only the best-characterised group. However, the mechanistic principles that are established by the mitochondrial proteins are applicable to the other proteins as well.

When the name cytochrome c is used in general discussions of proteins and electron transfer, the protein referred to is almost invariably mitochondrial class I cytochrome c; this is the convention used in the remainder of this article. Mitochondria are the subcellular organelles of eukaryotic organisms (organisms such as animals, plants, and fungi with complex cells containing a developed nucleus surrounded by a nuclear envelope) that carry out the final stages of metabolism and respiration (for further details, see Ref. [61]), and cytochrome c is one of the most stable, abundant, and easily isolated proteins from the respiratory electron transfer pathway. Hence, cytochrome c is the most thoroughly defined of the mitochondrial electron transfer proteins. Its main function is to transfer electrons from cytochrome c reductase to cytochrome c oxidase, both of which are complex, multi-headed enzymes that span the inner mitochondrial membrane. Cytochrome c performs this function in all eukaryotic organisms, but although its function is invariant, its structure is not. Sequence studies of the cytochromes c from different eukaryotic organisms have shown that at most only 26% of the amino acids are invariant (Fig. 10). Thus the sequence variation is a

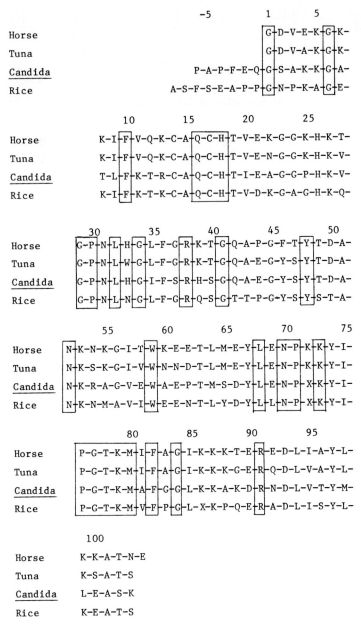

Fig. 10. The amino acid sequences of horse,[62] tuna,[63] *Candida*,[64,65] and rice[13,17,66] cytochromes *c*. The residues enclosed by boxes are invariant among the 84 mitochondrial cytochromes *c* that have been sequenced. The N-terminal residues are acetylated in the cytochromes from horse, tuna, and rice.

great help in defining which parts of the structure are essential for activity, although such discussions are beyond the scope of this article (instead, see Refs.[13, 14, and 35]). The amino acid sequences of the four cytochromes frequently referred to in the following sections are given in Fig. 10.

III. STRUCTURES AND PROPERTIES OF SELECTED INORGANIC REAGENTS

Most of the small-molecule inorganic reagents whose reactions with cytochrome c are discussed in this article are well characterised, and their properties relevant to this review are their size, shape, and charge, their oxidation–reduction potentials and electron self-exchange rates, and their magnetic properties. It is obvious why their structural and reactivity characteristics are important; their magnetic properties are important because of the possibility of using the inorganic reagent as an NMR probe (see Section IV). Wherland and Gray[2] have previously presented a catalogue of the properties of small-molecule reagents, but there is little overlap with the summary herein, which is presented in Table 3.

We have included a large number of reagents that have not been reacted with cytochrome c (compare Tables 3 and 5), both to present a generally useful compilation and in the expectation that future work will fill in some of the gaps. We have differed from Wherland and Gray in obtaining radii from crystal structure data by taking the distance from the metal ion to the most distant atom of the complex in the crystal and adding the covalent radius of that atom. Solvation is not taken into consideration by this procedure. Not all of the complexes are spherical (see later) and in these cases the value given represents the largest radial dimension. Values for k_{22} are given only when they have been experimentally determined. Wherland and Gray[2] report a number of values estimated from Marcus theory, but this procedure begs the question of whether the theory can be accurately applied. The magnetic properties given here are the number of unpaired electrons, an approximate designation of the g tensor as anisotropic or isotropic, and an estimate of the electron spin-lattice relaxation time. These latter two properties are usually determined by electron paramagnetic resonance (EPR) spectroscopy.

With only a few exceptions, the complexes of Table 3 fall into five broad groups: the anionic penta- and hexacyanides, the anionic oxalates, the anionic polyaminocarboxylates, the cationic penta- and hexammines, and the cationic complexes with relatively bulky aromatic ligands. The exceptions

TABLE 3

Structures and properties of small-molecule reagents

Complex[a]	Radius[b] (Å)	Magnetic properties[c]			E_m[d] (mV)	k_{22}[e] ($M^{-1}\,s^{-1}$)	Ref.
		S	g	T_{1e}			
$[Fe(CN)_6]^{3-}$	3.6	$\frac{1}{2}$	a	s	420 (1)	9.6×10^3 (1)	[67–72]
$[Fe(CN)_6]^{4-}$	3.6	0					
$[Co(CN)_6]^{3-}$	3.6	0			−830 (2)		[73]
$[Co(CN)_6]^{4-}$	3.6						
$[Ru(CN)_6]^{4-}$	3.6	0					
$[Cr(CN)_6]^{3-}$	3.6	$\frac{3}{2}$	i	l	−1280 (3)		[68]
$[Cr(CN)_6]^{4-}$							[74–76]
$[Mo(CN)_8]^{4-}$	3.9	0					[77, 78]
$[Fe(CN)_5(N_3)]^{3-/4-}$					240 (4)		[79]
$[Fe(CN)_5(imid)]^{2-/3-}$					350 (5)		[80]
$[Fe(CN)_5(py)]^{2-/3-}$					480 (5)		[80]
$[Fe(CN)_5(4\text{-}NH_2py)]^{2-/3-}$					350 (6)		[81]
$[Fe(CN)_5(CO)]^{2-/3-}$					1180 (5)		[80]
$[Fe(CN)_5(NH_3)]^{2-/3-}$	3.6				330 (4)		[79, 82]
$[Fe(CN)_5(PPh_3)]^{2-/3-}$					540 (4)		[79]
$[Co(ox)_3]^{3-/4-}$					570 (7)		[83]
$[Cr(ox)_3]^{3-}$	3.1	$\frac{3}{2}$	i	l			[84, 85]
$[Zr(ox)_4]^{4-}$		0					
$[IrCl_6]^{2-}$		$\frac{1}{2}$	i	s	1030 (8)		[86, 87]
$[IrCl_6]^{3-}$							
$[Mn(edta)(H_2O)]^{-}$	4.9				820 (9)		[88]
$[Mn(edta)(H_2O)]^{2-}$							[89, 90]
$[Fe(edta)(H_2O)]^{-}$	4.7	$\frac{5}{2}$		l	120 (10)		[2, 91–99]
$[Fe(edta)(H_2O)]^{2-}$		$\frac{4}{2}$		l			[97, 99]
$[Cr(edta)(H_2O)]^{-}$		$\frac{3}{2}$		l			[98]
$[Co(edta)]^{-}$		0			400 (2)	3×10^{-5} (2)	[73, 100, 101, 105]
$[Co(edta)]^{2-}$	4.4						[94, 102, 103]
$[Ga(edta)(H_2O)]^{-}$	4.4	0					[95, 104]
$[Fe(nta)(H_2O)(OH)]^{-}$		$\frac{5}{2}$		l			[73, 99]
$[Fe(dtpa)]^{2-}$		$\frac{5}{2}$		l			[73, 99]
$[Ru(NH_3)_6]^{3+}$	3.6	$\frac{1}{2}$		s	51 (11)	3.6×10^3 (3)	[2, 86,
$[Ru(NH_3)_6]^{2+}$		0					106, 107]
$[Co(NH_3)_6]^{3+}$	3.6	0			100 (8)		[87, 106]
$[Co(NH_3)_6]^{2+}$	3.4						[73]
$[Cr(NH_3)_6]^{3+}$		$\frac{3}{2}$	i	l			[73]
$[Pt(NH_3)_6]^{4+}$		0					
$[Ru(NH_3)_5(bm)]^{3+/2+}$					150 (2)		[108]
$[Ru(NH_3)_5(py)]^{3+/2+}$					273 (12)		[109]

(continued)

TABLE 3 (continued)

Complex[a]	Radius[b] (Å)	Magnetic properties[c] S	g	T_{1e}	$E_m{}^d$ (mV)	$k_{22}{}^e$ (M^{-1} s^{-1})	Ref.
$[Ru(bipy)_3]^{3+/2+}$					1260 (2)		[73]
$[Os(bipy)_3]^{3+/2+}$					884 (2)		[73]
$[Fe(bipy)_3]^{3+/2+}$					1096 (4)		[79]
$[Fe(bipy)(CN)_4]^{-/2-}$					550 (4)		[79]
$[Co(phen)_3]^{3+}$	5.2	0			⎱ 370 (13) ⎰	⎱ 4×10 (4) ⎰	[109–112]
$[Co(phen)_3]^{2+}$							
$[Co(5,6\text{-}Me_2phen)_3]^{3+/2+}$					430 (14)		[113]
$[Co(4,7\text{-}Me_2phen)_3]^{3+/2+}$					340 (14)		[113]
$[Co(5\text{-}Cl\text{-}phen)_3]^{3+/2+}$					420 (14)		[113]
$[Co(4,7\text{-}(PhSO_3)_2\text{-}$ phen$)_3]^{3-/4-}$					330 (14)		[113]
$[Cr(phen)_3]^{3+}$		$\frac{3}{2}$	1				[114, 115]
$[Fe(cp)_2]^{+}$	3.6				⎱ 513 (2) ⎰	⎱ $5.7 \pm 1 \times 10^6$ (5) ⎰	[116–118]
$[Fe(cp)_2]$							

[a] Ligand abbreviations used in this table and elsewhere: imid, imidazole; py, pyridine; 4-NH$_2$py, 4-aminopyridine; bm, benzimidazole; bipy, 4,4′-bipyridine; terpy, 2,2′,2″-terpyridine; phen, 1,10-phenanthroline; 5,6-Me$_2$phen, 5,6-dimethyl-1,10-phenanthroline; 4,7-Me$_2$phen, 4,7-dimethyl-1,10-phenanthroline; 5-Cl-phen, 5-chloro-1,10-phenanthroline; 4,7-(PhSO$_3$)$_2$-phen, 4,7-di(phenyl-4′-sulphonate)-1,10-phenanthroline; cp, cyclopentadiene; edta, ethylenediamine-N,N,N',N'-tetraacetic acid; nta, nitrilotriacetic acid; dtpa, diethylenetriamine-N,N,N',N',N''-pentaacetic acid; hedta, N-(2-hydroxyethyl)-ethylenediamine-N,N',N'-triacetic acid; dpta, 1,2-diaminopropane-N,N,N',N'-tetraacetic acid; edap, ethylenediamine-N,N'-diacetic acid-N,N'-dipropionic acid; pac, polyaminocarboxylate.

[b] The radii of metal ion complexes have been obtained from crystal structure data by taking the distance from the central metal ion to the most distant atom of the ligand and adding the covalent radius of that atom. Solvation is not taken into account by this procedure, and where the complex is not spherically symmetric, the radius given represents the largest radial dimension.

[c] S is the electron spin; g refers to the electronic g tensor and (a) and (i) indicate whether it is anisotropic or isotropic, respectively; T_{1e} is the electron spin-lattice relaxation time and (l) and (s) indicate whether it is long or short, respectively.

[d] Redox potentials measured under the following conditions: (1) pH 7.0, 25°C, $I = 0.1$ M (KCl); (2) unspecified; (3) 1 M KCN; (4) pH 5.5–7.0, 25°C, $I = 0.05$ M (KNO$_3$); (5) 25°C, $I = 1.0$ M (NaCl); (6) pH 8.1, 23°C, $I = 0.10$ M (LiClO$_4$); (7) pH 3.65, 25°C, $I = 1.0$ M (KCl); (8) standard potential; (9) pH 4.6, 25°C, $I = 0.2$ M (KCl); (10) pH 4–6, 20°C; (11) $I = 0.10$ M (NaBF$_4$); (12) pH 6.5, 25°C, $I = 0.10$ M (phosphate); (13) pH 6.5, 25°C, $I = 0.10$ M (phosphate and Na$_2$SO$_4$); (14) pH 7.0, 25°C, $I = 0.05$ M (NaCl).

[e] Self-exchange rates measured under the following conditions: (1) pH 7.0, 25°C, $I = 0.1$ M (KCl). This rate is markedly dependent upon pH and ionic strength and composition (see Ref. [70]). (2) pH 4.0, 74°C, $I = 0.2$ M (NaNO$_3$); (3) in ^{2}H$_2$O, pH* 7, 25°C, $I = 0.1$ M; (4) 25°C, $I = 0.1$ M; (5) a variety of solvents ranging from CD$_3$OD to CD$_3$COOD (see Ref. [118]). The self-exchange rates for a number of complexes have been estimated (see Ref. [2]), including those for $[Fe(edta)(H_2O)]^{-/2-}$ (3×10^4 M^{-1} s^{-1} at $I = 0.1$ M) and $[Co(ox)_3]^{3-/4-}$ (2.8×10^{-7} M^{-1} s^{-1} at $I = 0.1$ M).

are included because their reactions with cytochrome *c* have been studied (see Table 5). Each of the major groups contains at least one each of the following components: a redox reagent whose E_m is close to that of cytochrome *c* (compare Tables 2 and 3), a diamagnetic redox-inactive complex, and a paramagnetic complex with a relatively long electron relaxation time. These complementary classes of reagent allow small-molecule–protein reactions to be characterised in considerable detail, as shown in Section IV.

The penta- and hexacyanide group of reagents has been extensively used even though, as Wherland and Gray[2] pointed out, properties of these reagents are more dependent upon the medium than those of most other reagents (excluding the polyaminocarboxylates). Thus reactions between cytochrome *c* and the cyano complexes comprise a large part of the mechanistic discussion in Section VI.

The polyaminocarboxylate group of reagents has a particularly complex solution chemistry, and some of the reagents in Table 3 have not been fully characterised. A full consideration of this complexity is beyond the scope of the present article but some discussion of it is necessary. The edta ligand contains six potential coordinating atoms (four oxygens and two nitrogens), and, when coordination occurs, five-membered rings are formed. However, maximum stability does not always accompany sexadentate coordination, because of strain in the ligand, and quinquedentate coordination often occurs. Figure 11 illustrates a common sexadentate structure (I) and a common quinquedentate structure (II). Often a particular solution will contain a mixture of the two forms, with the degree of mixing being pH dependent.[90,94,98,119,120] Thus the edta complexes of Cr(III), Co(II), and

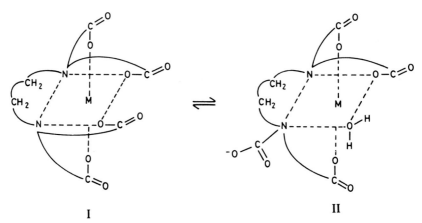

I   II

Fig. 11. Structural heterogeneity of edta complexes. The sexadentate structure I and quinquedentate structure II are two of the most common structures.

Co(III) are present in solution in the dynamic equilibrium of Fig. 11,[98,103] and approximately 80% of [Co(edta)]$^{2-}$ is present as structure I at 25°C, pH 5, and an ionic strength of 1 M.[94] The highest carboxyl pK_a for the Co(III) and Cr(III) complexes is approximately 3.0[98]; the Cr(III) complex also exhibits a pK_a of 7.5, which is due to the ionisation of coordinated H$_2$O.[98] Although a Cr(III) complex of structure type I (Fig. 11) has been isolated in the solid state, in the pH range 5–9 the complexes in solution are mainly of structure type II: [Cr(edta)(H$_2$O)]$^-$ and [Cr(edta)(OH)]$^{2-}$.[98] Two further complications exist with respect to the use of the cobalt complexes as redox reagents: there is a spin-state change accompanying electron transfer and in some reactions inner-sphere complexes may form. For example, [(edta)CoIII · (NC)FeII(CN)$_5$]$^{5-}$ may be formed on admixture of [Co(edta)]$^{2-}$ with [Fe(CN)$_6$]$^{3-}$.[121,122]

Crystal structure analyses of [Fe(edta)(H$_2$O)]$^-$ and [Mn(edta)(H$_2$O)]$^{2-}$ show the ligand to be sexadentate and H$_2$O to be coordinated to the metal ions (Fig. 12a).[89,91,92] These structures are largely maintained in solution,[90,93–95] although in the case of the iron complex there may be small populations of molecules containing seven-coordinate Fe(III) with two coordinated H$_2$O molecules and six-coordinate Fe(III) with no coordinated H$_2$O (structure I of Fig. 11). At lower pH in the solid state, one of the carboxyls is un-ionised, and [Fe(H · edta)(H$_2$O)] contains six-coordinate Fe(III) with an analogous structure to structure II of Fig. 11. The diamagnetic complex [Ga(H · edta)(H$_2$O)] has the same structure, suggesting that at higher pH [Ga(edta)(H$_2$O)]$^-$ resembles [Fe(edta)(H$_2$O)]$^-$. A recent report[97] indicated that [Fe(edta)(H$_2$O)]$^-$ is affected by some commonly used pH buffers. Its optical spectrum was affected by CO$_3$$^{2-}$ and HPO$_4$$^{2-}$ and, in the former case at least, this resulted from complexation between CO$_3$$^{2-}$ and [Fe(edta)(H$_2$O)]$^-$. The pK_a for the conversion of [Fe(edta)(H$_2$O)]$^-$ to [Fe(edta)(OH)]$^{2-}$ is the same as for the Cr(III) complex, 7.5,[94,96,97] and at higher pH dimeric species are formed.[96,123]

Although the structure of the iron(II)–edta complex is not known, it is generally taken to be analogous to the iron(III) complex (Fig. 12a). However, the pK_a of the coordinated H$_2$O is uncertain: 9.1 is usually given for this pK_a,[96] but more recent work[97] failed to detect an ionisation on [Fe(edta)(H$_2$O)]$^{2-}$ over the pH range 5–11.

Crystal structures of the ferric nta and dtpa complexes have not been determined. The depicted structures shown in Fig. 12 are those most consistent with pH studies,[96,124] spectroscopic investigations (Refs. [125 and 126] and J. Oakes, personal communication), and X-ray studies of a related compound.[127] The conversion of [Fe(nta)(H$_2$O)$_2$] to [Fe(nta)(H$_2$O)(OH)]$^-$ has a pK_a of 5.0.[124]

The crystal structure of $[La(edta)(H_2O)_3]^-$, which is also the dominant form in solution,[128] is illustrated in Fig. 12c to facilitate comparison between it and the iron complexes. Many of the lanthanides have short electron relaxation times, a property not prevalent among the complexes listed in Table 3, and this makes them exceedingly useful in NMR structure determinations (see Section IV).[129,130] What is required are lanthanide complexes with structures that closely match those of the redox probes, and although it might appear that the $[La(edta)(H_2O)_3]^-$ complexes satisfactorily achieve this, NMR experiments have shown this not to be the case.[99] Therefore, alternative lanthanide complexes are required.

Reactions between cytochrome *c* and the three groups of anionic complexes in Table 3 have been extensively studied, but reactions with the cationic complexes have not been thoroughly investigated. In part, this stems from the fact that cytochrome *c* is highly positively charged, and thus precursor complexes are formed more readily with anionic reagents. Nevertheless, sufficient studies have been made of the reactions of cytochrome *c* with all these different classes of reagents for their reaction mechanisms to be considered, and this is done in Section VI after a brief discussion of experimental methods and ionic strength dependences.

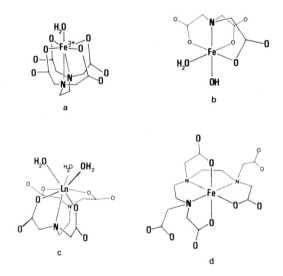

Fig. 12. The dominant solution structures of some NMR relaxation probes at pH 6. (a) $[Fe(edta)(H_2O)]^-$; (b) $[Fe(nta)(H_2O)(OH)]^-$; (c) $[Ln(edta)(H_2O)_3]^-$; (d) $[Fe(dtpa)]^{2-}$. Complex (c) is a relaxation probe when Ln = Gd.

IV. EXPERIMENTAL TECHNIQUES

A. Characterisation of an electron transfer reaction

To characterise an electron transfer reaction the following parameters should be determined:

1. The number of reaction sites on the protein.
2. The electron transfer rate constants at each site.
3. The association constants, and their attendant association and dissociation rate constants, at each site.
4. The activation enthalpy and entropy for each individual step in the reaction.
5. The structure(s) of the reactive complex(es), most importantly the distance between the immediate electron donor and acceptor sites.

Such an ideal characterisation has not been achieved for any electron transfer reaction involving a protein, partially because of the complications involved in studying large molecules and partially because discrete and identifiable reaction sites may not exist on the surface of a protein (see Section VI,D). Nevertheless, defined objectives are needed to work toward even if those objectives are not completely attainable.

Many experimental techniques have been used to characterise electron transfer reactions; the most informative of these are given in Table 4 and are

TABLE 4

Experimental characterisation of the reactions

Method	Property characterized
Rapid mixing and relaxation kinetics	Electron transfer rates
	Association constants
NMR spectroscopy	Electron transfer rates
	Association constants
	Number of interaction sites
	Structures of reactive complexes
Equilibrium dialysis	Number of interaction sites
	Association constants
Redox potentiometry	Number of interaction sites
	Association constants

described in this section. We have not attempted to describe the methodology of each technique, although we do illustrate the kinds of information they provide, but because of the stress we place on NMR we have discussed in some detail its application to mitochondrial cytochrome *c*.

B. NMR and mitochondrial cytochrome *c*

Like all spectroscopic techniques, NMR depends on the absorption and emission of energy associated with transitions between energy levels (in this case between nuclear energy levels) to produce a spectrum. The absorption peaks are characterised by a number of parameters, of which the most useful for our present purpose are the energy of absorption and the resonance linewidth. The energy of absorption is usually given relative to a standard

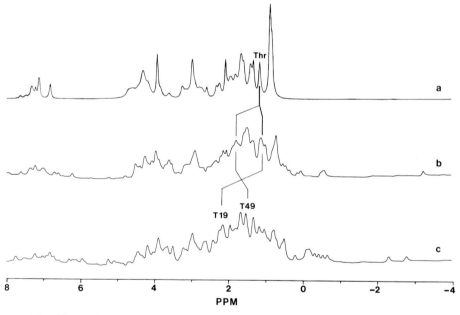

Fig. 13. Conformation-dependent shifts of NMR resonances of tuna cytochrome *c* at pH 7.0 and 25°C. (a) Simulated 300-MHz ^{1}H NMR spectrum in which all conformation-dependent shifts have been averaged to zero. The appearance of the spectrum depends only on the amino acid composition. (b) Experimental 300-MHz ^{1}H NMR spectrum of ferrocytochrome *c* showing the effects of diamagnetic conformation-dependent shifts. (c) Experimental 300-MHz ^{1}H NMR spectrum of ferricytochrome *c* showing the effects of both diamagnetic and paramagnetic conformation-dependent shifts. The positions of the methyl resonances of Thr 19 and Thr 49 are illustrated.

reference, as a chemical shift, on the parts per million (ppm) scale. Linewidths are measured in hertz.

Structural information is obtained from analysis of chemical shifts and linewidths. The importance of chemical shifts is illustrated in Fig. 13. This shows parts of the ^{1}H NMR spectra of tuna ferricytochrome c and tuna ferrocytochrome c, along with a spectrum simulated from the amino acid composition of tuna cytochrome c assuming the protein has no secondary or tertiary structure. The marked differences between the spectra are due to the highly ordered structure of the protein and the magnetic properties of low-spin Fe(III).[131]

The main problem with NMR of proteins is the assignment of resonances to specific nuclei within the protein, and for the kind of information required to properly characterise electron transfer reactions it is essential to have many resonance assignments. Fortunately, mitochondrial cytochrome c is one of the best NMR-characterised proteins, and over 40% of its CH ^{1}H NMR spectra have been firmly assigned to specific protons,[132–136] with the result that there is no part of the surface of cytochrome c that is further away than 6 Å from a group with an assigned resonance. As we shall see, this is of crucial importance in mapping reagent binding sites on the surface of cytochrome c. In addition to the assigned amino acid resonances, the heme resonances of ferricytochrome c are exceedingly useful for mechanistic studies. These resonances are strongly perturbed by the unpaired electron, with the result that most are shifted away from the main amino acid absorption region (Fig. 14).

The strength of the NMR method in studying electron transfer reactions is that spectra can be perturbed in a number of ways by redox reagents (see the following section) and by paramagnetic reagents. As Williams[137] pointed out, when the paramagnetic reagents are analogues of redox reagents, the binding sites for the redox reagents can be mapped. Figure 15 illustrates the general approach. The bound paramagnetic metal is at the centre of the coordinate system with a nucleus N at a distance r away. If the metal is a shift probe, the resonance of nucleus N will be shifted, and if

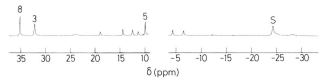

Fig. 14. The 300-MHz ^{1}H NMR spectrum of horse ferricytochrome c at pH 6.0 and 25°C, illustrating the heme resonance dispersion; 8, 3, and 5 refer to the methyl substituents at these positions on the porphyrin ring (see Fig. 1) and S refers to the methyl group of the methionine ligand (see Fig. 2).

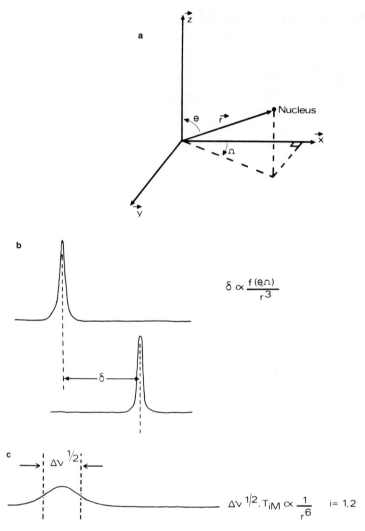

$$\delta \propto \frac{f(\theta,\Omega)}{r^3}$$

$$\Delta v^{1/2}, T_{iM} \propto \frac{1}{r^6} \qquad i = 1,2$$

Fig. 15. The effects of NMR shift and relaxation probes. (a) The coordinate system for the calculation of paramagnetic effects is centered on the unpaired electron and has the z axis aligned along the principal symmetry axis of the metal ion–protein complex. (b) If the electron relaxation time is short and the electron distribution anisotropic, the effect of probe binding is to shift the resonances of nearby nuclei from their unperturbed positions. (c) If the electron relaxation time is long, the dominant effect is on the relaxation times of nearby nuclei whose resonances are broadened. The perturbations to the chemical shift (δ) and relaxation times (T_{iM}, where $i = 1$ or 2, the spin-lattice and spin–spin relaxation times, respectively) are related to the geometry of the complex by the geometric relationships given above.

the metal is a relaxation probe, the resonance will be broadened. The chemical shift and linewidth changes can be related to the conformation around the bound metal by the relationships given in Fig. 15, but only when the perturbations are due to dipolar effects. When scalar effects are important, conformational analysis is not straightforward. The problems involved in this kind of analysis have been thoroughly discussed by Dwek[138,139] (and see Section VI,D). However, one point that is worth emphasising since it is sometimes overlooked is that the observed perturbations rely not only on the conformation around the binding site but also on the occupancy of the site. Thus, it is possible to observe a large effect on a resonance resulting from strong binding at a site far from the group whose resonance is observed, and this may be mistaken for weak binding at a nearby site. When there are many assignments or when the association constants are known, an accurate statement can usually be made concerning the origin of the effect.

The reason for the classification of magnetic properties in Table 3 is that it is generally possible to predict from them what NMR-perturbing properties a paramagnetic molecule has. A reagent with a long electron relaxation time is usually a relaxation probe, while a reagent with both a short electron relaxation time and an anisotropic g tensor is usually a shift probe. In practice, spherically symmetric transition metal complexes are not markedly anisotropic and are poor dipolar shift probes. Lanthanide complexes are much better dipolar shift probes and, therefore, lanthanide complexes similar to transition metal redox complexes are currently being developed for NMR (S. C. Tam, G. R. Moore, G. Williams, and R. J. P. Williams, unpublished data).

C. Rates of reaction

The most generally applicable techniques, and the ones which most readily yield accurate reaction rates, are rapid-mixing and relaxation kinetic techniques coupled with visible spectrophotometry. Optical spectra of cytochrome c, which are shown in Fig. 16, are markedly oxidation state dependent and have a high extinction coefficient that allows the redox compositions of dilute mixtures to be obtained. First-order kinetics are usually obtained for the reactions of proteins with redox reagents, but in some cases saturation kinetics are found (Fig. 17). Saturation kinetics are important because they may arise from precursor complex formation.

In simple reactions such as reaction (6) (see Section I)

$$\text{rate} = k[A^o][B^r] \qquad (7)$$

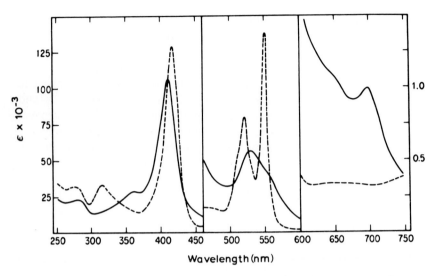

Fig. 16. Absorption spectra of horse cytochrome *c* at pH 7.0. Solid line, ferricytochrome *c*; dashed line, ferrocytochrome *c*. Center panel, use the ordinate on right × 20. (From Harbury and Marks.[140])

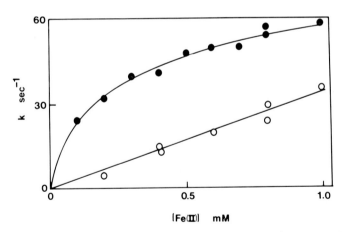

Fig. 17. The variation of first-order rate constants with $[Fe(CN)_6]^{4-}$ (FII) for the FII (in excess) reduction of 10^{-6} *M* horse ferricytochrome *c* at 25°C, pH 7.2 (Tris), and $I = 0.10$ *M* (NaCl). Rate constants (●) were obtained from a simple treatment (Ln absorbance change versus time), ignoring the fact that the reactions do not proceed to completion and they present the appearance of saturation kinetics. Rate constants (○) were obtained from a rigorous equilibration kinetics treatment that takes account of the fact that the reactions do not proceed to completion and they follow a simple first-order course. (From Butler *et al.*[141])

Under pseudo-first-order conditions with $\mathbf{B^r}$ in excess

$$k_{obs} = k[\mathbf{B^r}] \tag{8}$$

When precursor complex formation occurs under conditions of excess $\mathbf{B^r}$, as in reactions (9) and (10)

$$\mathbf{A^o} + \mathbf{B^r} \overset{K_{ass}}{\rightleftharpoons} \mathbf{A^o}{:}\mathbf{B^r} \tag{9}$$

$$\mathbf{A^o}{:}\mathbf{B^r} \overset{k_{et}}{\longrightarrow} \text{products} \tag{10}$$

curvature in the plot of k_{obs} against $[\mathbf{B^r}]$ results because

$$k_{obs} = K_{ass}k_{et}[\mathbf{B^r}]/(1 + K_{ass}[\mathbf{B^r}]) \tag{11}$$

Curvature in plots of k_{obs} against $[\mathbf{B^r}]$ may also arise from causes unconnected with complex formation,[141–143] but the reason for curvature can usually be determined. One particular cause of curvature, which has been misinterpreted,[144,145] results from the study of thermodynamically unfavourable reactions, and Butler *et al.*[141] pointed out that a rigorous kinetic treatment is essential for such reactions (Fig. 17).

The rapid kinetic methods can give information on complex formation by the competitive inhibition approach developed by Sykes and co-workers.[143] This method involves the use of a redox-inactive competitive inhibitor according to the reactive sequence

$$\mathbf{A^o} + \mathbf{B^r} \overset{k}{\longrightarrow} \text{products} \tag{12}$$

$$\mathbf{A^o} + \mathbf{I} \overset{K_I}{\rightleftharpoons} \mathbf{A^o}{:}\mathbf{I} \tag{13}$$

$$\mathbf{A^o}{:}\mathbf{I} + \mathbf{B^r} \overset{k_I}{\longrightarrow} \text{products} \tag{14}$$

where I is the inhibitor and k_I in many cases approaches zero. For this reaction sequence, the observed rate is dependent on [I] according to Eq. (15)

$$\text{rate} = k[\mathbf{A^o}]_t[\mathbf{B^r}] + K_I k_I[\mathbf{I}][\mathbf{A^o}]_t[\mathbf{B^r}]/(1 + K_I[\mathbf{I}]) \tag{15}$$

Thus, if I is a redox-inactive analogue of B, information concerning the binding of B may be obtained. Apart from this, highly charged redox-inactive complexes, which are not analogues of redox reagents, may be used to inhibit electron transfer.

NMR may be used to measure electron exchange rates, but, because the lifetimes of NMR excited states are much longer than the lifetimes of the corresponding states that give rise to optical absorption, NMR spectra often vary considerably with the redox state composition and are not simply

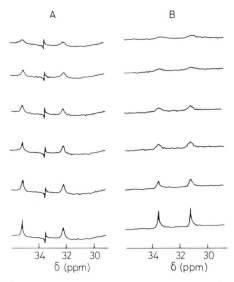

Fig. 18. $[Fe(CN)_6]^{4-}$-induced electron exchange broadening of ^1H NMR resonances of heme methyls 3 and 8 of (A) 0.41 mM horse ferricytochrome c and (B) 0.44 mM *Candida* ferricytochrome c; both in ^2H_2O at pH* 7.4 and 25°C, and both in the presence of varying concentrations of $K_4[Fe(CN)_6]$ at a constant ionic strength of 0.12 M (which was maintained by NaCl); with 0, 100, 200, 300, 600, and 1000 mol $[Fe(CN)_6]^{4-}$ per 100 mol cytochrome, from bottom to top, respectively. (From Eley *et al.*[148])

superimpositions of spectra of the fully oxidised and fully reduced species. These effects (described in Ref. [146]) are used for measuring electron self-exchange rates (Tables 2 and 3). Cross-reaction rates can also be measured, as first demonstrated by Stellwagen and Shulman[147] for the reaction of horse cytochrome c with $[Fe(CN)_6]^{3-/4-}$. One way of carrying out this experiment (Fig. 18) is to titrate $[Fe(CN)_6]^{4-}$ into a solution of ferricytochrome c at constant ionic strength. Heme methyl resonances broaden as the concentration of $[Fe(CN)_6]^{4-}$ increases. This is due to the reaction

$$CIII + FII \underset{}{\overset{K_1}{\rightleftharpoons}} CIII:FII \overset{k_2}{\rightarrow} CII:FIII \qquad (16)$$

where CIII is ferricytochrome c, FII is $[Fe(CN)_6]^{4-}$, and CIII:FII and CII:FIII are complexes of CIII and FII and of ferrocytochrome c and $[Fe(CN)_6]^{3-}$, respectively. Maximal broadening occurs when all the CIII is bound to hexacyanide, and under these conditions the broadening is equal to k_2/π. Analysis of the broadening as a function of the concentration of $[Fe(CN)_6]^{4-}$ allows both K_1 and k_2 to be determined (Fig. 19).[148]

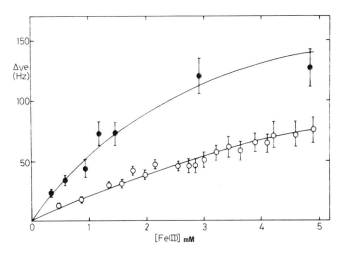

Fig. 19. Graph showing the electron exchange broadening of the heme methyl 8 resonance of horse (○) and *Candida* (●) ferricytochromes *c* as a function of the $[Fe(CN)_6]^{4-}$ concentration. The curves are theoretical lines calculated with the parameters $K_{ass} = 90\ M^{-1}$ and $k_{et} = 785\ s^{-1}$ for (○) and $K_{ass} = 285\ M^{-1}$ and $k_{et} = 770\ s^{-1}$ for (●). (From Eley *et al.*[148])

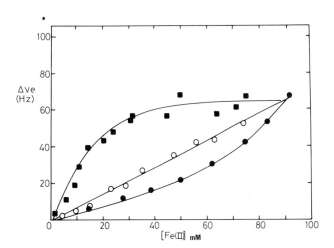

Fig. 20. Graph showing the induced broadening of the heme methyl 8 resonance of 8.5 mM horse ferricytochrome *c* (in ^2H_2O at pH* 7.0 and 27°C) as a function of the $Fe(CN)_6^{4-}$ concentration. ■, Variable ionic strength (final $I = 1.0\ M$); ○, constant $I = 1.0\ M$ maintained with Na cacodylate; ●, constant $I = 1.0\ M$ maintained with NaCl. (From Eley *et al.*[148])

Stellwagen and Shulman[147] described an analogous experiment at high cytochrome concentrations and uncontrolled ionic strengths, in which they obtained results similar to those shown in Fig. 20.[148] This type of curve has the appearance of saturation kinetics, although it does not result from the simple saturation of a binding site with a constant K_{ass}, but from apparent saturation resulting from a variable K_{ass} that is caused by variable ionic strength. When ionic strength is kept constant at a high value (Fig. 20), K_{ass} is small and there is no evidence for saturation kinetics.

In general, optical methods are preferable to NMR, because the latter is a relatively insensitive technique and requires higher protein concentrations than when optical spectra are monitored, i.e., $\geq 10^{-4}$ M compared with $\geq 10^{-6}$ M. This can cause problems in estimating and controlling ionic strength.

D. Complex formation association constants

The kinetic methods described in the previous section provide an accurate measurement of K_{ass}, but those methods are not always applicable and a number of equilibrium methods of determining K_{ass} are used. Probably the most widely applicable method is equilibrium dialysis.[149] This involves suspending a membrane bag containing a solution of the protein in a solution containing the small molecule. At equilibrium, the excess concentration of small molecule inside the bag is related to the extent of binding. The advantage of this method is that the concentration of protein, which is retained in the bag, can be held relatively low (the determining factors are the strength of binding and method of analysis for the small molecules), which allows both the ionic strength to be controlled and the analysis to be carried out using Scatchard plots.[150] The disadvantage of this method is that in cases of weak binding the analysis is not straightforward, but this is a problem with most methods. Given the overall simplicity of this method, it is surprising that there have been so few studies made of the binding of redox reagents to proteins by equilibrium dialysis. Figure 21, taken from Stellwagen and Cass,[151] shows the Scatchard plots obtained for the binding of $[Fe(CN)_6]^{3-/4-}$ to horse cytochrome c under a variety of conditions. Both the stoichiometry of binding and the association constants are readily obtained from these plots (see Table 6), which also illustrate the marked inhibitory effect that NaCl has on the binding.

Redox potentiometry can also give information on the association of redox reagents and proteins in cases in which there is a redox state difference

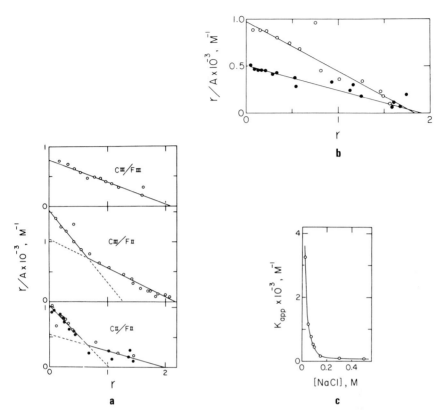

Fig. 21. $[Fe(CN)_6]^{3-/4-}$ binding to horse ferri- and ferrocytochrome c determined by equilibrium dialysis. (a) The abscissa on the Scatchard plots, r, represents the moles of $[Fe(CN)_6]^{3-/4-}$ bound per total moles of protein; the ordinate represents r divided by A, where A is the moles of unbound $[Fe(CN)_6]^{3-/4-}$. The intercept on the abscissa gives the number of binding sites and the intercept on the ordinate gives the association constants. The measurements were made at 25°C in 50 mM piperazine-N,N'-bis(2-ethanesulfonic acid) buffer at pH 7.0. (b) The binding of $[Fe(CN)_6]^{4-}$ to ferricytochrome c in the presence of 90 mM NaCl at pH 7.8 (○) and 1.0 M imidazole at pH 8.0 (●). [Compare with CIII/FII in (a).] The imidazole displaces Met 80 from its axial ligation and disrupts the structure by the exposed heme. (c) Effect of increasing NaCl concentration at variable ionic strength on the binding of $[Fe(CN)_6]^{4-}$ to ferricytochrome c. K_{app} is the apparent association constant. (From Stellwagen and Cass.[151])

to the association.[152] Thus, for the reaction

$$L + \text{oxid} + e^- \underset{E_{m_1}}{\rightleftharpoons} \text{red} + L$$
$$\big\Vert K_O \qquad\qquad \big\Vert K_R \qquad\qquad (17)$$
$$[\text{oxid-L}] + e^- \underset{E_{m_2}}{\rightleftharpoons} [\text{red-L}]$$

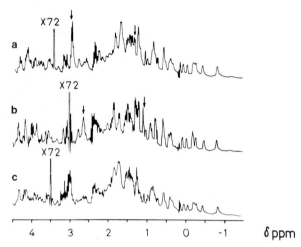

Fig. 22. The effect of metal hexacyanides on the 300-MHz ^{1}H NMR spectrum of *Candida* ferricytochrome c in ^2H_2O at pH* 7.2 and 25°C. (a) Cytochrome alone; (b) in the presence of 1000 mol $[Fe(CN)_6]^{3-}$ per 100 mol cytochrome; (c) in the presence of 1000 mol $[Co(CN)_6]^{3-}$ per 100 mol cytochrome c. X72 is the methyl resonance of trimethyllysine 72 (TML72) and ↓ indicates the positions of other resonances affected by $[Fe(CN)_6]^{3-}$. (From Eley *et al.*[154])

the measured midpoint potential at 25°C, E_m, is given by Eq. (18)

$$E_m = E_{m_1} + \frac{0.059}{n} \log \frac{1 + K_R[L]}{1 + K_O[L]} \qquad (18)$$

where oxid and red are the oxidised and reduced proteins, respectively, L is the small molecule, and K_R and K_O are association constants. The pH-dependent redox potentials given in Fig. 6 illustrate one application of this method (where [L] = [H$^+$]) but, despite attempts to apply this method to the binding of $[Fe(CN)_6]^{3-/4-}$ to cytochromes,[153] there are no clear examples of an E_m dependence resulting from the association of a redox reagent.

NMR can be used to measure association constants in two ways: from spectra of the protein perturbed by a paramagnetic small molecule, and from spectra of the small molecule perturbed by the protein. Figure 22 shows a region of the ^{1}H NMR spectrum of *Candida* ferricytochrome c in the absence and presence of $[Fe(CN)_6]^{3-}$.[154] This reagent is a contact shift probe,[155] shifting resonances of groups on the protein that binds it. Analysis of the shifts give K_{ass}, but this method has not yet been applied under conditions of controlled ionic strength. Figure 23 illustrates the effect of binding to cytochrome c on the ^{59}Co NMR resonance of $[Co(CN)_6]^{3-}$.[156] These data allow the association constants to be determined (see Table 6) and, as in Fig. 21, they illustrate the marked effect that NaCl has on the binding.

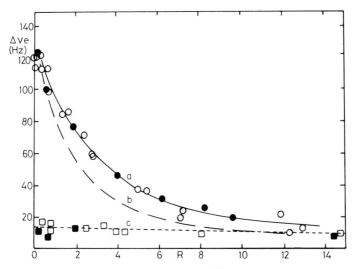

Fig. 23. Graph of the change in linewidth (Δve) of the ^{59}Co NMR resonance of $K_3[Co(CN)_6]$ as a function of the molar ratio (R) of $K_3[Co(CN)_6]$/cytochrome. $\bigcirc$ and $\bullet$, Experimental points for ferricytochrome c and ferrocytochrome c, respectively, in ^2H_2O at pH* 7.3, $I = 0.07\ M$ (no added salt), and 25°C; $\square$ and $\blacksquare$, experimental points for ferricytochrome c and ferrocytochrome c, respectively, in ^2H_2O at pH* 7.3, $I = 0.12\ M$ (KCl), and 25°C. The solid line (a) is a theoretical curve for two binding sites with K_{ass} of $2 \times 10^3\ M^{-1}$ and $1.5 \times 10^2\ M^{-1}$ and (b) is a theoretical curve for one binding site with K_{ass} of $2 \times 10^3\ M^{-1}$. Curve (c) is a least-squares fit, and for these data $K_{ass} \leq 1.5 \times 10^2\ M^{-1}$. (From Ragg and Moore.[156])

E.   Structures of reactive complexes

The NMR method of studying reactive complexes was briefly described in Section IV,B, and a discussion of results is given in Section VI,D. The only other method to give a firm indication of where the reaction takes place on the protein involves the study of chemically modified forms of the protein. The most generally useful approach uses proteins with single site modifications, usually with a single lysine residue modified,[157–159] but some workers have used less useful derivatives in which all the lysines have been modified.[160–163] The modified proteins may be gainfully studied by any of the techniques listed in Table 4, but undoubtedly rapid flow kinetics are the most useful methods, as Fig. 24 demonstrates for the oxidation of horse ferrocytochrome c with $[Fe(CN)_6]^{3-}$. This figure shows that the modification of some lysines causes a large decrease in activity, while modification of other lysines causes no decrease in activity. Similar results have been obtained with $[Co(phen)_3]^{3+}$ as oxidant, although rate enhancement rather than inhibition was found for some derivatives.[159]

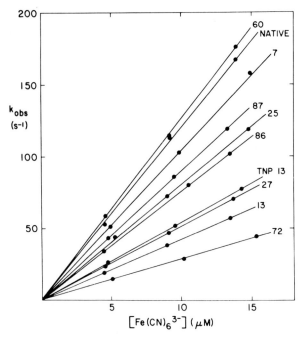

Fig. 24. Relationship between the observed pseudo-first-order rate constants (k_{obs}) and the concentration of [Fe(CN)$_6$]$^{3-}$ for the oxidation of native horse ferrocytochrome c and some of its carboxydinitrophenyl and trinitrophenyl (TNP) derivatives at 25°C in 5 mM Tris at pH 7.2 and at $I = 0.10$ M (NaCl). (From Butler *et al.*[159])

The modifications generally used are shown in Fig. 25[35,164–166]; they are a diverse collection ranging from large, negatively charged groups to small, neutral groups. The positively charged modifications have not been studied by kinetic methods, but this would be a valuable exercise because it should give an indication of the importance of steric factors in modifying the rate.

There is a general problem for the interpretation of rate data such as those shown in Fig. 24. Since

$$k_{obs} = K_{ass}k_{et} \qquad (\text{when} \quad K_{ass}[B^r] \ll 1) \qquad (19)$$

the rate difference between native cytochrome c and a derivative could arise from differences in K_{ass} or from differences in k_{et}. Usually it is assumed that the electron transfer pathway is unchanged by the modifications and, since the redox potentials of the singly modified derivatives are the same as for native cytochrome c,[165,166] that the intrinsic reactivity of the cytochrome

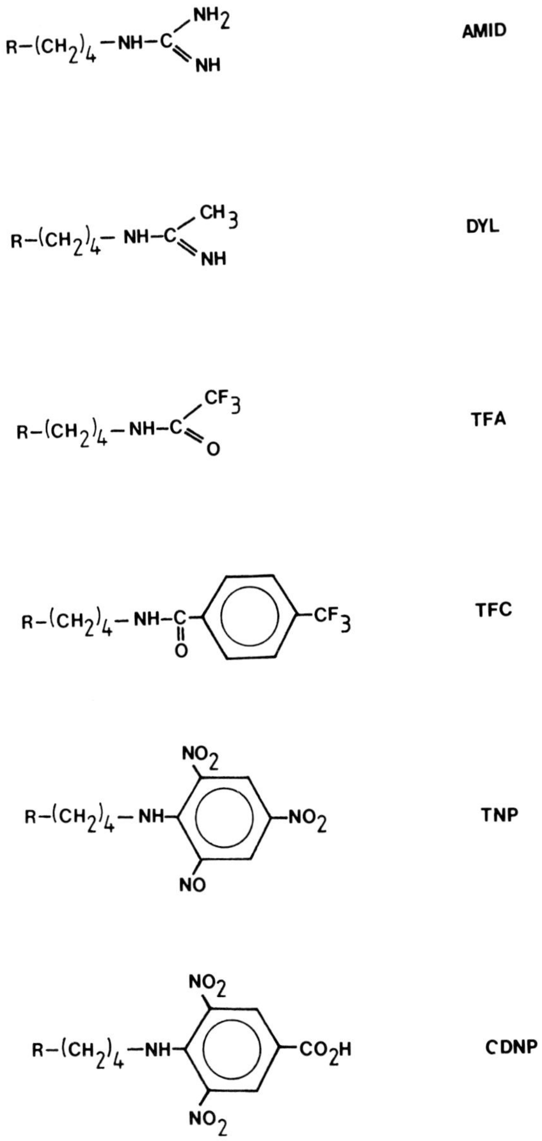

$$R-(CH_2)_4-NH-C\underset{\diagdown NH}{\overset{\diagup NH_2}{}} \qquad \textbf{AMID}$$

$$R-(CH_2)_4-NH-C\underset{\diagdown NH}{\overset{\diagup CH_3}{}} \qquad \textbf{DYL}$$

$$R-(CH_2)_4-NH-C\underset{\diagdown O}{\overset{\diagup CF_3}{}} \qquad \textbf{TFA}$$

$$R-(CH_2)_4-NH-\underset{O}{\overset{}{C}}-\!\!\!\!\bigcirc\!\!\!\!-CF_3 \qquad \textbf{TFC}$$

$$R-(CH_2)_4-NH-\!\!\!\!\bigcirc\!\!\!\!-NO_2 \qquad \textbf{TNP}$$

(with NO$_2$ above and NO below the ring)

$$R-(CH_2)_4-NH-\!\!\!\!\bigcirc\!\!\!\!-CO_2H \qquad \textbf{CDNP}$$

(with NO$_2$ above and NO$_2$ below the ring)

Fig. 25. Chemically modified lysine derivatives of cytochrome c (R): AMID, amidino; DYL, acetimidyl; TFA, trifluoroacetyl; TFC, (trifluoromethyl)phenylcarbamyl; TNP, trinitrophenyl; CDNP, carboxydinitrophenyl. At pH 7.0 AMID and DYL are positively charged modifications; TFA, TFC, and TNP are neutral modifications; and CDNP is a negatively charged modification.

remains unaffected. However, some of the lysines have key roles in stabilising the structure,[16,17,167] and in our view it may be a mistake to interpret observed rates simply in terms of the binding strength. Nevertheless, the methodology involving lysine modification has proved exceedingly useful for defining the general area of the surface of cytochrome c where reaction takes place, both for small reactants (Section VI) and for the larger, physiologically important reactants (Section VII).

The ionic strength dependence of reaction rates is sometimes interpreted to provide information about the reactive complexes, but this general approach is fraught with difficulties and is discussed separately in Section V.

F.   Summary of experimental methods

The experimental methods listed in Table 4 and described in this section characterise to a high order the reaction between a protein and a small redox reagent. As described, the overall procedure is at present applicable only to cytochrome c because other redox proteins have not been characterised to the same extent in terms of their NMR spectra or in terms of their chemically modified derivatives.

The kinetic methods undoubtedly probe only the reactive complex, but equilibrium methods probe both the reactive complex and the dead-end, nonreactive complexes. However, the equilibrium methods, especially the NMR method, are better at defining structures. The importance of non-reactive complexes can be gauged by the comparison of association constants determined by kinetic methods with those determined by equilibrium methods, and from the use of reagents of varying charge as kinetic inhibitors and as NMR probes.

V.   THE   PROBLEMS   OF   IONIC   STRENGTH
AND   COMPOSITION

The problems of ionic strength and composition are currently of major interest: variation of ionic strength, because it might yield mechanistic information,[2,168-172] and ionic composition, because of the possibility of specific ion effects.[148,156,173-176] The marked effects of variations in ionic strength and composition have already been alluded to in Figs. 7–9, 20, 21, and 23. In this section some of the consequences of specific ion effects are briefly considered and some of the approaches used to interpret the ionic strength dependence of reaction rates are described.

Ion binding has long been recognised as a problem in biochemical assays of cytochrome c activity,[35,177] but in general it has been disregarded as a problem in small-molecule studies. This is unfortunate because although Tris cacodylate, which does not bind to cytochrome c,[178] is used as a buffer in most biochemical work, chloride and phosphate are used in most small-molecule work, and both of these anions bind to cytochrome c at the front of the protein near the exposed heme edge (see Fig. 3).[13,35,36,151,161,179] One consequence of this is that the binding of reagents such as $[Fe(CN)_6]^{3-/4-}$ is strongly affected by the amount of chloride and phosphate present (Figs. 20, 21, and 23),[148,151,154,156] which means that the overall rates of reaction are also strongly affected. Figure 26 illustrates the latter point: k_{ox} for the oxidation of horse ferrocytochrome c by $[Fe(CN)_6]^{3-}$ varies by a factor of approximately 10 for the three sets of experiments.[180] Peterman and Morton[180] followed Margoliash's recommendation[35,178] and used Tris cacodylate as the supporting medium in many of their experiments, and other workers could with advantage follow this lead.

The variation of reaction rates as a function of ionic strength is well documented: with increasing ionic strength, the rate of reaction for like charged reactants increases while that of unlike charged reactants decreases. A quantitative description of this effect is important for two reasons.

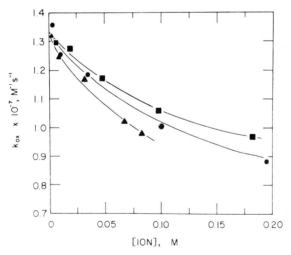

Fig. 26. Effect of chloride (●), phosphate (▲), and potassium (■) on the rate of oxidation of horse ferrocytochrome c by $[Fe(CN)_6]^{3-}$ at 24°C, pH 7.0, and constant ionic strength of 0.2 M (maintained with Tris cacodylate). The solid lines were calculated with k^i_{ox} (the oxidation rate for ion-bound cytochrome) of (●) 5.7 × 10⁶, (▲) 1 × 10⁵, and (■) 8.1 × 10⁶ M^{-1} s^{-1}. (From Peterman and Morton.[180])

First, in order to analyse the reaction energetics it is necessary to take account of the energy involved in formation of the precursor complex (see Section VI,F). This is not a simple problem, and Wherland and Gray[2,181] advocate an approach that assumes that only electrostatic interactions are important in complex formation. They suggest that the use of electrostatically corrected reaction rates in energetic calculations may circumvent the problem of assessing the energetics of complex formation. This approach is assessed in Section VI,F.

Second, the manner in which a reactant (either a small molecule or a protein) recognises the active site on the surface of a protein is a fundamental mechanistic problem, not only in electron transfer reactions but also in other biochemical processes. The ionic strength dependence of reaction rates is an important method of investigating this problem because, in principle, it allows the types of interactions that are most important to be specified. Thus, we can ask whether the net charge of the protein or the local charge at the interaction site principally determines the rate dependence on ionic strength, and whether the interactions are predominantly due to the protein monopoles (i.e., the individual charged groups on the protein) or to the overall dipole of the protein.

A full description of these topics is beyond the scope of the present article, because a clear consensus has not emerged from the workers involved in these studies, and only a brief account is presented here of the approaches currently in use for investigating the reactions of small-molecule redox reagents with proteins. An important point to note is that in the following discussion, and in Section VI,F, the implicit assumption is made that the ionic strength dependence of the rate reflects the ionic strength dependence of the association constant; the electron transfer rate constant is assumed to be independent of ionic strength. While this may be generally true, it has not been demonstrated for reactions of cytochrome c with small-molecule reagents; in the only relevant study of which we are aware, that of the reduction of cytochrome c by flavodoxin,[182] the electron transfer rate constant was slightly dependent on ionic strength.

The following three equations are the functions commonly used to describe ionic strength dependences:

$$\ln k_{obs} = \ln k_0 + 2Z_1Z_2\alpha\sqrt{I} \tag{20}$$

$$\ln k_{obs} = \ln k_0 + \frac{(2Z_1Z_2 + Z_2^2)\alpha\sqrt{I}}{1 + \kappa R_1} - \frac{Z_2^2\alpha\sqrt{I}}{1 + \kappa R_2} \tag{21}$$

$$\ln k_{obs} = \ln k_i - 3.576\left(\frac{e^{-\kappa R_1}}{1 + \kappa R_2} + \frac{e^{-\kappa R_2}}{1 + \kappa R_1}\right)\left(\frac{Z_1Z_2}{R_1 + R_2}\right) \tag{22}$$

where k_{obs}, k_0, and k_i are the observed rate constant at a given ionic strength I and the rate constant at zero ionic strength and infinite ionic strength, respectively; Z_1 and Z_2 are the charges on the reactants; R_1 and R_2 are the radii of the reactants; α is a constant (1.174 at 25°C); and $\kappa = 0.329I$ at 25°C. Equation (21) is the Brønsted–Debye–Hückel equation, which simplifies to Eq. (20) when κR_1, $\kappa R_2 \ll 1$. Equation (22) is the Wherland–Gray equation.

Gray and colleagues[2,109,163,181,183] carried out an extensive analysis of the applicability of these equations. Figure 27, taken from Cummins and Gray,[109] shows the ionic strength dependence of the rate of oxidation of horse cytochrome c by $[Ru(NH_3)_5(py)]^{3+}$ with the theoretical curves obtained from Eqs. (20)–(22), and it illustrates the kind of fit generally

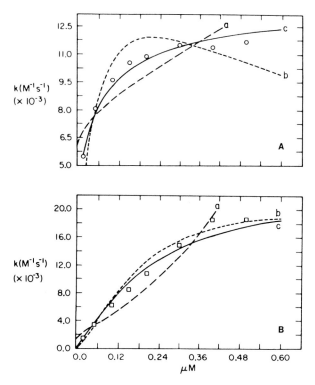

Fig. 27. Ionic strength dependence of the rate of oxidation of horse ferricytochrome c by $[Ru(NH_3)_5(py)]^{3+}$ at 25°C with (A) phosphate at pH 6.5 and (B) acetate at pH 5.3. The curves are theoretical fits using (a) Eq. (20), (b) Eq. (21), and (c) Eq. (22). (From Cummins and Gray.[109])

obtained for such data. Equation (20) does not fit the data, while Eqs. (21) and (22) both give a reasonable fit. The lack of fit for Eq. (20) is not surprising, since the assumption $\kappa R_1, \kappa R_2 \ll 1$ is not generally true for proteins. Moreover, Eqs. (20) and (21) are valid only at low ionic strength, although, as Koppenol[168] has pointed out, Eq. (22) is also strictly valid only at low ionic strength. In view of this low-ionic-strength restriction, it is surprising how well Eqs. (21) and (22) fit the data of Fig. 27, which cover a wide range of ionic strengths. These good fits result partially from the practice of treating the charge at the reaction site on cytochrome c as an adjustable parameter. Thus in Fig. 27, the charge of ferricytochrome c was taken to be (Fig. 27A) 6.5 and 1.8 for the curves of Eqs. (21) and (22), respectively; and (Fig. 27B) 11.1 and 6.2 for the curves of Eqs. (21) and (22), respectively. These values are significantly different from the charge $(+9)$ calculated from the amino acid sequence of horse ferricytochrome c[62] and from its pK_a values,[13] a difference that most probably results from a combination of specific ion effects (compare Fig. 27A with Fig. 27B) and inappropriate use of the equations at high ionic strength. An additional cause, and the one that is most important from a mechanistic viewpoint, is that the local charge at the protein reaction site, or a function of the local and net charge rather than the net charge itself, governs the ionic strength dependence of rate. Gray[2,109] made the reasonable point that Eqs. (21) and (22) are not so successful that they can be considered good models of the electrostatic interaction between proteins and small molecules, but, nevertheless, he favoured the view that the net charge on the protein is important. This viewpoint was also supported by Feinberg and Ryan[170] and Goldkorn and Schejter[184] from analyses of data similar to those of Fig. 27 for a variety of protein and reactant types. However, in energetic calculations Wherland and Gray generally use the computed charge from Eq. (22) rather than the net charge on the protein to obtain electrostatically corrected rate constants (see Section VI,F).

The recent work of Tam and Williams[185] and Chapman *et al.*[172] strongly supports the view that the charge at the active site determines the electrostatic work terms of the association reaction. These workers have shown that at $I = 0.1\ M$ the association constants for redox-inactive, spherically symmetric transition metal complexes binding to cytochromes, blue copper proteins, and Fe/S proteins are given by Bjerrum's equation when the protein charge is taken to be the charge at the binding site $(\pm 3\text{ or }4)$ rather than the net charge on the protein. Thus in order to establish precisely the effect of variable ionic strength, a thorough reexamination is needed of the overall rates of reaction for small-molecule reactions of cytochrome c at low ionic strength, with Tris cacodylate as the supporting buffer and in the absence of chloride and phosphate. At the same time, a more rigorous theoretical approach is required (see Refs. [168, 169, and 185]).

VI. NONPHYSIOLOGICAL REACTIONS OF CYTOCHROME *c*

A. Survey of reaction systems

The aim of the present article, as expressed in Section I, is to characterise the electron transfer function of cytochrome *c*; to do this we need to discuss well-defined reactions. Thus, although a wide range of small-molecule reactions of cytochrome *c* have been studied, not all of them will be considered in detail. Reactions with dithionite,[144,186,187] ascorbate (see Ref. [188], and references therein), catechols and quinols (see Refs. [189–191], and references therein), and free radicals such as the superoxide anion radical (see Refs. [162 and 192], and references therein) have been investigated, but, partially as a result of their complexity, which is due more to the small-molecule reagent than to the cytochrome, these reactions have not been investigated to the same extent as those involving transition metal complexes. Nevertheless, the energetics of some of the reactions involving organic reagents have proved to be amenable to analysis and therefore these reactions are considered later in this section.

The reactions we are most concerned with are those between cytochrome *c* and the reactants listed in Table 3, and these form the bulk of the discussion in this section. We also briefly describe two other reactions, both of which are potentially capable of providing detailed information about cytochrome *c* and both of which involve covalent labelling of the protein. These are the reactions with aquo Cr(II) and Ru(II) pentaammine complexes, and although neither of them has been thoroughly characterised, they are systems to note.

B. Covalently labelled cytochrome *c*

Hexaaquo Cr(II) was the first inorganic reagent to have its reaction site on cytochrome *c* investigated. On reduction of ferricytochrome *c* with Cr(II), the ferrocytochrome *c* produced contained a covalently bound Cr(III),[193–195] a consequence of the inner-sphere complexation of Cr(II) and inertness to aquation of Cr(III).[196] A unique 1 : 1 complex is formed. Two attempts have been made to identify the reaction site using different techniques, and these have produced different answers. Peptide mapping located Cr(III) bound to Asn 52 and Tyr 67,[197] a location which is consistent with a structure in which the bound Cr(III) is 6.8 Å away from the iron, 6.1 Å away from Met 80, and 5.4 Å away from part of the heme. This location is also consistent with the kinetic mechanism of Yandell *et al.*,[195] who showed that the rate-determining step in the Cr(II) reduction of ferricytochrome *c* was a breathing motion of the protein that allowed access to the Cr(II). However, this location was recently challenged by Petersen and Gupta,[198] who used NMR to deduce that the bound Cr(III) was 15–19 Å away from Met 80 and the heme. The discrepancy between these two structure studies[197,198] will be resolved only by further work. This covalent-labelling approach has recently been applied to blue copper proteins.[199,200]

An alternative approach involving covalently labelled cytochrome *c* has recently been developed by Gray and colleagues.[201,202] This approach makes use of chemically modified derivatives of horse cytochrome *c* with the electron transfer reagent bound to His 33. The parent complex is [Ru(NH$_3$)$_5$–his 33]–cytochrome *c*, and electron transfer, Ru(II)–Fe(III) → Ru(III)–Fe(II), takes place at a significant rate. The rationale behind this method is that the Ru–Fe separation may be obtained from the X-ray structure of horse cytochrome *c* so that the measured rates can be related directly to theoretical expectations. However, there are two complications. First, it is not clear that the Ru–Fe separation can be easily obtained. Although an X-ray structure of horse cytochrome *c* is available, it contains large errors in the positions of some side chains,[14] and although there are accurate structures of rice and tuna cytochromes *c* available,[16,17] neither of these proteins contains His 33 (Fig. 10). Also, it is known from NMR studies[203] that horse and tuna cytochromes *c* differ in the regions including residue 33, and that His 33 is not an immobile residue. Its side chain rapidly rotates,[203] but whether it does when the [Ru(NH$_3$)$_5$] moiety is bound is not known. Second, the ionisation of His 33 may influence the protein conformation around the heme edge. Burns and La Mar[204] have observed an unusual dynamic conformational process on horse cytochrome *c* that is influenced by a pK_a of 6.5–7.0, and His 33 is the only group known to have a pK_a in that region.[13] Nevertheless, despite the difficulty in accurately determining the structure, the study of covalently bound derivatives is important and promises to reveal some fundamental electron transfer properties of cytochrome *c*. A similar approach has recently been adopted by Hoffman and co-workers[205] with hemoglobin. These workers are studying hybrid hemoglobins containing two iron porphyrins and two zinc porphyrins with the metal–metal distances crystallographically defined.

C.   Reactions with transition metal complexes

Equilibrium constants, rate constants, and activation parameters for reactions of horse cytochrome *c* with transition metal complexes are given in

Tables 5 and 6, which include representative complexes of each of the main classes of reagents discussed in Section III. The best-characterised set of reactions are those involving the iron penta- and hexacyanides. This is partially due to historical development: the first study of the rate of reaction between a protein and a small-molecule inorganic redox reagent was the stopped-flow (SF) study carried out by Sutin and Christman[217] on the reaction of ferrocytochrome c with $[Fe(CN)_6]^{3-}$. This was followed by the work of Havsteen,[218] who studied the reaction of ferricytochrome c with $[Fe(CN_6)]^{4-}$ by a temperature-jump (TJ) method, and since this work there have been at least 30 papers dealing with various aspects of the reaction kinetics. Those papers relevant to the present article are listed in Table 6.

Table 6 is based on the reaction scheme proposed by Stellwagen and Shulman[147]:

$$\xrightarrow{\hspace{2cm} k_{red} \hspace{2cm}}$$

$$CIII + FII \underset{k_{-1}}{\overset{k_1}{\rightleftharpoons}} CIII:FII \underset{k_{-2}}{\overset{k_2}{\rightleftharpoons}} CII:FIII \underset{k_{-3}}{\overset{k_3}{\rightleftharpoons}} CII + FIII \qquad (23)$$

$$\xleftarrow{\hspace{2cm} k_{ox} \hspace{2cm}}$$

where CII and CIII are ferrocytochrome c and ferricytochrome c, respectively; FII and FIII are $[Fe(CN)_6]^{4-}$ and $[Fe(CN)_6]^{3-}$, respectively; and CIII:FII and CII:FIII are 1:1 adducts. $K_1 = k_1/k_{-1}$, $K_2 = k_2/k_{-2}$, $K_3 = k_3/k_{-3}$, and $K_{eq} = k_{ox}/k_{red}$.

The evidence in favour of this scheme is very strong, although the original work[147] and much of the subsequent supporting work[144,219,220] have been shown to contain errors (see Section IV,C). The main disagreement has been over the values of K_1 and K_3^{-1}, with values ranging from 15 to 10^4 M^{-1} having been reported (see Ref. [148], and references therein). Values of $<2 \times 10^2$ M^{-1} at $I = 0.12$ (NaCl) are now agreed upon by different techniques (Table 6).

The equilibrium dialysis results of Stellwagen and Cass[151] (Fig. 21 and Table 6) indicate that there are at least two binding sites for $[Fe(CN)_6]^{3-/4-}$ on horse cytochrome c, and this is supported by the NMR work of Eley et al.[154] and Williams et al.[221] (see Section VI,D); thus the scheme of reaction (23) may be an oversimplification. However, since all of the kinetic data given in Table 6 are derived from a scheme involving only one reaction site, additional kinetic information is needed to extend the scheme of reaction (23). Presumably, the reactions of Table 5 also proceed via a mechanism similar to that for $[Fe(CN)_6]^{3-/4-}$, and adducts are known to be formed between cytochrome c and some of the small-molecule reactants.[99,154,221] However, the kinetic competence of these adducts remains to be established, although given the proximity of the bound reactants to the heme (see Section VI,D) they are unlikely to be inactive. Nevertheless, there is no kinetic evidence to support the general application of reaction scheme (23).

TABLE 5

Overall rate constants and activation parameters for the oxidation and reduction of horse cytochrome c

Oxidant[a]	Reductant[a]	Conditions[b]	k_{ox} $(M^{-1}\,s^{-1})$	k_{red} $(M^{-1}\,s^{-1})$	$\Delta H^{\ddagger}$ $(kJ\,mol^{-1})$	$\Delta S^{\ddagger}$ $(J\,K^{-1}\,mol^{-1})$	Ref.
$[Co(ox)_3]^{3-}$	CII	pH 7.0; 25°C; $I = 0.5$ (P)	5.5				[206]
$[IrCl_6]^{2-}$	CII	pH 7.1; 25°C; $I = 0.1$ (NaCl)	3.7×10^9				[207]
CIII	$[Fe(edta)(H_2O)]^{2-}$	pH 7.0; 25°C; $I = 0.1$ (NaCl)		2.6×10^4	25.1	-75.2	[183]
CIII	$[Fe(edta)(H_2O)]^{2-}$	pH 7.2; 24°C; $I = 0.074$ (NaCl)		1.7×10^4			
CIII	$[Fe(dpta)]^{2-}$			9×10^3			[208]
CIII	$[Fe(edap)]^{2-}$			1.7×10^4			
CIII	$[Fe(dtpa)]^{3-}$	pH 7.2; 24°C; $I = 0.065$ (NaCl)		1.7×10^4			
CIII	$[Fe(hedta)(H_2O)]^{-}$			9×10^2			
CIII	$[Fe(nta)(H_2O)_2]^{-}$			1.6×10^2			
CIII	$[Ru(NH_3)_6]^{2+}$	pH 7.0; 25°C; $I = 0.1$ (Tris HCl)		3.8×10^4	12.0	-117.0	[209]

Complex		Conditions				Ref.
[Ru(NH$_3$)$_5$(py)]$^{3+}$	CII	pH 5.0; 25°C; I = 0.1 (A)	6 × 10^3		33.4	[109]
CIII	[Ru(NH$_3$)$_5$(bm)]$^{2+}$	pH 6.1; 25°C; I = 1.0 (NaCl)	4.7 × 10^5		−58.5	[108]
[Os(bipy)$_3$]$^{3+}$	CII	pH 7.1; 25°C; I = 0.1 (NaCl)	2.4 × 10^7			[207]
[Fe(phen)(CN)$_4$]$^-$	CII		4.4 × 10^8			
[Fe(phen)$_2$(CN)$_2$]$^+$	CII		8.5 × 10^8			
[Fe(bipy)(CN)$_4$]$^-$	CII	pH 7.0; 25°C; I = 0.1 (KCl)	1.6 × 10^8			[210]
[Fe(bipy)$_2$(CN)$_2$]$^+$	CII	pH 7.0; 25°C; I = 0.1 (KCl)	1.9 × 10^8			[210]
[Co(phen)$_3$]$^{3+}$	CII	pH 7.0; 25°C; I = 0.5 (P)	3.3 × 10^3			[206]
	CII	pH 7.0; 25°C; I = 0.1 (P)	1.5 × 10^3	47.2	−25.1	
[Co(5-Clphen)$_3$]$^{3+}$	CII		1.3 × 10^2	40.1	−66.9	
[Co(5,6-Me$_2$phen)$_3$]$^{3+}$	CII	pH 7.0; 25°C; I = 0.1 (P)	2.7 × 10^2	51.8	−25.1	[211]
[Co(4,7-Me$_2$phen)$_3$]$^{3+}$	CII		2.8 × 10	61.0	−12.5	
[Co(4,7-(PhSO$_3$)$_2$phen$_3$]$^{3-}$	CII		2.9 × 10^3	53.5	0.0	
[Fe(cp)$_2$]$^+$	CII	pH 7.0; 25°C; I = 0.5 (P)	6.2 × 10^6			[117]

[a] CII and CIII are ferro- and ferricytochrome c, respectively.

[b] I, Ionic strength (molar); P, phosphate; A, acetate.

49

TABLE 6. Rate constants, equilibrium constants, and activation parameters

Reactant	Method	Conditions	K_{eq}	$\dfrac{k_{ox}}{(M^{-1}\,s^{-1})}$
FII and FIII	SF	pH 7.2; 25°C; $I = 0.1$ (NaCl)	2.6×10^2	9.1×10^6
FII and FIII	SF	pH 7.0; 25°C; $I = 0.1$ (KCl)	2.35×10^2	8.0×10^6
FIII	SF	pH 7.2; 25°C; $I = 0.1$ (NaCl)		1.2×10^7
FIII	SF	pH 7.0; 24°C; $I = 0.2$ (NaCl)		7.1×10^6
FIII	SF	pH* 6.6; 24°C; $I = 0.2$ (NaCl)		3.1×10^6
FII	NMR	pH 7.4; 25°C; $I = 0.12$ (NaCl)		
FII	NMR	pH* 7.4; 25°C; $I = 0.12$ (NaCl)		
Binding	NMR	pH 7.3; 25°C; $I = 0.12$ (KCl)		
Binding	ED	pH 7.0; 25°C; 50 mM (pips)		
		pH 7.8; 25°C; 300 mM (NaCl)		
FII and FIII	TJ	pH 7.0; 22°C; $I = 0.18$ (Na$_2$SO$_4$)	3.1×10^2	8.1×10^6
FIII	PR	pH 7.0; 25°C; $I = 0.2$ (Na$_2$SO$_4$)		4.1×10^6
FII and FIII	TJ	pH 7.2; 25°C; $I \sim 0.1$ (P)	2.2×10^2	5.7×10^6
FII and FIII	TJ	pH* 7.2; 25°C; $I \sim 0.1$ (P)	1.2×10^2	3.1×10^6
FIII	CF	pH 7.0; 23°C; $I = 0.07$ (P)		2.5×10^7
FIII	TJ	pH* 7.0 (?)		
FII and FIII	TJ/PR	pH 7.1; 25°C; $I = 0.02$ (?)		
[Fe(CN)$_5$(4-NH$_2$py)]$^{3-}$	SF	⎫		
[Fe(CN)$_5$(imid)]$^{3-}$	SF	⎬ pH 7.2; 25°C; $I = 0.1$ (NaCl)		
[Fe(CN)$_5$(NCS)]$^{3-}$	SF	⎭		1.3×10^6
[Fe(CN)$_5$(NH$_3$)]$^{2-}$	SF	pH 8.0; 25°C; $I = 0.1$ (NaCl)		2.75×10^6
[Fe(CN)$_5$(PPh$_3$)]$^{2-}$	SF	⎫		3.0×10^7
[Fe(CN)$_5$(CNS)]$^{3-}$	SF	⎬ pH 7.0; 25°C; $I = 0.1$ (KCl)		1.0×10^7
[Fe(CN)$_5$(NH$_3$)]$^{2-}$	SF	⎪		2.5×10^6
[Fe(CN)$_5$(N$_3$)]$^{3-}$	SF	⎭		9.0×10^5

[a] The table is arranged around reaction scheme (23); the reactants are defined in the notes to Table 3 and reaction (23); the methods are SF, stopped-flow; ED, equilibrium dialysis; TJ, temperature jump; and PR, pulse radiolysis; pH and pH* are the relevant measurements in H$_2$O and ^{2}H$_2$O, respectively; I is the

for the reactions of horse cytochrome c with iron penta- and hexacyanides[a]

k_{red} $(M^{-1}s^{-1})$	K_1 (M^{-1})	k_2 (s^{-1})	k_{-2} (s^{-1})	K_3^{-1} (M^{-1})	$\Delta H^{\ddagger}$ (kJ mol^{-1})	$\Delta S^{\ddagger}$ (J K^{-1} mol^{-1})	Ref.
3.5×10^4	$<2 \times 10^2$			$<2 \times 10^2$			[141]
3.4×10^4					4.6	−96.1	[2, 210]
							[108]
							[180]
5.1×10^4	1.3×10^2	3.85×10^2					[148]
7.1×10^4	9×10	7.85×10^2					
			6.6×10^4	1.3×10^2			[148, 156]
	$\begin{cases} 9.6 \times 10^2 \\ 3.1 \times 10^2 \end{cases}$						
	$\begin{cases} 1 \times 10^2 \\ 1 \times 10^2 \end{cases}$						[151]
2.6×10^4							[54]
							[212]
2.6×10^4							[213]
2.6×10^4							
							[214]
		2×10^2	5×10^4				[215]
		3.3×10^2	4.6×10^4				[216]
6.7×10^5							
4.2×10^5							[141]
					5.0	−83.6	
					5.0	−92.0	[210]
					10.0	−87.8	
					12.1	−87.8	

ionic strength (molar); P is phosphate. Stasiw and Wilkins[79] report that reactions with $[Fe(CN)_5(CNS)]^{4-}$ are complicated by the ligand and exchange NCS→ and SCN→ in the corresponding Fe(III) complex. The formulations given above for the Fe(III) form of this complex are those from the original papers.

TABLE 7

Equilibrium and rate constants for reactions of *Candida* cytochrome c^a

Reactant	Method	Conditions	k_{ox}	k_{red}	K_1	k_2	Ref.
FIII	SF	pH 7.2; 25°C $I = 0.1$ (NaCl)	2.1×10^7				[108]
FII	NMR	pH* 7.4; 25°C $I = 0.12$ (NaCl)		2.9×10^5	2.85×10^2	7.7×10^2	[148]
$[Ru(NH_3)_5(bm)]^{2+}$	SF	pH 6.1; 25°C; $I = 1.0$ (NaCl)		1×10^6			[108]
$[Co(phen)_3]^{3+}$	SF	pH 7.2; 25°C; $I = 0.1$ (NaCl)	2.7×10^3				[108]

^{a} See footnote to Table 6 for an explanation of abbreviations and the arrangement of the table.

Before discussing the energetics of the reactions of Tables 5 and 6 (Section VI,F), we describe the structures of the reactant:cytochrome adducts (Section VI,D). Characterisation of these structures has been considerably aided by studies of mitochondrial cytochromes that are related to horse cytochrome c but that have different amino acid sequences. The most important of these are the cytochromes from tuna, rice, and *Candida* (Fig. 10). We do not expect large kinetic differences between these cytochromes and horse cytochrome c, an expectation borne out by experiments for tuna and *Candida* cytochromes c.[108,145,148,222] *Candida* cytochrome c, which is the most divergent of the four shown in Fig. 10, is the one that, after horse cytochrome c, has been most thoroughly investigated. Table 7 summarises relevant kinetic data for this protein, and a comparison of these data with those of Tables 5 and 6 emphasises the similarity between the different cytochromes. The small kinetic differences are discussed in Section VI,F.

D. Structures of precursor and product complexes

The structures of complexes of cytochrome c with $[Cr(CN)_6]^{3-}$, $[Cr(ox)_3]^{3-}$, $[Fe(edta)(H_2O)]^-$, $[Fe(dtpa)]^{2-}$, and $[Fe(nta)(H_2O)(OH)]^-$ have been defined by the NMR method as described in Section IV,B, and the complexes of cytochrome c with $[Fe(CN)_6]^{3-}$, $[Fe(CN)_5(4-NH_2py)]^{2-}$, $[Fe(CN)_5(imid)]^{2-}$, and $[Co(phen)_3]^{3+}$ have been studied by kinetic investigations of chemically modified derivatives of horse cytochrome c as described in Section IV,E. There is good agreement between the two methods on the structure of the complex common to both studies, that of ferrocytochrome c with $[M(CN)_6]^{3-}$, and it is this complex we consider first.

1.   Complexes with $[Fe(CN)_6]^{3-}$ and $[Co(ox)_3]^{3-}$

$[Cr(CN)_6]^{3-}$ and $[Cr(ox)_3]^{3-}$, NMR relaxation probes that are analogous to $[Fe(CN)_6]^{3-}$ and $[Co(ox)_3]^{3-}$, each bind to the same regions of cytochrome *c* with about the same binding strength.[154] There is one binding region, however, that is different for the two reagents, but this region is probably not a kinetically important region, as we discuss later.

As described in Section IV,B, we need to determine which NMR resonances of cytochrome *c* are affected by the relaxation probe and which are unaffected. The former come from groups relatively close to the bound reagent, while the latter come from groups relatively far away. The obvious starting place for such a structure definition is the well-resolved heme resonances of ferricyto-chrome *c* (Fig. 14). Two studies of the effect of $[Cr(CN)_6]^{3-}$ upon these resonances have been reported.[154,223] Both of these studies showed that the resonance of heme methyl 3 is specifically affected while those of heme methyl 5 and 8 are unaffected. Unfortunately, these data are insufficient

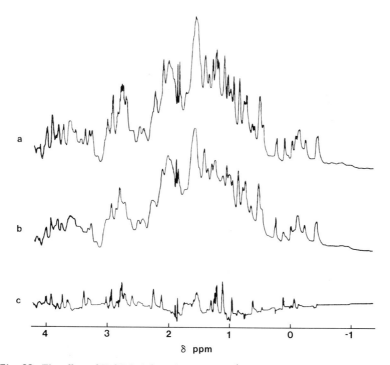

Fig. 28. The effect of $K_3[Cr(ox)_3]$ on the 470-MHz ^1H NMR spectrum of horse ferricyto-chrome *c* in ^2H_2O at pH* 7.1 and 25°C. (a) Cytochrome alone; (b) in the presence of 41 mol $[Cr(ox)_3]^{3-}$ per 100 mol cytochrome; (c) the difference spectrum (a) minus (b). (From Eley *et al.*[154])

to define the binding sites, and the suggested locations given by Hopfield and Ugurbil[223] should be regarded with caution, especially as their ionic strength was uncontrolled, which means that their association constants and site occupancies (essential data for the site location procedure) are uncertain. The problem of lack of data can be partially overcome by using the assigned amino acid resonances. Difference spectroscopy is then required for the relaxation effects to be observed.[76,154] An example of this method is shown in Fig. 28, which illustrates the aliphatic regions of spectra of horse ferricytochrome c in the absence and presence of $[Cr(ox)_3]^{3-}$, and the resultant paramagnetic difference spectrum (PDS). The subtraction removes all the resonances unaffected by the probe and leaves a PDS containing those resonances that have been specifically affected.

Most of the resonances in the PDS of Fig. 28 have been assigned to specific nuclei; these are shown in Fig. 29, which compares both the aromatic and aliphatic regions of the PDS for $[Cr(CN)_6]^{3-}$ and $[Cr(ox)_3]^{3-}$ binding to horse ferricytochrome c. It is immediately clear from this comparison that the two reagents behave almost the same toward cytochrome c. The difference between them is reflected by resonances of Phe 36 and Ile 95, the former of which is more strongly affected by $[Cr(CN)_6]^{3-}$ while the latter is more strongly affected by $[Cr(ox)_3]^{3-}$. These differences result from a small dif-

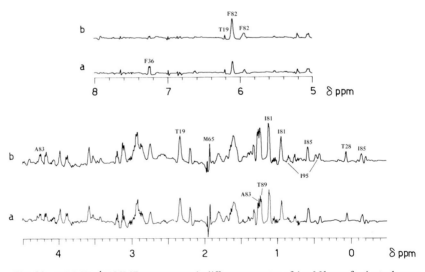

Fig. 29. 470-MHz ^{1}H NMR paramagnetic difference spectra of 4 mM horse ferricytochrome c in ^2H_2O at pH* 7.5 and 25°C, corresponding to the addition of (a) 18 mol $[Cr(CN)_6]^{3-}$ per 100 mol cytochrome and (b) 16 mol $[Cr(ox)_3]^{3-}$ per 100 mol cytochrome. Resonance labels: F, phenylalanine; T, threonine; I, isoleucine; A, alanine; and M, methionine. The numbers refer to residue positions (see Fig. 10). (From Eley.[224])

ference in a binding site at the back of the molecule (according to the orientation of Fig. 3) that is ≥ 20 Å away from the heme. All the remaining assigned resonances of Fig. 29 come from binding sites on the top or front of cytochrome c, two of which are relatively close to the heme. The titration of ferrocytochrome c with $[Cr(CN)_6]^{3-}$ and $[Cr(ox)_3]^{3-}$ defines the precursor complexes, and the results of NMR titrations were entirely compatible with those of Fig. 29.[154,224]

Similar NMR experiments with tuna cytochrome c have been done.[224] Using the X-ray structure of tuna cytochrome c[16,225] as a guide to the solution structure, we have identified the top and front-face binding sites shown in Fig. 30.[99,224] Together with the back site, the sites of Fig. 30 account for all the perturbations of assigned NMR resonances. The numbering of the sites in Fig. 30 was chosen to correspond to the numbering adopted by Osheroff *et al.*[174] for the location of phosphate binding sites: phosphate binds at sites 1 and 2.

Site 1 is positioned at the top left of cytochrome c and includes residues 65 and 89. Lysines 5, 86, 87, and 88 may provide the binding groups. Site 2 lies to the right of the heme on the front right edge and includes the amino acids Val 11, Ala 15, and Thr 19. Lysines 7, 25, and 27 may provide the binding groups. Site 3 lies on the front face to the left of the heme crevice. It is binding at this site that is predominantly responsible for the perturbation of the resonance of heme methyl 3. Resonances of the amino acids Ile 81, Phe 82, and Ala 83 are affected by binding at this site. Lysines 13, 72, and 86 may provide the binding groups. Resonances of Val 28 are affected by binding at both sites 2 and 3 and resonances of Ile 85 by binding at both sites 1 and 3. Further description of these binding sites, including the important reagent–heme distances and binding strengths, is deferred until Section VI, D, 4, when the general binding properties of a highly charged protein surface are discussed. However, it can be stated now that sites 1 and 3 have a higher affinity for simple substitution-inert anions than does site 2.

Kinetic studies with four different chemically modified derivatives of horse cytochrome c, the carboxydinitrophenyl, trifluoroacetyl, (trifluoro-methyl)phenylcarbamyl, and trinitrophenyl (CDNP, TFA, TFC, and TNP, respectively) derivatives (Fig. 25), have led to the definition of the kinetically competent $[Fe(CN)_6]^{3-}$ binding sites.[157–159] An example of the variation of rate observed for different derivatives is shown in Fig. 24, and in Fig. 31 the kinetic results are summarised on a ribbon diagram of cytochrome c together with the NMR results. Although there are small differences among the various kinetic studies (see Section VI,D,4), they all point to site 3 being the kinetically most important site, with site 2 being of lesser importance. Kinetically, site 1 is relatively unimportant. The major importance of site 3 and unimportance of site 1 were also demonstrated by kinetic experiments

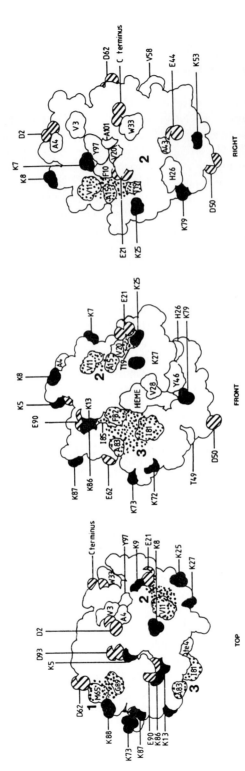

Fig. 30. Representation of the three major anion binding sites on cytochrome c. The space-filling diagrams for tuna cytochrome c illustrate the positions of negatively charged groups (striped) and positively charged groups (solid). Important residues with assigned NMR resonances are shown in outline. Top, front and right refer to views of the conventional orientation of cytochrome c (see Fig. 3). The stippled residues are those affected by low concentrations of probes of the type $[Cr(CN)_6]^{3-}$ and $[Cr(ox)_3]^{3-}$, and the numbering scheme refers to binding sites defined in the text. The anionic Cr(III) complexes bind preferentially to sites 1 and 3. $[Fe(edta)(H_2O)]^-$ binds preferentially at the lower end of site 2, close to the acidic residue Glu 21, where it also affects the resonances of Val 20 and His 26.

with $[Fe(CN)_5(4\text{-}NH_2py)]^{3-}$ and $[Fe(CN)_5(imid)^{3-}$,[159] but the importance of site 2 was not investigated with these reactants.

Lysine resonances are not particularly easy to distinguish by NMR, especially when there are 19 lysines, as there are for horse cytochrome c (Fig. 10). However, when the lysine is converted into trimethyllysine (TML) or substituted by arginine, resonances can often be distinguished, in the former case by ${}^1H$ NMR and in the latter case by ${}^{13}C$ NMR. Three of the lysines of horse and tuna cytochromes c are so replaced: Lys 13 becomes Arg 13 in *Candida* cytochrome c, Lys 72 becomes TML 72 in rice and *Candida* cytochromes c, and Lys 86 becomes TML 86 in rice cytochrome c (Fig. 10). These substitutions allow us to determine the importance of these residues by NMR, even though the NMR spectra of rice and *Candida* cytochromes c are not as thoroughly characterised as are those of horse and tuna cytochromes c.[226]

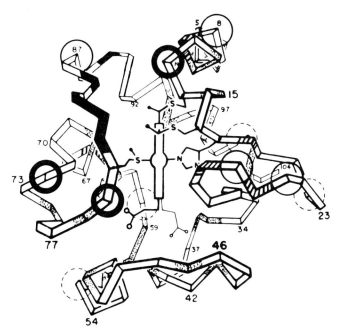

Fig. 31. Ribbon diagram of the polypeptide chain folding of horse cytochrome c, illustrating a comparison of the sites for metal hexacyanides as determined by NMR and kinetic investigations of chemically modified cytochromes. Residues for which NMR assignments are known are indicated by darkened, striped, and stippled ribbon regions signifying respectively, resonances that are strongly, weakly, and unaffected by $[Cr(CN)_6]^{3-}$. Regions encircled by heavy, light, and dashed lines illustrate the approximate positions of some of the lysine residues that have been shown by studies of chemically modified cytochrome c to have, respectively, either a large, weak, or no effect upon the rates of reaction of cytochrome c with $[Fe(CN)_6]^{3-/4-}$. (From Williams *et al.*[221])

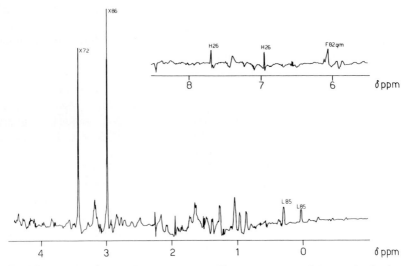

Fig. 32. 470-MHz ^{1}H NMR paramagnetic difference spectrum of 2 mM rice ferricyto-chrome c in ^{2}H$_2$O at pH* 7.0 and 25°C, corresponding to 14 mol [Cr(ox)$_3$]$^{3-}$ per 100 mol cytochrome. (From Eley.[224])

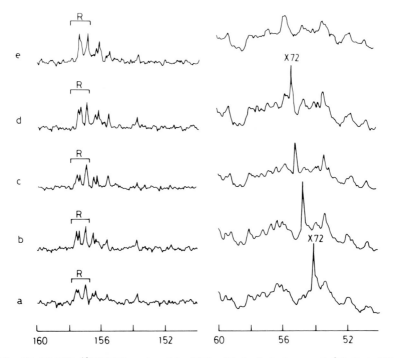

Fig. 33. 75-MHz ^{13}C NMR spectra of 6 mM *Candida* ferricytochrome c in ^{2}H$_2$O at pH* 7.0 and 25°C. X72 is the methyl resonance of TML72 and R are the quarternary carbon resonances of the four arginine residues (13, 38, 54, and 91). (a) Cytochrome alone; (b)–(d) with 200, 700, and 1700 mol [Fe(CN)$_6$]$^{3-}$ per 100 mol cytochrome; (e) with 1700 mol [Fe(CN)$_6$]$^{3-}$ and 200 mol [Cr(CN)$_6$]$^{3-}$ per 100 mol cytochrome. (From Eley.[224])

Figure 32 shows the PDS for $[Cr(ox)_3]^{3-}$ binding to rice ferricytochrome *c*.[224] Resonances that are indicative of each of the sites 1–3 are present, showing that all of the sites exist on rice cytochrome *c*. However, there appears to be a difference in the exact location of site 2, since resonances of His 26 are strongly affected (these are only weakly affected in horse and tuna cytochromes *c*). The important information to be gained from Fig. 32 is that resonances of both TML 72 and TML 86 are strongly affected and, from a titration of the paramagnetic effect, it is clear that they are affected by sites with different binding strengths: the TML 86 site is the stronger.

The ^{1}H NMR spectra of Fig. 32 confirm that TML 72 of *Candida* ferricytochrome *c* binds $[Fe(CN)_6]^{3-}$, and a similar result is shown by the ^{13}C NMR spectra of Fig. 33.[224] However, Arg 13 does not bind $[Fe(CN)_6]^{3-}$, contrary to the expectation based on the chemical modification experiment (Figs. 24 and 31). This is not due to *Candida* cytochrome *c* lacking a binding site (see Section VI,D,4).

2. Complexes with $[Fe(edta)(H_2O)]^-$, $[Fe(dtpa)]^{2-}$ and $[Fe(nta)(H_2O)(OH)]^-$

Complexes of cytochrome *c* with $[Fe(edta)(H_2O)]^-$, $[Fe(dtpa)]^{2-}$, and $[Fe(nta)(H_2O)(OH)]^-$ have been characterised by ^{1}H NMR experiments similar to those described in the previous section. In addition to being redox reagents, all of these Fe(III) complexes are NMR relaxation probes (Tables 3 and 5).

The PDS for the addition of $[Fe(edta)(H_2O)]^-$ to tuna ferricytochrome *c* shown in Fig. 34 contains only a few resonances, but at higher probe concentrations other resonances are perturbed.[99,221] These perturbations show that the three regions of cytochrome *c* that bind the anionic Cr(III) complexes, sites 1–3 (Fig. 30), also bind $[Fe(edta)(H_2O)]^-$, but a comparison of Fig. 34 with Fig. 29 shows that the relative binding strengths of these sites are modified. Thus the relative affinity of site 2, which includes Ala 15, Thr 19, and Val 20, toward $[Fe(edta)(H_2O)]^-$ is increased relative to the affinity of sites 1 and 2 compared with the binding of $[Cr(CN)_6]^{3-}$ and $[Cr(ox)_3]^{3-}$. Analogous experiments with tuna ferrocytochrome *c* have produced comparable results.[99] In these experiments, the addition of $[Fe(edta)(H_2O)]^-$ probably resulted in the formation of a fraction of $[Fe(edta)(H_2O)]^{2-}$, since excess ascorbic acid was present to maintain the cytochrome in its fully reduced state.

Figure 30 illustrates the location of the $[Fe(edta)(H_2O)]^-$ binding sites on the surface of cytochrome *c*. A change in the relative broadening of certain resonances indicates that site 2 binds $[Fe(edta)(H_2O)]^-$ at a slightly different location to its binding site for $[Cr(ox)_3]^{3-}$.

The decrease in binding strength at sites 1 and 3 is consistent with the decrease in charge on going from $[Cr(ox)_3]^{3-}$ to $[Fe(edta)(H_2O)]^-$, but this simple electrostatic effect is overshadowed by other effects at site 2. Inspection of Fig. 30 suggests a reason for the relative increase in the binding strength of site 2. The presence of Glu 21 in the centre of this site indicates that the carboxylate group of this residue may be able to interact with the partially exposed Fe(III) of $[Fe(edta)(H_2O)]^-$. Presumably the interaction is accompanied by the expulsion of an existing ligand on $[Fe(edta)(H_2O)]^-$ so that the sevenfold coordination is maintained. Thus this appears to be an

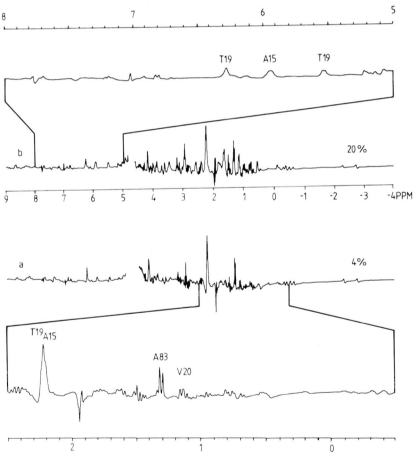

Fig. 34. 300-MHz ^{1}H NMR paramagnetic difference spectra of 4 mM tuna ferricytochrome *c* in ^{2}H$_2$O at pH* 5.5 and 27°C, corresponding to 4 mol $[Fe(edta)(H_2O)]^-$ and 20 mol $[Fe(edta)$-$(H_2O)]^-$ per 100 mol cytochrome. (From Williams.[99])

example of the conformational heterogeneity discussed in Section III, and as partially illustrated by Fig. 11, for edta complexes.

NMR experiments with $[Fe(dtpa)]^{2-}$ and $[Fe(nta)(H_2O)(OH)]^-$ binding to tuna ferricytochrome c yielded results similar to the experiments described above, but they did not show the strong binding to site 2 that is characteristic of $[Fe(edta)(H_2O)]^-$. Instead, these complexes demonstrated behavior intermediate between the simple, octahedral Cr(III) complexes and the asymmetric $[Fe(edta)(H_2O)]^{3-}$, and they showed approximately the same affinity for all three binding sites. Compared with $[Cr(ox)_3]^{3-}$ this results from a lower affinity for sites 1 and 3, due to the reduction in the total charge, and a slightly increased affinity for site 2. The different response of these Fe(III) complexes at site 2 compared to $[Fe(edta)(H_2O)]^-$, and their different structures (Fig. 12), support the view that a caboxylate ligand to the Fe(III) of $[Fe(edta)(H_2O)]^-$ is displaced on binding.[99] This displacement is facilitated by the strain in the coordinated edta ligand which is absent in the coordinated nta and dtpa ligands.

3.   Complex with $[Co(phen)_3]^{3+}$

The precursor complex ferrocytochrome $c:[Co(phen)_3]^{3+}$ has been investigated by the kinetic study of singly modified TNP and CDNP derivatives of horse cytochrome c (Fig. 25).[158,159] Instead of the modifications being accompanied by a decrease in rate, as with the $[Fe(CN)_6]^{3-}$ reactions (Fig. 24), they were accompanied by an increase in rate, which is consistent with the respective charges of the inorganic reactants and the lysine-modified protein in the activated complex. Butler *et al.*[159] found that the order of increasing bimolecular rate constants for the $[Co(phen)_3]^{3+}$ reactions was CDNP-27 > -72 > -13 > -25 > TNP-13 = CDNP-7 > -87 > -86 = -60 > native cytochrome c, where the numbers refer to the modified lysine (see Fig. 37 for a diagrammatic representation of these data).

These data indicate that a reaction site involving Lys 27 is the most important site, with a site involving Lys 13 and Lys 72 of lesser importance. Lys 86 and Lys 87 are relatively unimportant. This pattern fits in surprisingly well with the anionic reaction sites previously described (Fig. 30). The apparent preference of $[Co(phen)_3]^{3+}$ for a region similar to anion site 2 may, as with $[Fe(edta)(H_2O)]^-$, be associated with the presence of Glu 21. After the introduction of a negative charge onto Lys 25 or Lys 27, there will be two negatively charged groups at site 2, and these should produce a greater electrostatic attraction for $[Co(phen)_3]^{3+}$ than will site 3, which contains only one negative charge on Lys 13 or Lys 72. Also, there are more lysines on the left side of the heme than on the right side (Fig. 30), which causes greater electrostatic repulsion for $[Co(phen)_3]^{3+}$ interacting at sites 1 and 3 compared to its interaction at site 2.

The interaction of positively charged reagents at the front face of cytochrome c cannot easily be studied by NMR because the lifetime of the reactant complex is very short compared to the molecular correlation time.[138] Thus, although $[Cr(NH_3)_6]^{3+}$ and $[Cr(en)_3]^{3+}$ have been observed to bind to cytochrome c by NMR (M. N. Robinson, G. R. Moore, and G. Williams, unpublished data), their binding sites are relatively far from the heme and are unlikely to be kinetically important. However, it may be possible to observe significant relaxation effects caused by $[Cr(phen)_3]^{3+}$ in NMR spectra of CDNP-modified cytochromes, and this would test the suggestion that Glu 21 is involved in the binding.

4. The binding surface on cytochrome c

In the following discussion some general aspects of the binding of ions to a charged protein surface are considered and the structural studies of the reactive complexes are summarised. Physiological implications of this discussion are considered in Section VII.

Ion binding occurs at the surface of cytochrome c, and therefore it is important to know what the structure of the surface looks like. However, because of the dynamic properties of the surface, it is far harder to define precisely the surface structure of a protein than it is to define its internal structure. Indeed, since many of the surface groups possess a number of conformations that are almost energetically equivalent, it is a gross simplification to describe the surface as though there were a unique structure. With cytochrome c, most attention has been focused on the lysines, all of which are located on the surface, because it is these groups that are responsible for binding the physiological oxidoreductases (see Section VII).

The lysines of cytochrome c are distributed in a highly asymmetric manner[41,227,228]; the majority are clustered on the front face near the heme (Fig. 30). There are few carboxylates on the front face. Moore and Pettigrew[13] have described the binding surface of cytochrome c in detail, and have pointed out that only 7 of the 16 lysines of tuna cytochrome c are crystallographically well defined.[16] Of the remaining 9 lysines, some have blurred images in the electron density maps and some have different positions in what should be identical X-ray structures. Four of the 7 well-defined lysines are involved in interactions with other residues that presumably reduce their conformational flexibility. The surface carboxylates are generally well defined, probably because their side chains are shorter than that of lysine. Thus, in summary, Asp 2, 50, and 93, Glu 21, 66, 69, and 90, and

Lys 13, 27, 53, 55, 79, 86, and 99 are well defined, and the C-terminal carboxyl, Asp 62, Glu 44, and Lys 5, 7, 8, 25, 39, 72, 73, 87, and 88 are poorly defined. Both TML residues of rice cytochrome c are well defined,[17] and are in slightly different positions than those of their unmethylated counterparts in tuna cytochrome c.

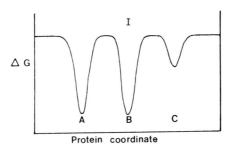

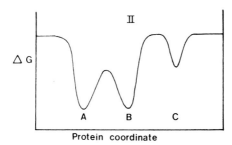

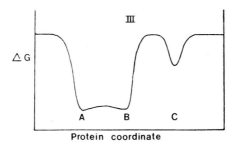

Fig. 35. Three hypothetical examples of ion binding to a protein. ΔG is the binding energy; the protein coordinate represents the location of the binding sites on the protein. The stoichiometry is three ions/protein (see text for further discussion).

Bearing in mind the conformational flexibility of lysine, and the fact that its side chain from α-C to ε-N is about 10 Å long when fully extended, we can ask whether there are discrete anion binding sites on the surface of cytochrome c, as is suggested by the representation of NMR results given in Fig. 30. Figure 35 is a diagrammatic representation of the problem. There are three binding sites on the front face of the protein, two of about equal affinity and one of lesser affinity. Figure 35 represents the situations when all three binding sites are distinct (case I), and when one binding site is distinct but the other two sites are either indistinct (case III) or are only partially distinct (case II). In case III the two sites have almost completely merged and it is incorrect to describe them as separate sites, but for the sake of simplicity we shall do so.

In case I, for an ion to move between sites A and B it would have to first dissociate from the protein, but in the other cases it need not fully dissociate from the protein in order to move between sites, i.e., the ion could migrate on the protein surface. This might be facilitated by, for example, the two sites sharing a binding group. Which case best describes the binding of anions to cytochrome c must be determined. At present we do not have sufficient data to provide a complete answer to this question, but the most probable situation is as follows. Site 2, on the right side of the heme (Fig. 30), resembles site C in Fig. 35, while sites 1 and 3 (Fig. 30) resemble sites A and B in Fig. 35, case I or II. Thus there may be some movement of bound anions between sites 1 and 3 but not to the extent indicated by Fig. 35, case III.

There is a considerable amount of data relating to the binding of [Fe-$(CN)_6]^{3-/4-}$ to cytochrome c that should be considered alongside the NMR data. The most significant data indicate that there are only two relatively strong binding sites on horse cytochrome c (Section IV,D and Table 6). This is not inconsistent with the structural model proposed in Fig. 30, if the binding site near Phe 36 and Ile 95 identified by NMR (Section VI,D,1) is a weak site that is not detected by the equilibrium methods described in Section IV,D, and if sites 1 and 3 are negatively cooperative such that only one of them can be occupied on a given cytochrome c molecule and that they have about equal probabilities of being occupied. Such negative cooperativity could result from electrostatic repulsion between anions bound in sites 1 and 3 or from the two sites sharing a lysine that acts as a binding group in both sites.

It is important to obtain the distances between the bound reactants and the heme, but in view of the foregoing discussion this clearly requires a number of assumptions. Also, there is the spectroscopic problem of not having NMR pseudocontact shift probes that are analogues of the relaxation probes. Nevertheless, we have carried out a preliminary structure determination for the binding of $[Cr(CN)_6]^{3-}$ and $[Cr(ox)_3]^{3-}$ at site 3 of tuna ferricytochrome c.[224] To do this we assume the site can be described as in Fig. 35, case I. The family of structures obtained for the complex are illustrated in Fig. 36.

Figure 36 represents the solutions to a computer search for the structure of the complex as a series of contour diagrams of the distance between the centre of $[Cr(CN)_6]^{3-}$ and the methyl group of Ala 83, which is the closest group to the bound anion in the complex. The X-ray structure of tuna ferricytochrome c[16] was used for this computer search. The most probable position of the centre of the bound anion is at the coordinates $x = 6.1\ (\pm 0.3)$, $y = 11.1\ (\pm 0.3)$, and $z = 3.1\ (\pm 0.7)$, and from this position the Cr–Fe distance is 20 ± 1 Å. The imprecise definition of the binding site may be an accurate representation of the location of the bound but mobile anion, or it could be a result of errors in the analysis. This will become clearer as the structure is refined, but there are two features of this model that are unlikely to change significantly upon refinement and that are important to note.

First, the distance between the bound anion and the protein surface is quite large. After taking into account the radius of the anion, the closest approach is 8–9 Å. At first sight this appears surprising: why should the anion be restrained from approaching closer to the surface? Part of the answer may be that the approach of the anion to the surface induces an electrostatic image force. Such an effect was discussed by Perutz[229] in connection with the analogous problem of the approach of diffusible electrolytes to a protein, and he concluded that ions cannot penetrate a layer of water about one molecule thick on the surface of the protein. This accounts for 3–4 Å of the distance between the anion and cytochrome c, and solvation of the anion may also contribute to the large separation. The large anion–protein separation is consistent with the low steric requirements of the association reaction shown by NMR studies of the binding of Cr(III) complexes of substituted malonates, which have approximate maximal radii of 3–8 Å. These complexes bind to sites 1, 2, and 3 with approximately the same strength as $[Cr(ox)_3]^{3-}$.[99]

Second, the distances between the anion (taking into account its radius) and the ε-N of all the lysines were determined, and all were found to be rather long. The ε-N positions were taken from the X-ray coordinates, and even though some of them are not well defined, coordinates for each have been published.[16] The closest lysines (with their distances) are Lys 13 (~ 12.0 Å), Lys 72 (16.0 Å), Lys 86 (~ 13.5 Å), and Lys 87 (16.5 Å). These measurements are encouraging in that lysines 13 and 72 are considered to be kinetically the most important,[157–159] but the actual distances are far too large for them to form a strong binding site. However, we have already noted that the positions of some of the lysinces are ill defined and Lys 72 falls into this category, although Lys 13 is well defined and, moreover, forms a salt bridge

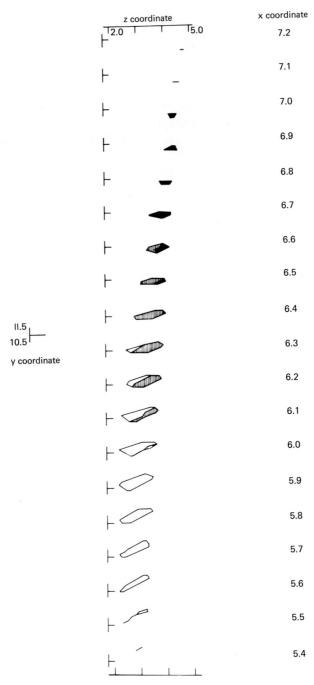

Fig. 36. Solutions of computer search for binding at site 3 of tuna ferricytochrome *c*. The solutions are presented as a series of contour diagrams referenced to the protein X-ray coordinates. Ala 83 CH_3 is the protein group closest to the bound ion; the contour distinctions refer to the distance from this group: (black) ≤ 11.0 Å; (striped) ≤ 11.5 Å; (clear) ≤ 12.0 Å. (From Eley.[224])

with Glu 90 in both the solid and the solution states[16,167] (see Fig. 30). Thus, if this lysine is an anion ligand, the salt bridge must be broken. The ^{13}C NMR spectra of *Candida* cytochrome *c* (Fig. 33) are relevant to the problem of the role of residue 13 because they show that Arg 13, which is involved in a salt bridge with Asp 90, does not bind $[Cr(CN)_6]^{3-}$ or $[Fe(CN)_6]^{3-}$. However, the geometry of site 3 on this protein is slightly different than that of site 3 on horse and tuna cytochromes *c*,[224] which is probably a reflection of the trimethylation of Lys 72 and the substitution of Val 11 (horse and tuna) by Lys 11 (*Candida*), and it is not clear that this is a satisfactory comparison. Thus, in order to clarify the role of Lys 13, it is necessary to determine both K_{ass} and k_{et} for the reactions between the modified cytochrome and $[Fe(CN)_6]^{3-/4-}$.

The structures of the complexes with anions bound at sites 1 and 2 have not been defined by NMR in the same manner as has the site 3 complex. This is because there are fewer groups in and around binding site 1 with assigned resonances, and because of uncertainty over the occupancy of site 2. Nevertheless, because of their importance in considerations of electron transfer mechanisms, we have estimated the range of minimum distances between anions bound at these sites and the heme.[224] These distances are summarised in Table 8 along with the corresponding measurements for binding at site 3. Detailed investigations of the binding of $[Fe(edta)(H_2O)]^-$ and its analogues should lead to a clearer definition of binding-site 2. A quantitative analysis of such binding has not yet been carried out.

TABLE 8

Distance between bound Cr(III) complexes and the heme
of tuna ferricytochrome c^a

Site	Anion–Fe(III) distance (Å)	Anion–nearest heme proton distance (Å)
1	15–21	13–19 (methyl 1)
2	13–21	10–18 (meso β)
3	17–19	11–13 (methyl 3 and thioether 4 methyl)

^{a} The distances take account of the anion radius (see Table 3). The distances for site 3 were obtained from the structures depicted in Fig. 36. Those for sites 1 and 2 were obtained as follows.[224] The limits of sites 1 and 2 were determined by considering which resonances were completely unaffected by anion binding; this resulted in the definition of two regions of space that encompassed the binding sites. The surfaces of those regions of space closest to the protein were taken to define the closest approach of the bound anion, and the minimum and maximum distances were measured for anions located in these parts of space. Thus the range of values given for sites 1 and 2 represent the range of *minimum* heme–anion separations. There are other equally plausible possibilities, with the anion not placed at the position of closest approach to the protein, and these all give larger heme–anion separations.

The kinetic studies of chemically modified cytochromes are summarised in Fig. 37, which records the results of two different studies of the $[Fe(CN)_6]^{3-}$ reaction: a study of the $[Co(phen)_3]^{3+}$ reaction and a study of the reduction of ferricytochrome c by O_2^-.[157–159,192] The latter reaction was investigated with CDNP-modified cytochrome c. The representation of the reactivity patterns in Fig. 37 is similar to that used by Rieder and Bosshard[230] in their differential chemical modification study of cytochrome c complexed to its oxidoreductases, and it is different from the representations customarily used in kinetic studies (see, for example, Refs. [159 and 228]). We prefer the representation of Fig. 37, both because it presents the results without modifying them according to a particular concept of the interaction and because reactivity contours are not used. Reactivity contours occasionally resemble energy diagrams and they are then misleading.

Although we do not draw quantitative structural conclusions from Fig. 37, there is one structural feature we wish to emphasize. The reactivity pattern for negatively charged reagents is not centered at a common point on the protein surface. As mentioned in Section V, there are two general schools of thought concerning the electrostatic interactions between cytochrome c and its electron transfer partners. One of these is that local electrostatic effects at the reaction site are most important, and the other is that the overall charge and its distribution on the protein is most important. In the context of the latter suggestion, the dipole moment of cytochrome c has been widely discussed,[168,169,228] and Koppenol et al.[228,231] have calculated that the positive end of this dipole crosses the protein surface on the heme face approximately midway between Lys 13 and Lys 72, as they are represented in Fig. 37. Leaving aside the complications inherent in calculating the dipole moment of a structure containing a number of mobile groups, such as the surface of cytochrome c, we note that neither the reactivity pattern for O_2^- (Fig. 37) nor the binding of $[Fe(edta)(H_2O)]^-$ (Fig. 30) fit into a scheme in which the dipole moment is assumed to be dominant. Furthermore, on the basis of the chemical evidence reported by Osheroff et al.,[174] the binding of carbonate and phosphate also do not appear to be connected with the overall dipole moment. In fact, the carbonate binding site closely resembles the O_2^- reactivity pattern. Thus, at least for some small-molecule reagents, local effects are more important than the overall dipole moment. These local effects are predominantly electrostatic in origin, but Butler et al.[192] suggest that in addition to these, unspecified steric effects might also be important for the O_2^- reaction.

The binding of octahedral complexes such as $[Fe(CN)_6]^{3-}$ is more complicated to analyse. Clearly, local effects are important because there is more than one binding site (Fig. 30). However, one of the strongest binding sites (site 3) appears to be centred close to the spot at which the positive end of the

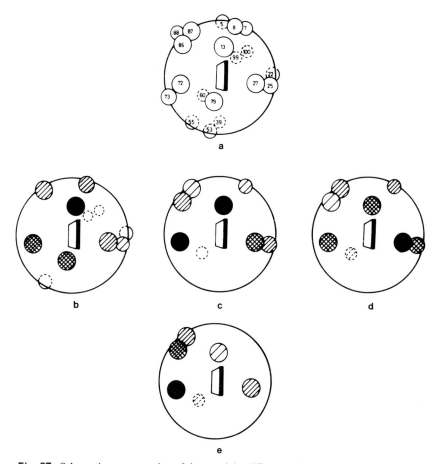

Fig. 37. Schematic representation of the reactivity differences between chemically modified derivatives of horse cytochrome *c* and the native protein. The singly modified derivatives studied are represented in each reactivity diagram, (b)–(d), by the position of the lysine marker. The order of importance of the lysines is solid > crosshatched > narrow stripes > wide stripes > clear = native. (a) Approximate location of lysine ε-NH$_2$ groups of horse cytochrome *c*. The protein is viewed from the same orientation as in Fig. 3, with the heme viewed edge on (central slab). The molecule is cut by an imaginary plane just behind the heme and perpendicular to the heme plane. The location of lysine ε-NH$_2$ groups in front of and behind the imaginary plane is indicated by closed and dashed circles, respectively. (b) Reaction of [Fe(CN)$_6$]$^{3-}$ with TFA and TFC cytochrome *c*.[157] (c) Reaction of [Fe(CN)$_6$]$^{3-}$ with TNP and CDNP cytochrome *c*.[159] (d) Reaction of [Co(phen)$_3$]$^{3+}$ with TNP and CDNP cytochrome *c*.[159] (e) Reaction of O$_2^-$ with CDNP cytochrome *c*.[192]

dipole crosses the protein surface. This is near Ile 81 and Phe 82,[228,231] a position that should be compared with the location of site 3 given in Fig. 30. NMR studies of CDNP-Lys 7 or CDNP-Lys 8 horse cytochrome c will help to determine whether the overall dipole is important for anion binding at site 3, because these modifications cause a substantial shift of the dipole axis without modifying groups that might be locally important.[228]

5. Summary of complex formation

We have described extensively the binding surface located on the front face of cytochrome c because all the available evidence indicates that this area encompasses the electron transfer site(s) for both physiological and nonphysiological reactants (as has been discussed in this article and Refs. [1–3, 13, 35, and 41]). Important properties of this surface are that it contains a highly asymmetric and mobile distribution of charge that is not well defined by the crystallographic structure. The close approach of small ions may be hindered by solvation and by electrostatic image forces, but, once bound, ions may migrate on the protein surface. Such motion will have important mechanistic consequences. In general, the binding of small ions is governed by local forces rather than by the overall charge and dipole of the protein.

E. Reduction of ferricytochrome c with organic hydroxy compounds

The redox reactions of ascorbic acid, catechols, and quinols are complicated because they involve both protons and electrons (Fig. 38), but the energetics of their reactions with cytochrome c are more easily analysed than are the reactions of cytochrome c with transition metal complexes (see Section VI,F). This is because the redox properties of quinols and catechols are easily varied by a suitable choice of aromatic ring substituents. Rich and Bendall[189,190] and Wilson and colleagues[188,191] have exploited this capacity for variation in their studies with cytochrome c. The former workers investigated the reaction of 13 substituted p-benzoquinols and the latter workers studied a range of 21 dihydroxy compounds including ascorbate, catechols, and quinols. The reactions of these compounds with cytochrome c do not proceed by an inner-sphere mechanism.[188,191]

The ascorbate reaction is important because this is the only organic reactant that has been studied with CNDP-modified cytochromes. From the

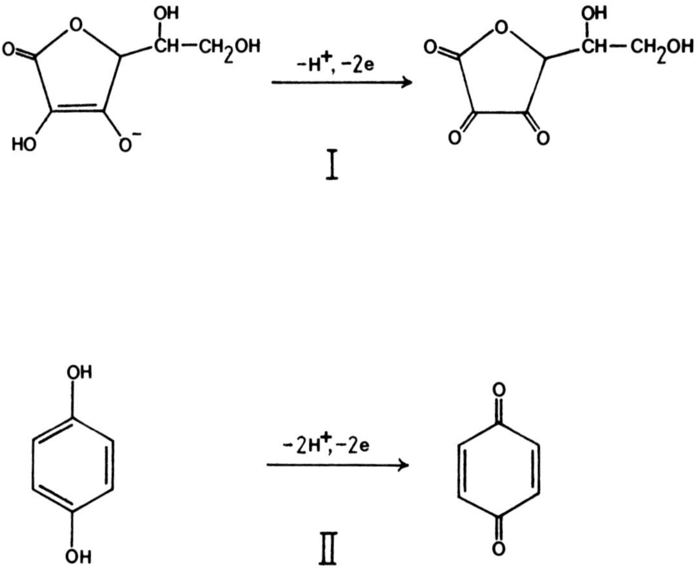

I

II

Fig. 38. Redox reactions of (I) ascorbate and (II) *p*-benzoquinol. The pK_a of the acidic hydroxyl of ascorbic acid is 4.2.

observed variations of reaction rates Konig *et al.* (cited as a footnote in Ref. [192]) concluded that the reactivity profile was similar to that found for $[Fe(CN)_6]^{3-}$ and CDNP-cytochrome *c* (Fig. 37). Thus, as other workers have assumed,[188] the reaction site for ascorbate reduction is at the exposed heme edge of cytochrome *c*, and presumably the catechols and quinols also react at this site. However, Myer *et al.*,[232] disagreed with this location of the reaction site. These workers suggested that ascorbate binds to the buried Arg 38 on the back of cytochrome *c*, but their experiments were rather complicated (they involved measuring the kinetics of reduction of urea-denatured cytochrome *c*) and we do not find them convincing.

Although the quinone/quinol couple contains a large number of potential reductants, Rich and Bendall[189] controlled their experimental conditions so that the anionic quinol (QH⁻) was the major reductant. The reduction of cytochrome *c* was then accompanied by the formation of the semiquinone (QH·):

$$QH^- + CIII \xrightarrow{k_q} QH\cdot + CII$$

By measuring the rate k_q and the redox potential of the couple QH·/QH⁻ for a range of quinols, Rich and Bendall demonstrated that the parameters were simply correlated according to Marcus' theory (see Fig. 39). This is discussed further in Section VI,F.

The reactions with catechols (Ct) also appear to be relatively simple, considering the number of potential reductants. Again, deprotonated species are the major reductants, but, because of the variations in pK_a values in the series of compounds studied by Saleem and Wilson,[191] both monoanionic (Ct^-) and dianionic species are important. Nevertheless, Saleem and Wilson were able to determine the rate of reduction of cytochrome c by Ct^- and they showed that this correlated well with the substituent electronic parameter σ on a Hammett plot.

F. Energetics of reactions

1. The Marcus theory

We shall follow other workers[1−3] and discuss the energetics of the electron transfer reactions of cytochrome c within the framework of the Marcus theory (see Refs. [233–236], and references therein). This theory considers the electron transfer event occurring within a precursor complex in which the donor and acceptor groups are close together. Other theoretical approaches based on electron tunnelling have been developed (see Refs. [237–239], and references therein) because it has become clear that in many biochemical reactions the donor and acceptor centres are relatively far apart (see Ref. [238] and Section VII). There is a direct relationship between the rate of electron tunnelling and the distance between redox centres, but there is not an explicit distance dependence to the rate of transfer given by Marcus theory. Thus, when an experimentally determined rate is the same as the rate calculated from Marcus theory for an adiabatic reaction, the transfer distance is not an important parameter, but when the experimental rate is less than the calculated rate, the distance between redox centres may be an important rate-determining parameter and electron tunnelling may then be a more appropriate formalism to analyse the reaction. We do not consider these concepts in the present article. However, it is pertinent to note that the charge-transfer band observed by Potasek and Hopfield[240] in their study of the reaction between cytochrome c and $[Fe(CN)_6]^{3-/4-}$, and cited by them as evidence for vibronically coupled electron tunnelling,[237] has been recently shown by Austin and Hopfield[215,241] to come from a binuclear cyano-bridged species formed between $[Fe(CN)_6]^{3-}$ and $[Fe(CN)_6]^{4-}$ and not to be associated with electron transfer involving cytochrome c.

The foundation of the Marcus theory[233,234] is that the motion of an electron is too fast to be coupled to the much slower nuclear motions of vibration and rotation. When an electron is transferred between two molecules in a radiationless transition, the transfer is adiabatic and the reactant and product activated complexes must be of equal energy. The frequency of electron transfer is then controlled by a Franck–Condon transmission coefficient that depends on the overlap and density of electronic, vibrational, and rotational states in the reactant and product complexes. Since the protein–redox reagent complex has a large number of low-energy vibrational and rotational states, most of which are independent of the precise location of the electron, the value of the transmission coefficient will depend mainly on the degree of overlap of the electronic energy levels of the two redox centres. Marcus theory therefore begins with the precursor complex in its ground vibrational state and it requires simple harmonic distortion to occur until a geometry is attained in which the electron transfer may be achieved iso-energetically. The rate constant is then given by

$$k = pZe^{-\Delta G^*_{12}} \tag{24}$$

where p is the probability of electron transfer in the activated complex, Z is the collision frequency between two uncharged reactants, and ΔG^*_{12} is the free energy of activation for the electron cross-exchange reaction. Z is generally taken to be 10^{11} M^{-1} s^{-1} and for an adiabatic reaction $p = 1$. For our purposes the most useful form of the Marcus theory is the correlation equation that relates an electron cross-exchange reaction (referred to by the subscript 12) to its component self-exchange reactions (referred to by the subscripts 11 and 22). This equation is

$$\Delta G^*_{12} = \Delta G^*_{11}/2 + \Delta G^*_{22}/2 + (\Delta G^{\circ}_{12}/2)(1 + \alpha) \tag{25}$$

where

$$\alpha = \Delta G^{\circ}_{12}/4(\Delta G^*_{11} + \Delta G^*_{22}) \tag{26}$$

Unless ΔG°_{12}, the overall cross-reaction free energy change, is large, $\alpha \ll 1$ and can be disregarded. With this limitation,

$$\Delta G^*_{12} = \tfrac{1}{2}(\Delta G^*_{11} + \Delta G^*_{22} + \Delta G^{\circ}_{12}) \tag{27}$$

$$\Delta S^*_{12} = \tfrac{1}{2}(\Delta S^*_{11} + \Delta S^*_{22} + \Delta S^{\circ}_{12}) \tag{28}$$

$$\Delta H^*_{12} = \tfrac{1}{2}(\Delta H^*_{11} + \Delta H^*_{22} + \Delta H^{\circ}_{12}) \tag{29}$$

In its rate form, Eq. (25) becomes

$$k_{12} = (k_{11}k_{22}K_{12}f_{12})^{1/2} \tag{30}$$

where

$$\log f_{12} = (\log K_{12})^2/[4 \log(k_{11}k_{22}/Z^2)] \tag{31}$$

For reactions considered here, $f \approx 1$.

ΔG^*, ΔH^*, and ΔS^* are related to the experimentally derived quantities (denoted by the superscript $\ddagger$) as follows:

$$\Delta G^{\ddagger} = \Delta G^* RT \ln(hZ/KT) \tag{32}$$

$$\Delta S^{\ddagger} = \Delta S^* + R \ln(hZ/KT) - \tfrac{1}{2}R \tag{33}$$

$$\Delta H^{\ddagger} = \Delta H^* - \tfrac{1}{2}RT \tag{34}$$

where K is the Boltzmann constant and h is Planck's constant.

Equations (24)–(34) adequately describe many electron transfer reactions between small molecules (see Refs. [1 and 242], and references therein); the question that we address is, do these equations adequately describe electron transfer reactions involving cytochrome c? This problem has been tackled by a number of workers, most notably by Sutin[3] and by Gray and colleagues,[2,109,181] and it is clear that in many cases this theoretical approach is valid for cytochrome c. However, there are a number of assumptions made in the derivation of these equations that can produce problems. The most serious of these assumptions is that the work terms required to bring together the three sets of reactants cancel each other. Sutin[3] described two cases important for reactions of cytochrome c in which they do not cancel.

First, when a reaction involves similarly charged reactants, the formation of a precursor complex has to overcome unfavourable electrostatic interactions; these contribute to the work terms for all self-exchange reactions involving charged reactants. Thus when a cross-exchange reaction involves similarly charged reactants, there are unfavourable work terms in all three reactions and they may reasonably be assumed to cancel. However, when the cross-exchange reaction involves dissimilarly charged reactants, precursor complex formation is aided by attractive electrostatic interactions. Therefore, in these cases the work terms do not cancel, and the reaction will proceed faster than predicted by Eq. (30).

Second, when the cross-reaction is between a hydrophilic and a hydrophobic reactant, there are unfavourable nonelectrostatic interactions that are not present in their constituent self-exchange reactions. Thus the cross-reaction proceeds more slowly than predicted by Eq. (30).

Various approaches have been adopted to overcome these work problems. In one approach, the measured rate constants are corrected for the work that is required to form the complexes; in a second approach, a range of similar reactions are compared that vary only in one term, usually ΔG°_{12}. Work term corrections are difficult because the magnitude of the nonelectrostatic work cannot be determined and consequently it is usually disregarded in energetic calculations. However, although in discussing the surface of cytochrome c we have emphasised its high charge, especially around the heme crevice, it is important to note that approximately 55% of the surface area accessible to a sphere of radius 1.4 Å is apolar.[23] Corrections for electrostatic work can be made by modifying Eq. (30) to include the stability constants of the three precursor complexes (P_{11}, P_{22}, and P_{12}) and the stability constant of the cross-reaction successor complex (P_{21}).[234] Thus,

$$k_{12} = (P_{12}P_{21}k_{11}k_{22}K_{12}/P_{11}P_{22})^{1/2} \tag{35}$$

when $f = 1$. However, these stability constants are rarely known. An alternative method that has been advocated by Wherland and Gray[2] is to estimate the electrostatic work involved in complex formation from the ionic strength dependence of the reaction rate. The problems inherent in this approach have already been discussed in Section V.

The second approach, the comparison of related reactions, is a relatively simple approach that is not of general application to inorganic reagents because only rarely can their redox potentials be changed without the attendant alteration of some other reactivity characteristic.

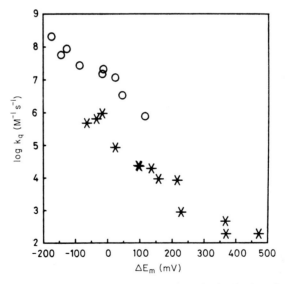

Fig. 39. Marcus plot of rate constant against ΔE_m for quinol reduction of cytochrome c (∗) and solubilised plant cytochrome f (○) according to the reaction $QH^- + heme^{3+} \xrightarrow{k_q} QH\cdot + heme^{2+}$. Both sets of data follow Eq. (36) and have slopes of $1/i18$. The displacement of the two lines occurs because the k_{11} and/or work terms are different for the two proteins. (From Rich.[190])

2.   Reactions with quinols

The redox potentials of *p*-benzoquinol can be readily altered by varying the nature of the ring substituents. Rich and Bendall[189,190] have exploited this to show that there is a simple correlation between k_{12} and ΔG°_{12} for a series of reactions between cytochrome *c* and substituted *p*-benzoquinols. Equation (30) can be restated thus at 25°C

$$\log k_{12} = (\Delta E_m/118) + \tfrac{1}{2}(\log k_{11} + \log k_{22}) \tag{36}$$

when $f = 1$. Provided that both $\log k_{11} + \log k_{22}$, and the interaction work terms, are constant over the range of reactions, a plot of $\log k_{12}$ against ΔE_m should give a straight line of slope 1/118. Figure 39, covering a ΔE_m range of 600 mV and a k_{12} range of 10^4, shows that the reactions of cytochrome *c* with 13 substituted *p*-benzoquinols fit the theoretical expectations very well.

3.   Reactions with positively charged
     transition metal complexes

Reactions involving positively charged transition metal complexes should be among the simplest to consider because the electrostatic repulsion terms can be assumed to cancel each other. Sutin[3] calculated kinetic parameters for a number of reactions of cytochrome *c* and some of his results are given in Table 9 along with the calculated rate for the reaction of ferrocytochrome *c* with $[Fe(cp)_2]^+$. The observed rates and activation parameters are given in Table 5.

It can be seen from Table 9 that there is good agreement between the calculated and observed values, especially for the $[Co(phen)_3]^{3+}$ and $[Ru(NH_3)_5(bm)]^{2+}$ reactions. This has been attributed to the fact that both of these reagents contain π systems that can interact directly with the exposed heme.[2,3] Cummins and Gray[109] also claimed that the reaction of $[Ru(NH_3)_5(py)]^{3+}$ with ferrocytochrome *c* fits into the same general pattern (Table 5). However, calculation of k_{12} for this reaction using the values given in Tables 2 and 3 and taking k_{22} to be 3.38×10^2 M^{-1} s^{-1} (this was calculated, not measured[109]) yields a value of 8.2×10^2 M^{-1} s^{-1} (at $I = 0.1$ M) as compared to the observed value of 6×10^3 M^{-1} s^{-1}. Thus either there is a real mechanistic difference, which seems unlikely, or the calculated value of k_{22} is in error.

McArdle *et al.*[211] extended the study of positively charged reagents to a variety of $[Co(phen)_3]$ complexes containing modified 1,10-phenanthrolines

TABLE 9

Calculated rate constants for reactions of cytochrome c
with positively charged reactants[a]

Reaction	k_{12}^{calc}	$k_{12}^{calc}/k_{12}^{obs}$	$\Delta H_{calc}^{\ddagger}$ (kJ mol^{-1})	$\Delta S_{calc}^{\ddagger}$ (J K^{-1} mol^{-1})
$[Ru(NH_3)_6]^{2+}$ + CIII	1.2×10^5	3.1	29.3	-58.5
$[Ru(NH_3)_5(bm)]^{2+}$ + CIII	6.3×10^4	1.1		
CII + $[Co(phen)_3]^{3+}$	2.0×10^3	1.3	46.4	-25.1
CII + $[Fe(cp)_2]^+$	3.3×10^7	5.3		

[a] From Sutin,[3] except k_{12}^{calc} for the reaction with $[Fe(cp)_2]^+$, which was calculated from data given by Pladziewicz and Carney[117] and Gupta *et al.*[37]

(Table 5). However, since the self-exchange rates of these reactants are unknown, and the use of calculated self-exchange rates may introduce large errors, calculations such as those of Table 9 cannot be carried out. Also, on a simple plot of log k_{12} against ΔE_m the data are scattered. Therefore, it is not possible to simply demonstrate that these reactions are adiabatic and, since some of the rates are rather low (Table 5), it is possible that some are nonadiabatic. These reactions have been extensively discussed[2,211] and attention has been drawn to the variation of activation parameters. This variation, together with the variation in rate, is taken to indicate that there are sizable steric effects in these reactions that are absent from the reactions of the smaller $[Co(phen)_3]^{3+}$.

4. Reactions with negatively charged transition metal complexes

The reasons that these reactions are the easiest to study from a structural aspect also make them the hardest to analyse from an energetic viewpoint. The electrostatic attractive forces lead, in some cases, to stable precursor complex formation and, in all cases, to work terms that do not cancel in the simple correlation expressions. Sutin[3] has amply demonstrated this by calculating k_{12} for a variety of reactions involving iron penta- and hexa-cyanides using Eq. (30) and disregarding work terms. In all cases he found the calculated k_{12} to be less than the observed k_{12}, with the difference ranging from 10 to 10^3. Equation (35) could be used, but the stability constants are not generally well established. Those for $[Fe(CN)_6]^{3-/4-}$ binding to cytochrome c are known (Table 6), and if they are taken to be $10^2 \ M^{-1}$, with P_{11} and P_{22} both arbitrarily set at 1 M^{-1}, the calculated k_{12} is in reasonable agreement with the observed k_{12}.

NMR studies indicate that $[Co(ox)_3]^{3-}$ and $[Fe(CN)_6]^{3-}$ have similar binding sites and binding strengths. If we make the reasonable assumption that a comparison of their rates leads to a cancellation of work terms, the energetics of these two reactions can be compared. To do this we need to know the respective k_{22} values and these have either been measured ([Fe(CN)$_6$]$^{3-}$, Table 3) or calculated ([Co(ox)$_3$]$^{3-}$, 2.8×10^{-7} M^{-1} s^{-1}[206]). Taking these values of k_{22}, the ratio of k_{12}^{calc} for $[Fe(CN)_6]^{3-}/[Co(ox)_3]^{3-}$ is 8×10^3, compared to the ratio of k_{12}^{obs} of about 10^6. The discrepancy of 10^2–10^3 could arise because k_{22} for $[Co(ox)_3]^{3-}$ has been grossly underestimated or because there is a real reactivity difference, with $[Co(ox)_3]^{3-}$ reacting nonadiabatically.

A straightforward comparison is not as simple with $[Fe(edta)(H_2O)]^{2-}$ because its binding to cytochrome c is different from that of $[Cr(ox)_3]^{3-}$ and $[Fe(CN)_6]^{3-}$. Nevertheless, assuming that k_{22} for $[Fe(edta)(H_2O)]^-$ is 3×10^4 M^{-1} s^{-1} (this is not a direct measurement[2]) and ignoring electrostatic corrections, k_{12}^{calc} is 9×10^4 M^{-1} s^{-1}. This is greater than k_{12}^{obs} (Table 5), which indicates that either k_{22} has been wrongly estimated, the unfavourable work terms in the cross-reaction outweigh the attractive electrostatic work terms, or the reaction is nonadiabatic. The possibility of high, unfavourable work terms during electron transfer is significant given the possibility that, on binding to cytochrome c, the inner coordination sphere of $[Fe(edta)(H_2O)]^{-/2-}$ is significantly perturbed.

Wei and Ryan[208] analysed reactions of cytochrome c with various iron polyaminocarboxylate (pac) complexes and reported reasonably good Marcus correlations in most cases. They assumed that k_{22} for the iron complexes was the same and used ionic strength dependences to correct the observed cross-reaction rates for electrostatic interactions, and then determined theoretical k_{12} ratios from the relationship

$$\frac{k_{12}(Fe \cdot pac)}{k_{12}(Fe \cdot dtpa)^{3-}} = \left(\frac{K_{12}(Fe \cdot pac)}{K_{12}(Fe \cdot dtpa)^{3-}} \right)^{1/2} \tag{37}$$

The calculated and observed (in parentheses) k_{12} ratios are $[Fe(dtpa)]^{3-}$, 1.0 (1.0); $[Fe(edta)(H_2O)]^{2-}$, 1.6 (2.2); $[Fe(dpta)]^{2-}$, 1.3 (1.2); $[Fe(edap)]^{2-}$, 2.2 (1.9); $[Fe(hedta)(H_2O)]^-$, 1.1 (0.17); $[Fe(nta)(H_2O)_2]^-$, 0.6 (0.04). The good agreement for some of the reagents is striking. The poor agreement for the hedta and nta complexes is acribed by Wei and Ryan[208] to nonelectrostatic interactions between the redox reagent and the protein. However, the pK_a of coordinated H_2O in the ferric states of these reagents is lower than those of the other reagents, i.e., 4–5 as compared to 7.5–10.[96,208] Also, for the hedta complex,[96] and probably for the nta complex as well, the pK_a is increased to 9 on reduction to the ferrous state, and therefore the electron transfer is accompanied by a proton transfer. Because of these considerations, simple Marcus analysis ignoring the proton is not valid.

Summarising the analysis of the energetics of reactions involving negatively charged reactants, it is clear that when electrostatic work terms can be eliminated, there is reasonably good agreement between experimentally measured k_{12} values and theoretical expectations for a variety of different reagents. However, nonelectrostatic work terms appear to be important in some cases. To remove uncertainty at some parts of the analysis, experimental determination of k_{22} for $[Co(ox)_3]^{3-/4-}$ and $[Fe(edta)(H_2O)]^{-/2-}$ is necessary.

5. Reactions of *Candida* cytochrome *c*

Creutz and Sutin[108] studied a range of reactions of *Candida* cytochrome *c*, including chemical modification and reduction with inner-sphere reagents, and in general they found small reactivity differences. Their results with outer-sphere electron transfer reagents are summarised in Table 7. Comparison of these data with those of Tables 5 and 6 reveals that the reactions of *Candida* cytochrome *c* are slightly faster than the corresponding reactions of horse cytochrome *c*. The reactions of horse cytochrome *c* with [Co-(phen)$_3$]$^{3+}$ and [Ru(NH$_3$)$_5$(bm)]$^{2+}$ fit very well with the Marcus correlation equation (Table 9),[3] and the reaction of horse cytochrome *c* with [Fe-(CN)$_6$]$^{3-}$ fits reasonably well if certain allowances are made for the electrostatic interactions. Therefore, the question is posed, why do horse and *Candida* cytochromes *c* have different reactivities, given that they have almost the same redox potential (Table 2)? The situation is further complicated because, as Creutz and Sutin[108] noted, the measured k_{11} values for horse and *Candida* cytochromes *c* are different. Gupta[40] has shown that k_{11} for *Candida* cytochrome *c* is a factor of 10 slower than for horse cytochrome *c*, a difference that enhances the difference in cross-reaction rates.

The apparent anomaly of enhanced *Candida* cytochrome *c* reactivity can be largely resolved within the framework of the Marcus theory by comparing the reactant interaction sites on the surfaces of the two proteins. This is easily done for [Fe(CN)$_6$]$^{3-}$. The NMR measurements described in Section VI,D indicate that the location of the main binding sites on the two proteins is slightly different, though both are by the exposed heme. Also, NMR electron exchange experiments (see Figs. 18 and 19) have shown that K_{ass} for [Fe(CN)$_6$]$^{4-}$ binding to ferricytochrome *c* is different for the two proteins. When this is taken into account, and using Eq. (35) with the assumption that P_{11} is the same for both proteins, the ratio of calculated k_{12} (*Candida/horse*) is 1.0, which compares favourably with the observed ratio of 1.7.

The reactivity differences for the positively charged reactants cannot be so clearly analysed. $[Co(phen)_3]^{3+}$ was shown by chemical modification studies to interact predominantly on the right side of the front surface (see Fig. 37), and we suggested this might be because Glu 21 was located on this part of the surface. *Candida* cytochrome *c* also contains Glu 21 and, compared to horse cytochromes *c*, this part of the protein surface is more able to interact with positively charged reagents because lysines 22 and 25 of horse cytochrome *c* are replaced by uncharged residues in *Candida* cytochrome *c* (see Fig. 10). Thus small differences in the binding surface of cytochrome *c* are transformed into faster rates by variation of electrostatic work terms.

6. The Wherland–Gray reactivity series

The analytical approach adopted by Wherland and Gray[2,181] has already been mentioned several times. Their procedure aims to place all the energetic uncertainties into one parameter (k_{11}) and then to analyse the variation of this parameter for a variety of reactions in mechanistic terms. We denote self-exchange rates calculated by their procedure with the superscript "est." Work terms for the electrostatic interactions are calculated using data from ionic strength dependences or using the net protein charge, and a correction is made for the differences between reaction driving forces by assuming they are adiabatic and using Eq. (30). This requires that the respective k_{22} values be known, and, as described earlier, despite the uncertainty some are calculated from small-molecule reactions.

k_{11}^{est} ($I = 0.1\ M$) for the reactions of cytochrome *c* with $[Fe(CN)_6]^{3-}$, $[Co(ox)_3]^{3-}$, and $[Fe(edta)(H_2O)]^{2-}$, respectively, are 1.2×10^4, 4.6, and $6.2\ M^{-1}\ s^{-1}$. For comparison, k_{11}^{est} for $[Co(phen)_3]^{3+}$ and $[Ru(NH_3)_6]^{2+}$ are 7.1×10^2 and $1.6 \times 10\ M^{-1}\ s^{-1}$.[2,206] These values should be compared to the experimentally determined k_{11} ($I = 0.1\ M$) of $1.2 \times 10^3\ M^{-1}\ s^{-1}$.[37] The reactivity order of $[Fe(CN)_6]^{3-} > [Co(phen)_3]^{3+} > [Ru(NH_3)_6]^{2+} > [Fe(edta)(H_2O)]^{2-} \sim [Co(ox)_3]^{3-}$ is interpreted by Gray and colleagues in terms of the mechanistic preferences of the reactants: $[Fe(CN)_6]^{3-}$ has a cylindrical π-orbital set, $[Co(phen)_3]^{3+}$ has an extensive π-orbital network but a more rigid ligand type that introduces specific orientation and steric factors, and $[Ru(NH_3)_6]^{2+}$ is hydrophilic, like $[Co(ox)_3]^{3-}$ and $[Fe(edta)(H_2O)]^{2-}$, but it possesses relatively expanded d_π orbitals that allow more efficient electron transfer than is obtained with the other hydrophilic reagents. Variation within the series, and differences from k_{11}, are thus interpreted as a failure to account for nonelectrostatic interactions and as an indicator of nonadiabaticity. However, it is difficult to see why the reactivity differences are so large, especially for the positively charged reactants with which reasonably good Marcus correlation is obtained with the measured k_{11} (Table 9). There is, therefore, a fundamental difference between the analysis of Sections VI,F,3–5 and the Wherland–Gray analysis, and we believe that it is the latter procedure, which has greater scope for the introduction of errors, that is at fault.

Wherland and Gray have done similar analyses with other protein types, and have shown that most follow a reactivity trend similar to cytochrome c (though k_{11}^{est} for $[Fe(CN)_6]^{3-}$ and $[Co(phen)_3]^{3+}$ are occasionally switched). A particularly important point is that the k_{11}^{est} ratio $[Fe(edta)(H_2O)]^{2-}/[Co(ox)_3]^{3-}$ is similar for most of the proteins studied. Because of this, k_{11}^{est} for these hydrophilic reagents is suggested to be an indicator of the reactivity of the protein independent from that of the small-molecule reactants.[206] Mauk *et al.*[242] have gone further than this and proposed a method, using Hopfield's expression,[237] for calculating electron tunnelling distances from values of k_{11}^{est} that are referenced to k_{11}^{est} for $[Fe(edta)(H_2O)]^{2-}$ (however, see comments by De Vault in Ref. [239]). Clearly, if the conformation of $[Fe(edta)-(H_2O)]^{2-}$ is perturbed on binding to cytochrome c (see Section VI,D), then this entire procedure is undermined.

7. Summary of reaction energetics

Marcus theory accounts satisfactorily for the energetics of the reactions of cytochrome c with a variety of types of small-molecule reagents when interaction work terms can be cancelled or incorporated into the analysis. This is demonstrated by the preceding analysis of the reactions with quinones, $[Co(phen)_3]^{3+}$, $[Ru(NH_3)_6]^{2+}$, $[Ru(NH_3)_5(bm)]^{2+}$, and $[Fe(CN)_6]^{3-}$. Uncertainty in the application of simple Marcus theory to a number of reactions may be due to nonadiabaticity or to insufficient characterisation of the cross-reaction considered and its component self-reactions.

The energetics of electrochemical reactions of cytochrome c have been analysed with the Marcus theory and the free-energy profile calculated for the reaction at a gold electrode modified with 4,4′-bipyridyl.[243,244] Although these results are not considered in detail in the present article, there are two important points relevant to our main themes. First, the reactions appear to take place on the front face of cytochrome c. Second, a relatively high binding energy (~ 30 kJ mol^{-1}) is required to partly overcome the high activation free energy (~ 60 kJ mol^{-1}). This binding energy represents a rate enhancement of approximately 10^5. Thus this electrochemical reaction has marked similarities with other reactions of cytochrome c, such as those with $[Fe(CN)_6]^{3-/4-}$ and its physiological oxidoreductases (see following section).

The application of Marcus theory to protein–protein reactions can also be considered. Leaving aside the fact that some interprotein redox centre separations are too great for classical outer-sphere electron transfer, a large problem in these analyses is determining work term corrections. As we shall see, all the physiological reactions of cytochrome c involve precursor complex formation between highly charged partners. Nevertheless, some experimental tests have been made, though not between cytochrome c and its physiological partners, and reasonable agreement has been reported for some reactions.[211,245]

G. Summary of nonphysiological reactions and reactivity characteristics of cytochrome *c*

The nonphysiological electron transfer reactions of cytochrome c have a number of common features, of which the most important is that small-molecule reagents react at the exposed heme edge with a rate that is generally given by the Marcus theory for an adiabatic process. Thus the rate of each particular reaction may be expressed in terms of the overall driving force ΔG_{12}°, the equilibrium constant for the formation of a precursor complex, and the reactivity characteristics of each individual reactant. ΔG_{12}° is determined by the redox potentials of the small-molecule reagents and cytochrome c, and the structural and electronic influences on these have been extensively discussed elsewhere.[13,67,98,246] Factors influencing the binding energy for the formation of a precursor complex include the charge and size of the small-molecule reagent, the ionic strength of the solution, and the presence of inhibitors. All of these factors are currently being investigated.[168–174,185] We have emphasized the importance of controlling the ionic composition (Section V) and have described procedures for determining the structures and strengths of the precursor complexes (Sections IV and VI). Finally, the reactivity characteristics of the individual reactants must be considered. Those of the small-molecule reagents have already been discussed (Section III and Refs. [1, 2, 246, and 247], and references therein) and in the following discussion we summarise only some of the reactivity characteristics of cytochrome c.

Two of the central structural features of cytochrome c are its exposed heme edge and asymmetric charge distribution, both of which are mechanistically important. In view of the location of small-molecule reaction sites on both sides of the exposed heme (see Section VI,D), it is interesting to note that EPR and NMR studies[131,135,136,248] have shown that the partially filled Fe(III) d_{yz} orbital has a node in the plane of the porphyrin ring and is directed toward the protein surface above and below the heme plane between heme methyl 3 and thioether 4 at the top of the heme (according to the conventional orientation in Figs. 3 and 30). The first excited state, in which the orbital vacancy is placed in d_{xz}, is approximately 7.1 kJ mol^{-1} (~ 3 RT) higher in energy, and it directs the unpaired electron above and below the heme plane toward the bottom of the heme. Although a full molecular orbital description of the heme is not available, the highest occupied molecular orbital must preserve the symmetry of d_{yz} and d_{xz} and have a nodal plane in the plane of the heme. This leads to the interesting possibility that for relatively low-energy electron transfer to occur, it is only necessary to get the electron donor/acceptor close to the exposed heme and not to a particular part of the exposed heme. A full molecular orbital description of the heme will assess this possibility.

The flexibility of the protein structure is an important determinant of reactivity characteristics. One manifestation of this flexibility is the redox state conformation change. The energetics of this conformation change will be reflected in the protein self-exchange rate[233,234] although, in general, such exchange rates are more easily measured than interpreted. Thus, the self-exchange rates of horse and *Candida* cytochrome *c* differ by an order of magnitude despite the structural similarities (Table 2). This may be a reflection of work term differences or of a difference in the energetics of the conformation changes. Warshel and Churg[249] have attempted to estimate the protein reorganisation energy using an *ab initio* computational method and the X-ray coordinates of tuna cytochrome *c*. However, the value they obtained (1.2 kJ mol^{-1}) is less than RT (2.5 kJ mol^{-1}), which is inconsistent with the observation of two distinct structural states by X-ray crystallography and NMR. In view of this, and because NMR studies suggest that the redox state conformational charge in solution is not well defined by the X-ray structure,[99] an experimental determination of the reorganisation energy is required.

Other important roles for protein flexibility are in ion binding, which was discussed in Section VI,D, and in the formation of isoenergetic activated reactant complexes. The low-energy vibrational modes of the protein (and solvent) that affect the instantaneous redox potential are important because it is through molecules in these vibrational states that electron transfer occurs. A general description of the mobility of cytochrome *c* has been presented[250] and it is interesting to note that different regions of the front face possess different mobilities, with one of the most mobile regions being at the top of the heme near Phe 82 and heme methyl 3.

Finally, as is usually the case with solution reactions, the solvent plays an important role. This is demonstrated by the ^{1}H/^{2}H isotope effects (Fig. 8

and Table 6) and the large negative activation entropies (Tables 5 and 6). The solvent is involved in the formation of both the association precursor complex and the isoenergetic electron transfer complexes. It is clear from the experimentally determined distances between the small molecules and the protein surface and from the low steric requirements of the association reaction (Section VI,D,4) that a solvent shell at least one molecule thick is maintained around the protein in the association precursor complex. For electron transfer to occur, either this solvent shell must be rearranged to allow access of the small molecule to the heme, or the electron must tunnel through the barrier.

VII. PHYSIOLOGICAL REACTIONS OF CYTOCHROME *c*

The physiological reactions of cytochrome *c* have been reviewed from several viewpoints[13,35,177,251,252] and so we shall discuss only those aspects of the reactions that are addressed directly by the studies of non-physiological reactions as described in previous sections.

The most important of the physiological oxidoreductases of cytochrome *c* are the mitochondrial enzymes cytochrome *c* reductase (called bc_1) and cytochrome *c* oxidase (called aa_3), but other physiological reactants have been investigated. These are sulphite oxidase, yeast cytochrome b_2 (called b_2), cytochrome b_5 (called b_5), and yeast cytochrome *c* peroxidase (called CcP). The latter two proteins are important because their X-ray structures have been determined,[253,254] thus allowing the structures of possible precursor complexes to be investigated by computer modelling.[255,256]

Cytochrome *c* is reduced by bc_1, b_5, b_2, and sulphite oxidase, and it is oxidised by aa_3 and CcP. One common feature of these reactions is that they involve the formation of relatively strong protein–protein complexes. These are formed by the association of one oxidoreductase molecule per cytochrome *c* molecule, and their association constants are markedly ionic strength dependent; for example, K_{ass} for the cytochrome $c:b_5$ complex is $4 \times 10^6 \, M^{-1}$ at $I = 0.001 \, M$ and $8 \times 10^4 \, M^{-1}$ at $I = 0.01 \, M$ (at 25°C, pH 7, phosphate).[257] This has obvious parallels with the binding of small ions to cytochrome *c* and, again, opinion is divided between the local charge model[258,259] and the overall dipole model[171,228,231] to account for the ionic strength effect.

The binding sites on cytochrome *c* for its oxidoreductases have been determined by kinetic studies of chemically modified cytochromes (see Refs. [228, 251, and 252], and references therein) and by the determination of the chemical reactivities of the lysine residues of cytochrome c in both its free and complexed forms.[230,260] The results of these studies are summarised in Fig. 40 and they reveal that not only do the physiological reactants all react on the front face of cytochrome c, as proposed by Salemme *et al.*,[41] but also that they all appear to have a preference for the top left side. It is this preference for the top left side, which encloses the positive end of the protein dipole, that led Koppenol and Margoliash[228] to advocate the importance of the overall dipole to complex formation. However, as was

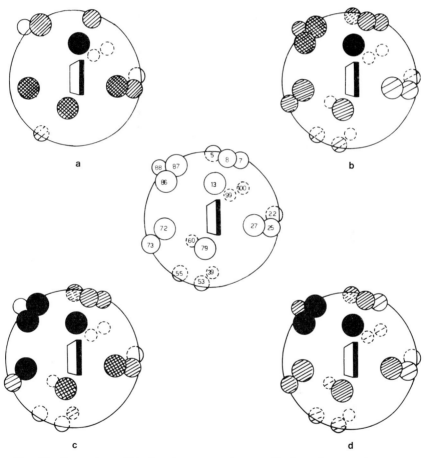

Fig. 40. Comparison of binding sites on cytochrome c for its various reaction partners. See Fig. 37 for key to representation. The diagram summarises a considerable amount of work that has been fully described (see Refs. [13, 35, 228, 251, and 252], and references therein). (a) Cytochrome b_5; (b) cytochrome c oxidase; (c) cytochrome c peroxidase; (d) cytochrome c reductase.

pointed out in Section VI,D,4, the overall dipole does not provide the dominant interactions for binding small molecules, and since these reactants approximate point charges much better than do proteins, it seems to us unlikely that the overall dipole provides the dominant interactions for protein complexation.

The best-characterised complex is the cytochrome c:CcP complex. Poulos and Kraut[256] proposed a model for this complex based on general electrostatic considerations by a procedure similar to Salemme's[255] in his work with the cytochrome c:b_5 complex. The overall dipole of the proteins was not considered, and all that was done was to search the surfaces of the proteins to find complementary patches of charged residues that could be docked together without causing unfavourable steric interaction. Poulos and Kraut[256] found that the two proteins could be oriented so that lysines 13, 27, 72, and 87 on cytochrome c could interact with aspartates 37, 79, 216, and 34 on CcP, and that with such a complex the two hemes were oriented with respect to each other as shown in Fig. 41. The chemical modification studies depicted in Fig. 40 and preliminary chemical experiments with CcP[261,262] are consistent with the proposed model.

The relative heme orientations in the proposed cytochrome c: CcP com-

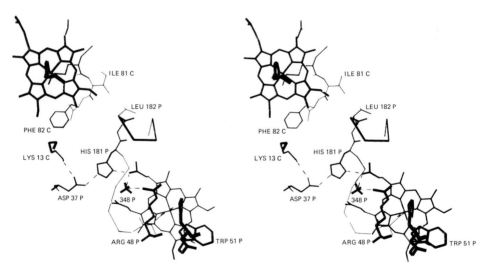

Fig. 41. Stereoscopic view of the interheme region of the proposed cytochrome c: cytochrome c peroxidase (CcP) complex. Cytochrome c residues are labelled with a C and CcP residues are labelled with a P. The broken lines represent hydrogen bonds and ionic interactions. (From Poulos and Finzel.[256])

plex are shown in Fig. 41.[256] There are two notable features of this complex. First, the hemes are almost parallel, with a perpendicular separation between their planes of 6.8 Å. Second, the heme edges are separated by a distance of 16.5 Å. The full significance of these features is not understood, although it is clear that the interheme separation is too great for classical outer-sphere electron transfer. In Salemme's model of the cytochrome $c:b_5$ complex[255] the two hemes are also almost parallel, although the distance between the heme edges is only 8.4 Å. However, it is premature to conclude that parallel hemes are always necessary for electron transfer, because the four hemes of the class III cytochromes c_3 are not parallel (see Fig. 5).

Along with the proposed model of the cytochrome $c:b_5$ complex,[255] which is consistent with chemical modification studies (Fig. 40) and preliminary NMR experiments,[263] the proposed model of the cytochrome c:CcP complex provides a good framework for the development of ideas and experiments concerning hemoprotein-mediated electron transfer. Based on the small-molecule interactions of cytochrome c we can ask the following questions.

1.   What are the dominant electrostatic interactions that govern complex formation?

2.   How important are nonelectrostatic interactions?

3.   Are electrostatic image forces important in the formation of a productive complex?

4.   What happens to the solvation shells around each protein on complex formation?

5.   Do the proteins within the complex have independent mobility so that as one moves against the other the possibility of correct alignment, and hence the probability of electron transfer, is increased?

A number of reported studies have begun to address these questions but, with the exception of the first question, there are insufficient data with which to assess them. The roles of nonelectrostatic interactions are particularly difficult to assess. That in some cases of protein association such interactions are important is illustrated by the dimeric class II cytochromes (see Fig. 4), in which the subunit interface is composed almost entirely of hydrophobic residues. The forces responsible for the stability of protein complexes have been assessed from a thermodynamic viewpoint (see Refs. [264 and 265], and references therein) and extensive reorganisation of the solvation sphere has been recognised to be a dominant process in many cases. In fact, Salemme[255] pointed out in his study of the cytochrome $c:b_5$ complex that water would be excluded from the interface, and Mauk *et al.*[257] have shown that the formation of this complex is accompanied by a large entropic change and a small enthalpic change, a result indicative of extensive solvent reorganisation.

The recent characterisation of active, covalently linked cytochrome c:CcP complexes[261,266,267] may allow some of the problems outlined above to be tackled. Thus the complex reported by Pettigrew and Seilman[267] appears to consist of CcP covalently bound at the front face of cytochrome c, yet the cross-linked cytochrome c can be reduced by free cytochrome c. Pettigrew and Seilman considered that this result, in conjunction with the result of Hill and Nicholls[268] that cytochrome c in a noncovalent complex with aa_3 can be reduced by N,N,N',N'-tetramethyl-p-phenylenediamine (TMPD) but not by ascorbate, implied that there were separate routes for electron entry into cytochrome c. This is consistent with the view that the entire exposed heme edge of cytochrome c is capable of transferring electrons (see Section VI,G), with some reagents having a preference for one part of the front face while other reagents have a preference for another part. However, there is also the problem of the difference in charge between free and complexed cytochrome c. This problem has been studied by Petersen and Cox[269] with cytochrome c complexed to phosvitin. One molecule of phosvitin, a highly charged protein of molecular weight 35,000 that is about 50% phosphoserine, can bind at least 20 molecules of cytochrome c at pH 7 (see Ref. [177], and references therein). Although such complexes are not physiologically important they have been widely studied, and Petersen and Cox[269] found that in their complex, the reactivity of cytochrome c toward negatively charged reagents such as ascorbate and $[Fe(CN)_6]^{3-}$ was strongly inhibited, while its reactivity toward uncharged reagents such as TMPD was little affected and its reactivity toward $[Co(phen)_3]^{3+}$ was greatly stimulated. Thus there is a marked charge effect, but the fact remains that cytochrome c complexed with a strongly acidic protein, which presumably binds at the front face of cytochrome c, is capable of accepting electrons from a variety of reagents.

The fact that complexed cytochrome c is reactive raises the question of whether there is a unique complex, as is generally assumed, or a variety of complexes for each of the oxidoreductases. This is clearly related to the question of mobility within the complex, a question that is directly posed by the work of Yoshimura *et al.*[270] on the rate of electron transfer between horse and *Candida* cytochromes c bound to phosvitin. These workers found that in the phosvitin–cytochrome c complex, the rate of electron transfer between the two cytochromes was $2 \times 10^6 \ M^{-1} \ s^{-1}$ compared to the rate for the cytochromes free in solution, $2\text{--}3 \times 10^4 \ M^{-1} \ s^{-1}$ (at 20°C in 20 mM Tris HCl at pH 7.4), and they proposed that the rate enhancement was due to an increase in the frequency of productive collisions.

Finally, there is a methodological point concerning the small-molecule reactions of cytochrome *c* that is relevant to its physiological reactions. The interactions between cytochrome *c* and its oxidoreductases can be studied by using small-molecule reagents to compete with the oxidoreductase for the cytochrome. Provided the interaction of the small molecule with cytochrome *c* has been characterised, information concerning the protein–protein interaction can be obtained. This approach has recently been used to study the cytochrome $c:b_5$ interaction by NMR[262] and by kinetic methods,[271] with promising results that are generally consistent with Salemme's model.[255]

Acknowledgments

We are indebted to Professor R. J. P. Williams, F.R.S., for numerous discussions on the subject of this review and for his critical reading of the manuscript. The work was supported by the Science and Engineering Research Council (SERC) and the Medical Research Council. G.R.M. gratefully acknowledges the award of an SERC Advanced Fellowship and G.W. thanks the Fellows of Merton College for a senior scholarship.

References

[1] Bennett, L. E. *Prog. Inorg. Chem.* **1973**, *18*, 1–176.

[2] Wherland, S.; Gray, H. B. *In* "Biological Aspects of Inorganic Chemistry"; Addison, A. W.; Cullen, W. R.; Dolphin, D.; James, B. R., Eds.; Wiley: New York, 1977, pp. 289–368.

[3] Sutin, N. *Adv. Chem. Ser.* **1977**, *162*, 156–172.

[4] Keilin, D. "The History of Cell Respiration and Cytochrome"; Cambridge Univ. Press: London and New York, 1966.

[5] Bielka, H.; Horecker, B. L.; Jakoby, W. B.; Karlson, P.; Keil, B.; Liebecq, C.; Lindberg, B.; Webb, E. C. "Enzyme Nomenclature 1978"; Academic Press: New York, 1979.

[6] Xavier, A. V.; Czerwinski, E. W.; Bethge, P. H.; Mathews, F. S. *Nature (London)* **1978**, *275*, 245–247.

[7] Meyer, T. E.; Kamen, M. D. *Adv. Protein Chem.* **1982**, *35*, 105–212.

[8] Mathews, F. S.; Argos, P.; Levine, M. *Cold Spring Harbor Symp. Quant. Biol.* **1971**, *36*, 387–395.

[9] Brunori, M.; Wilson, M. T. *TIBS* **1982**, *7*, 295–299.

[10] Weber, P. C.; Salemme, F. R.; Mathews, F. S.; Bethge, P. H. *J. Biol. Chem.* **1981**, *256*, 7702–7704.

[11] Ambler, R. P. *In* "From Cyclotrons to Cytochromes"; Robinson, A. B.; Kaplan, N. O., Eds.; Academic Press: New York, 1980, pp. 263–279.

[12] Sawyer, L.; Jones, C. L.; Damas, A. M.; Harding, M. M.; Gould, R. O.; Ambler, R. P. *J. Mol. Biol.* **1981**, *153*, 831–835.

[13] Moore, G. R.; Pettigrew, G. W. "Cytochromes *c*"; Springer-Verlag: New York, 1984, in preparation.

[14] Dickerson, R. E.; Timkovich, R. (1975) *In* "The Enzymes", Vol. 11; Boyer, P., Ed.; 3rd ed., Academic Press: New York, 1975, pp. 397–547.

[15] Tanaka, N.; Yamane, T.; Tsukihara, T.; Ashida, T.; Kakudo, M. *J. Biochem. (Tokyo)* **1975,** *77,* 147–162.

[16] Takano, T.; Dickerson, R. E. *J. Mol. Biol.* **1981,** *153,* 79–115.

[17] Ochi, H.; Hata, Y.; Tanaka, N.; Kakudo, M.; Sakurai, T.; Aihara, S.; Morita, Y. *J. Mol. Biol.* **1983,** *166,* 407–418.

[18] Matsuura, Y.; Takano, T.; Dickerson, R. E. *J. Mol. Biol.* **1982,** *156,* 389–409.

[19] Korszun, Z. R.; Salemme, F. R. *Proc. Natl. Acad. Sci. U.S.A.* **1977,** *74,* 5244–5247.

[20] Salemme, F. R.; Freer, S. T.; Xuong, N. H.; Alden, R. A.; Kraut, J. *J. Biol. Chem.* **1973,** *248,* 3910–3921.

[21] Timkovich, R.; Dickerson, R. E. *J. Biol. Chem.* **1976,** *251,* 4033–4046.

[22] Ambler, R. P.; Meyer, T. E.; Kamen, M. D.; Schichman, S. A.; Sawyer, L. *J. Mol. Biol.* **1981,** *147,* 351–356.

[23] Stellwagen, E. *Nature (London)* **1978,** *275,* 73–74.

[24] Wüthrich, K. *Proc. Natl. Acad. Sci. U.S.A.* **1969,** *63,* 1071–1078.

[25] Moore, G. R.; Williams, R. J. P.; Chien, J. C. W.; Dickinson, L. C. *J. Inorg. Biochem.* **1980,** *12,* 1–15.

[26] Weber, P. C.; Howard, A.; Xuong, N. H.; Salemme, F. R. *J. Mol. Biol.* **1981,** *153,* 399–424.

[27] Emptage, M. H.; Xavier, A. V.; Wood, J. M.; Alsaadi, B. M.; Moore, G. R.; Pitt, R. C.; Williams, R. J. P.; Ambler, R. P.; Bartsch, R. G. *Biochemistry* **1981,** *20,* 58–64.

[28] Moore, G. R.; McClune, G. J.; Clayden, N. J.; Williams, R. J. P.; Alsaadi, B. M.; Ångström, J.; Ambler, R. P.; Van Beeumen, J.; Tempst, P.; Bartsch, R. G.; Meyer, T. E.; Kamen, M. D. *Eur. J. Biochem.* **1982,** *123,* 73—80.

[29] Reed, C. A.; Mashiko, T.; Bentley, S. P.; Kastner, M. E.; Scheidt, W. R.; Spartalian, K.; Lang, G. *J. Am. Chem. Soc.* **1979,** *101,* 2948–2958.

[30] Weber, P. C. *Biochemistry* **1982,** *21,* 5116–5119.

[31] Higuchi, Y.; Bando, S.; Kusunoki, M.; Matsuura, Y.; Yasuoka, N.; Kakudo, M.; Yamanaka, T.; Yagi, T.; Inokuchi, H. *J. Biochem. (Tokyo)* **1981,** *89,* 1659–1662.

[32] Haser, R.; Pierrot, M.; Frey, M.; Payan, F.; Astier, J. P.; Bruschi, M.; Le Gall, J. *Nature (London)* **1979,** *282,* 806–810.

[33] Pierrot, M.; Haser, R.; Frey, M.; Payan, F.; Astier, J. P. *J. Biol. Chem.* **1982,** *257,* 14341–14348.

[34] Lemberg, R.; Barrett, J. "Cytochromes"; Academic Press: New York, 1973.

[35] Ferguson-Miller, S.; Brautigan, D. L.; Margoliash, E. *In* "The Porphyrins", Vol. 7; Dolphin, D., Ed.; Academic Press: New York, 1979, pp. 149–240.

[36] Margalit, R.; Schejter, A. *Eur. J. Biochem.* **1973,** *32,* 492–499.

[37] Gupta, R. K.; Koenig, S. H.; Redfield, A. G. *J. Magn. Reson.* **1972,** *7,* 66–73.

[38] Margalit, R.; Schejter, A. *FEBS Lett.* **1970,** *6,* 278–280.

[39] Moore, G. R.; Harris, D. A.; Leitch, F. A.; Pettigrew, G. W. *Biochim. Biophys. Acta* **1984,** *764,* 331–342.

[40] Gupta, R. K. *Biochim. Biophys. Acta* **1973,** *292,* 291–294.

[41] Salemme, F. R.; Kraut, J.; Kamen, M. D. *J. Biol. Chem.* **1973,** *248,* 7701–7716.

[42] Pettigrew, G. W.; Bartsch, R. C.; Meyer, T. E.; Kamen, M. D. *Biochim. Biophys. Acta* **1978,** *503,* 509–523.

[43] Smith, G. M. *Biochemistry* **1979,** *18,* 1628–1634.

[44] Moore, G. R.; Pettigrew, G. W.; Pitt, R. C.; Williams, R. J. P. *Biochim. Biophys. Acta* **1980,** *590,* 261–271.

[45] Keller, R. M.; Wüthrich, K.; Pecht, I. *FEBS Lett.* **1976,** *70,* 180–184.

[46] Leitch, F. A.; Moore, G. R.; Pettigrew, G. W. *Biochemistry* **1984**, *23*, 1831–1838.

[47] Perini, F.; Kamen, M. D.; Schiff, J. A. *Biochim. Biophys. Acta* **1964**, *88*, 74–90.

[48] Wood, F. E.; Cusanovich, M. A. *Arch. Biochem. Biophys.* **1975**, *168*, 333–342.

[49] Goldkorn, T.; Schejter, A. *Arch. Biochem. Biophys.* **1976**, *177*, 39–45.

[50] Keller, R. M.; Wüthrich, K.; Schejter, A. *Biochim. Biophys. Acta* **1977**, *491*, 409–415.

[51] Barakat, R.; Strekas, T. C. *Biochim. Biophys. Acta* **1982**, *679*, 393–401.

[52] Der Vartanian, D. V.; Xavier, A. V.; Le Gall, J. *Biochimie* **1978**, *60*, 321–325.

[53] Moura, J. J. G.; Santos, H.; Moura, I.; Le Gall, J.; Moore, G. R.; Williams, R. J. P.; Xavier, A. V. *Eur. J. Biochem.* **1982**, *127*, 151–155.

[54] Brandt, K. G.; Parks, P. C.; Czerlinski, G. H.; Hess, G. P. *J. Biol. Chem.* **1966**, *241*, 4180–4185.

[55] Davis, L. A.; Schejter, A.; Hess, G. P. *J. Biol. Chem.* **1974**, *249*, 2624–2632.

[56] Pettigrew, G. W.; Meyer, T. L.; Bartsch, R. G.; Kamen, M. D. *Biochim. Biophys. Acta* **1975**, *430*, 197–208.

[57] Kreishman, G. P.; Anderson, C. W.; Su, C.-H.; Halsall, H. B.; Heineman, W. R. *Bioelectrochem. Bioenerg.* **1978**, *5*, 196–203.

[58] Eddowes, M. J.; Hill, H. A. O. *Biosci. Rep.* **1981**, *1*, 521–532.

[59] Sokol, W. F.; Evans, D. H.; Niki, K.; Yagi, T. *J. Electroanal. Chem.* **1980**, *108*, 107–115.

[60] Bianco, P.; Haladjian, J. *Electrochim. Acta* **1981**, *26*, 1001–1004.

[61] Lehninger, A. L. "Principles of Biochemistry"; Worth Publ.: New York, 1982.

[62] Margoliash, E.; Smith, E. L.; Kreil, G.; Tuppy, H. *Nature (London)* **1961**, *192*, 1125–1127.

[63] Kreil, G. *Hoppe-Seyler's Z. Physiol. Chem.* **1963**, *334*, 154–166; *ibid.* **1965**, *340*, 86–87.

[64] Narita, K.; Titani, K. *J. Biochem. (Tokyo)* **1968**, *63*, 226–241.

[65] Lederer, F. *Eur. J. Biochem.* **1972**, *31*, 144–147.

[66] Mori, E.; Morita, Y. *J. Biochem. (Tokyo)* **1980**, *87*, 249–266.

[67] Chadwick, B. M.; Sharpe, A. G. *Adv. Inorg. Chem. Radiochem.* **1966**, *8*, 84–176.

[68] Baker, J. M.; Bleaney, B.; Bowers, K. D. *Proc. Phys. Soc. London, Ser. B* **1956**, *69*, 1205–1215.

[69] Hanania, G. I. H., Irvine, D. H.; Eaton, W. A.; George, P. *J. Phys. Chem.* **1967**, *71*, 2022–2030.

[70] Campion, R. J.; Deck, C. F.; King, P.; Wahl, A. C. *Inorg. Chem.* **1967**, *6*, 672–681.

[71] Figgis, B. N.; Gerloch, M.; Mason, R. *Proc. R. Soc. London, Ser. A* **1969**, *309*, 91–118.

[72] Pierrot, M.; Kern, R.; Weiss, R. *Acta Crystallogr.* **1966**, *20*, 425–428.

[73] Buckingham, D. A.; Sargeson, A. M. *In* "Chelating Agents and Metal Chelates"; Dwyer, F. P.; Mellor, D. P., Eds.; Academic Press: New York, 1964, pp. 237–282.

[74] Lewis, W. B.; Morgan, L. O. *Transition Met. Chem. (N.Y.)* **1968**, *4*, 33–112.

[75] Gudel, H. U.; Stucki, H.; Ludi, A. *Inorg. Chim. Acta* **1973**, *7*, 121–124.

[76] Campbell, I. D.; Dobson, C. M.; Williams, R. J. P.; Xavier, A. V. *J. Magn. Reson.* **1973**, *11*, 172–181.

[77] Griffith, W. P. *Compr. Inorg. Chem.* **1973**, *1*, 105–195.

[78] Hoard, J. L.; Nordsieck, H. H. *J. Am. Chem. Soc.* **1939**, *61*, 2853–2863.

[79] Stasiw, R.; Wilkins, R. G. *Inorg. Chem.* **1969**, *8*, 156–157.

[80] Toma, H. E.; Creutz, C. *Inorg. Chem.* **1977**, *16*, 545–550.

[81] Hrepic, N. V.; Malin, J. M. *Inorg. Chem.* **1979**, *18*, 409–413.

[82] Tullberg, A.; Vannerberg, N. G. *Acta Chem. Scand., Ser. A* **1974**, *A28*, 340–346.

[83] Hin-Fat, L.; Higginson, W. C. E. *J. Chem. Soc. A* **1967**, 298–301.

[84] Yager, T. D.; Eaton, G. R.; Eaton, S. S. *Inorg. Chem.* **1979**, *18*, 725–727.

[85] Van Neikerk, J. N.; Schoening, F. R. L. *Acta Crystallogr.* **1952**, *5*, 196–202.

[86] Griffiths, J. H. E.; Owen, J.; Ward, I. M. *Proc. R. Soc. London, Ser. A* **1953**, *219*, 526–542.

[87] Dean, J. A. (Ed.) "Lange's Handbook of Chemistry"; McGraw-Hill: New York, 1973, pp. 6-1–6.16.

[88] Tanaka, N.; Shirakaski, T.; Ogina, H. *Bull. Chem. Soc. Jpn.* **1958**, *38*, 1515–1517.

[89] Richards, S.; Pedersen, B.; Silverton, J. V.; Hoard, J. L. *Inorg. Chem.* **1964**, *3*, 27–33.

[90] Oakes, J.; Smith, E. G. *J. Chem. Soc., Faraday Trans. 2* **1981**, *77*, 299–308.

[91] Lind, M. D.; Hamor, M. J.; Hamor, T. A.; Hoard, J. L. *Inorg. Chem.* **1964**, *3*, 34–43.

[92] Stezowski, J. J.; Countryman, R.; Hoard, J. L. *Inorg. Chem.* **1973**, *12*, 1749–1754.

[93] Garbett, K.; Lang, G.; Williams, R. J. P. *J. Chem. Soc. A* **1971**, 3433–3436.

[94] Oakes, J.; Smith, E. G. *J. Chem. Soc., Faraday Trans. 1* **1983**, *79*, 543–552.

[95] Lang, G.; Aasa, R.; Garbett, K.; Williams, R. J. P. *J. Chem. Phys.* **1971**, *55*, 4539–4548.

[96] Gustafson, R. L.; Martell, A. E. *J. Phys. Chem.* **1963**, *67*, 576–582.

[97] Bull, C.; McClune, G. J.; Fee, J. A. *J. Am. Chem. Soc.* **1983**, *105*, 5290–5300.

[98] Garvan, F. L. *In* "Chelating Agents and Metals Chelates"; Dwyer, F. P.; Mellor, D. P., Eds.; Academic Press: New York, 1964, pp. 283–333.

[99] Williams, G. Ph.D. Thesis, University of Oxford, **1983**.

[100] Adamson, A. W.; Vorres, K. S. *J. Inorg. Nucl. Chem.* **1956**, *3*, 206–214.

[101] Im, Y. A.; Busch, D. H. *J. Am. Chem. Soc.* **1961**, *83*, 3357–3362.

[102] Weakliem, H. A.; Hoard, J. L. *J. Am. Chem. Soc.* **1959**, *81*, 549–555.

[103] Wilkins, R. G.; Yelin, R. *J. Am. Chem. Soc.* **1967**, *89*, 5496–5497; *Inorg. Chem.* **1968**, *7*, 2667–2669.

[104] Kennard, C. H. L. *Inorg. Chim. Acta* **1967**, *1*, 347–354.

[105] Busch, D. H.; Bailar, J. C. *J. Am. Chem. Soc.* **1953**, *75*, 4574–4575.

[106] Stynes, H. C.; Ibers, J. A. *Inorg. Chem.* **1971**, *10*, 2304–2308.

[107] Meyer, T. J.; Taube, H. *Inorg. Chem.* **1968**, *7*, 2369–2379.

[108] Creutz, C.; Sutin, N. *J. Biol. Chem.* **1974**, *249*, 6788–6795.

[109] Cummins, D.; Gray, H. B. *J. Am. Chem. Soc.* **1977**, *99*, 5158–5167.

[110] Khare, G. P.; Eisenberg, R. *Inorg. Chem.* **1970**, *9*, 2211–2217.

[111] Baker, B. R.; Basolo, F.; Neumann, H. M. *J. Phys. Chem.* **1959**, *63*, 371–377.

[112] Farina, R.; Wilkins, R. G. *Inorg. Chem.* **1968**, *7*, 514–518.

[113] McArdle, J. V.; Coyle, C. L.; Gray, H. B.; Yoneda, G. S.; Holwerda, R. A. *J. Am. Chem. Soc.* **1977**, *99*, 2483–2489.

[114] Cookson, D. J.; Hayes, M. T.; Wright, P. E. *Nature (London)* **1980**, *283*, 682–683.

[115] Handford, P. M.; Hill, H. A. O.; Lee, R. W.-K.; Henderson, R. A.; Sykes, A. G. *J. Inorg. Biochem.* **1980**, *13*, 83–88.

[116] Bernstein, T.; Herbstein, F. H. *Acta Crystallog., Sect. B* **1968**, *B24*, 1640–1645.

[117] Pladziewicz, J. R.; Carney, M. J. *J. Am. Chem. Soc.* **1982**, *104*, 3544–3545.

[118] Yang, E. S.; Chan, M.-S.; Wahl, A. C. *J. Phys. Chem.* **1980**, *84*, 3094–3099.

[119] Shimi, I. A. W.; Higginson, W. C. E. *J. Chem. Soc.* **1958**, 260–263.

[120] Funaki, Y.; Harada, S.; Okumiya, K.; Yasunaga, T. *J. Am. Chem. Soc.* **1982**, *104*, 5325–5328.

[121] Adamson, A. W.; Gonick, E. *Inorg. Chem.* **1963**, *2*, 129–132.

[122] Rosenheim, L.; Speiser, D.; Haim, A. *Inorg. Chem.* **1974**, *13*, 1571–1575.

[123] Schugar, H. J.; Hubbard, A. T.; Anson, F. C.; Gray, H. B. *J. Am. Chem. Soc.* **1969**, *91*, 71–77.

[124] Martell, A. W. (Ed.) "Stability Constants of Metal-Ion Complexes", Spec. Publ. No. 17. Am. Chem. Soc.: Washington, D.C., 1964.

[125] Sievers, R. E.; Bailar, J. C. *Inorg. Chem.* **1962**, *1*, 174–182.

[126] Erickson, L. E.; Ho, F. F.-L.; Reilley, C. N. *Inorg. Chem.* **1970**, *9*, 1148–1153.

[127] Whitlow, S. H. *Inorg. Chem.* **1973**, *12*, 2286–2289.

[128] Lind, M. D.; Byungkook Lee; Hoard, J. L. *J. Am. Chem. Soc.* **1965**, *87*, 1611–1612.

[129] Inagaki, F.; Miyazawa, T. *Prog. Nucl. Magn. Reson. Spectrosc.* **1981**, *14*, 67–111.

[130] Williams, R. J. P. *Struct. Bonding (Berlin)* **1982**, *50*, 79–119.

[131] Clayden, N. J.; Moore, G. R.; Williams, G.; Williams, R. J. P. *J. Mol. Biol.* **1984**, submitted for publication.

[132] Moore, G. R.; Williams, R. J. P. *Eur. J. Biochem.* **1980**, *103*, 493–550.

[133] Williams, G.; Moore, G. R.; Robinson, M. N.; Williams, R. J. P.; Porteous, R.; Soffe, N. *J. Mol. Biol.* **1984**, submitted for publication.

[134] Moore, G. R.; Robinson, M. N.; Williams, G.; Williams, R. J. P. *J. Mol. Biol.* **1984**, submitted for publication.

[135] Redfield, A. G.; Gupta, R. K. *Cold Spring Harbor Symp. Quant. Biol.* **1971**, *36*, 405–411.

[136] Keller, R. M.; Wuthrich, K. *Biochim. Biophys. Acta* **1978**, *533*, 195–206.

[137] Williams, R. J. P.; Moore, G. R.; Wright, P. E. *In* "Biological Aspects of Inorganic Chemistry"; Addison, A. W.; Cullen, W. R.; Dolphin, D.; James, B. R., Eds. Wiley: New York, 1977, pp. 369–401.

[138] Dwek, R. A. "NMR in Biochemistry"; Oxford Univ. Press: London and New York, 1973.

[139] Morris, A. T.; Dwek, R. A. *Q. Rev. Biophys.* **1977**, *10*, 421–484.

[140] Harbury, H. A.; Marks, R. H. L. *Inorg. Biochem.* **1973**, *2*, 902–954.

[141] Butler, J.; Davies, D. M.; Sykes, A. G. *J. Inorg. Biochem.* **1981**, *15*, 41–53.

[142] Yoneda, G. S.; Holwerda, R. A. *Bioinorg. Chem.* **1978**, *8* 139–159.

[143] Chapman, S. K.; Davies, D. M.; Watson, A. D.; Sykes, A. G. *ACS Symp. Ser.* **1983**, *211*, 177–197.

[144] Miller, W. G.; Cusanovich, M. A. *Biophys. Struct. Mech.* **1975**, *1*, 97–111.

[145] Ohno, N.; Cusanovich, M. A. *Biophys. J.* **1981**, *36*, 589–605.

[146] Martin, M. L.; Delpuech, J.-J.; Martin, G. J. "Practical NMR Spectroscopy"; Heyden: London, 1980.

[147] Stellwagen, E.; Shulman, R. G. *J. Mol. Biol.* **1973**, *80*, 559–573.

[148] Eley, C. G. S.; Ragg, E.; Moore, G. R. *J. Inorg. Biochem.* **1984**, in press.

[149] Van Holde, K. E. "Physical Biochemistry"; Prentice-Hall: Englewood Cliffs, New Jersey, 1971, pp. 43–44.

[150] Cantor, C. R.; Schimmel, P. R. "Biophysical Chemistry"; Freeman: San Francisco, 1980, pp. 849–886.

[151] Stellwagen, E.; Cass, R. D. *J. Biol. Chem.* **1975**, *250*, 2095–2098.

[152] Dutton, P. L.; Wilson, D. F. *Biochim. Biophys. Acta* **1974**, *346*, 165–212.

[153] Pettigrew, G. W.; Leitch, F. A.; Moore, G. R. *Biochim. Biophys. Acta* **1983**, *725*, 409–416.

[154] Eley, C. G. S.; Moore, G. R.; Williams, G.; Williams, R. J. P. *Eur. J. Biochem.* **1982**, *124*, 295–303.

[155] Huang, Z.-X.; Moore, G. R. *J. Magn. Reson.* **1983**, *52*, 505–510.

[156] Ragg, E.; Moore, G. R. *J. Inorg. Biochem.* **1984**, in press.

[157] Ahmed, A. J.; Millett, F. *J. Biol. Chem.* **1981**, *256*, 1611–1615.

[158] Butler, J.; Davies, D. M.; Sykes, A. G.; Koppenol, W. H.; Osheroff, N.; Margoliash, E. *J. Am. Chem. Soc.* **1981**, *130*, 469–471.

[159] Butler, J.; Chapman, S. K.; Davies, D. M.; Sykes, A. G.; Speck, S. H.; Osheroff, N.; Margoliash, E. *J. Biol. Chem.* **1983**, *258*, 6400–6404.

[160] Le Bon, T. R.; Cassatt, J. C. *Biochem. Biophys. Res. Commun.* **1977**, *76*, 746–750.

[161] Morton, R. A.; Breskvar, K. *Can. J. Biochem.* **1977**, *55*, 146–151.

[162] Ilan, Y.; Shafferman, A.; Feinberg, B. A.; Lau, Y.-K. *Biochim. Biophys. Acta* **1979**, *548*, 565–578.

[163] Holwerda, R. A.; Read, R. A.; Scott, R. A.; Wherland, S.; Gray, H. B.; Millett, F. *J. Am. Chem. Soc.* **1978**, *100*, 5028–5033.

[164] Wallace, C. J. A.; Offord, R. E. *Biochem. J.* **1979**, *179*, 169–182.

[165] Smith, H. T.; Staudenmayer, N.; Millett, F. *Biochemistry* **1977**, *16*, 4971–4978.

[166] Brautigan, D. L.; Ferguson-Miller, S.; Tarr, G. E.; Margoliash, E. *J. Biol. Chem.* **1978**, *253*, 140–148.

[167] Osheroff, N.; Borden, D.; Koppenol, W. H.; Margoliash, E. *J. Biol. Chem.* **1980**, *255*, 1689–1697.

[168] Koppenol, W. H. *Biophys. J.* **1980**, *29*, 493–508.

[169] Van Leeuwen, J. W.; Mofers, F. J. M.; Veerman, E. C. I. *Biochim. Biophys. Acta* **1981**, *635*, 434–439.

[170] Feinberg, B. A.; Ryan, M. D. *J. Inorg. Biochem.* **1981**, *15*, 187–199.

[171] Van Leeuwen, J. W. *Biochim. Biophys. Acta* **1983**, *743*, 408–421.

[172] Chapman, S. K.; Sinclair-Day, J. D.; Sykes, A. G.; Tam, S. C.; Williams, R. J. P. *Chem. Commun.* **1983**, 1152–1154.

[173] Osheroff, N.; Koppenol, W. H.; Margoliash, E. *In* "Frontiers of Biological Energetics", Vol. 1; Dutton, P. L.; Leigh, J. S.; Scarpa, A., Eds.; Academic Press: New York, 1978, pp. 439–449.

[174] Osheroff, N., Brautigan, D. L.; Margoliash, E. *Proc. Natl. Acad. Sci. U.S.A.* **1980**, *77*, 4439–4443.

[175] Smith, L.; Davies, H. C.; Nava, M. E. *Biochemistry* **1980**, *19*, 1613–1617.

[176] Brooks, S. P. J.; Nicholls, P. *Biochim. Biophys. Acta* **1982**, *680*, 33–43.

[177] Nicholls, P. *Biochim. Biophys. Acta* **1974**, *346*, 261–310.

[178] Barlow, G. H.; Margoliash, E. *J. Biol. Chem.* **1966**, *241*, 1473–1477.

[179] Andersson, T.; Thulin, E.; Forsen, S. *Biochemistry* **1979**, *12*, 2487–2493.

[180] Peterman, B. F.; Morton, R. A. *Can. J. Biochem.* **1979**, *57*, 372–377.

[181] Wherland, S.; Gray, H. B. *Proc. Natl. Acad. Sci. U.S.A.* **1976**, *73*, 2950–2954.

[182] Simondsen, R.; Weber, P. C.; Salemme, F. R.; Tollin, G. *Biochemistry* **1982**, *21*, 6366–6375.

[183] Hodges, H. L.; Holwerda, R. A.; Gray, H. B. *J. Am. Chem. Soc.* **1974**, *96*, 3132–3137.

[184] Goldkorn, T.; Schejter, A. *J. Biol. Chem.* **1979**, *254*, 12562–12566.

[185] Tam, S. C.; Williams, R. J. P. *J. Chem. Soc.* **1984**, in press.

[186] Creutz, C.; Sutin, N. *Proc. Natl. Acad. Sci. U.S.A.* **1973**, *70*, 1701–1703.

[187] Lambeth, D. O.; Palmer, G. *J. Biol. Chem.* **1973**, *248*, 6095–6103.

[188] Al-Ayash, A. I.; Wilson, M. T. *Biochem. J.* **1979**, *177*, 641–648.

[189] Rich, P. R.; Bendall, D. S. *Biochim. Biophys. Acta* **1980**, *592*, 506–518.

[190] Rich, P. R. *Faraday Discuss. Chem. Soc.* **1982**, *74*, 349–364.

[191] Saleem, M. M. M.; Wilson, M. T. *Biochem. J.* **1982**, *201*, 433–444.

[192] Butler, J.; Koppenol, W. H.; Margoliash, E. *J. Biol. Chem.* **1982**, *257*, 10747–10750.

[193] Kowalsky, A. *J. Biol. Chem.* **1969**, *244*, 6619–6625.

[194] Dawson, J. W.; Gray, H. B.; Holwerda, R. A.; Westhead, E. W. *Proc. Natl. Acad. Sci. U.S.A.* **1972**, *69*, 30–33.

[195] Yandell, J. K.; Kay, D. P.; Sutin, N. *J. Am. Chem. Soc.* **1973**, *95*, 1131–1137.

[196] Taube, H.; Myers, H.; Rich, R. *J. Am. Chem. Soc.* **1953**, *75*, 4118–4119.

[197] Grimes, C. J.; Piszkiewicz, D.; Fleischer, E. B. *Proc. Natl. Acad. Sci. U.S.A.* **1974**, *71*, 1408–1412.

[198] Petersen, R. L.; Gupta, R. K. *FEBS Lett.* **1979**, *107*, 427–430.

[199] Chapman, S. K. Ph.D. Thesis, University of Newcastle, **1983**.

[200] Farver, O.; Pecht, I. *Isr. J. Chem.* **1981**, *21*, 13–17.

[201] Yocom, K. M.; Shelton, J. B.; Shelton, J. R.; Schroeder, W. A.; Worosila, G.; Isied, S. S.; Bordingnon, E.; Gray, H. A. *Proc. Natl. Acad. Sci. U.S.A.* **1982,** *79,* 7052–7055.

[202] Winkler, J. R.; Nocera, D. G.; Yocom, K. M.; Bordingnon, E.; Gray, H. B. *J. Am. Chem. Soc.* **1982,** *104,* 5798–5800.

[203] Moore, G. R.; Williams, R. J. P. *Eur. J. Biochem.* **1980,** *103,* 533–541.

[204] Burns, P. D.; La Mar, G. N. *J. Am. Chem. Soc.* **1979,** *101,* 5844–5846.

[205] McGourty, J. L.; Blough, N. V.; Hoffman, B. M. *J. Am. Chem. Soc.* **1983,** *105,* 4470–4472.

[206] Holwerda, R. A.; Knaff, D. B.; Gray, H. B.; Clemmer, J. D.; Crowley, R.; Smith, J. M.; Mauk, A. G. *J. Am. Chem. Soc.* **1980,** *102,* 1142–1146.

[207] Holzwarth, J. F. *Faraday Discuss. Chem. Soc.* **1982,** *74,* 393.

[208] Wei, J.-F.; Ryan, M. D. *J. Inorg. Biochem.* **1982,** *17,* 237–246.

[209] Ewall, R. X.; Bennett, L. E. *J. Am. Chem. Soc.* **1974,** *96,* 940–942.

[210] Cassatt, J. C.; Marini, C. P. *Biochemistry* **1974,** *13,* 5323–5328.

[211] McArdle, J. V.; Yocum, K.; Gray, H. B. *J. Am. Chem. Soc.* **1977,** *99,* 4141–4145.

[212] Ilan, Y.; Shafferman, A.; Stein, G. *J. Biol. Chem.* **1976,** *251,* 4336–4345.

[213] Ilan, Y.; Shinar, R.; Stein, G. *Biochim. Biophys. Acta* **1977,** *461,* 15–24.

[214] McCray, J. A.; Kihara, T. *Biochim. Biophys. Acta* **1979,** *548,* 417–426.

[215] Chang, A. M.; Austin, R. *Biophys. J.,* **1984,** in press.

[216] Ilan, Y.; Shafferman, A. *Biochim. Biophys. Acta* **1979,** *548,* 161–165.

[217] Sutin, N.; Christman, D. R. *J. Am. Chem. Soc.* **1961,** *83,* 1773–1774.

[218] Havsteen, B. H. *Acta Chem. Scand.* **1965,** *19,* 1227–1231.

[219] Cusanovich, M. A. *In* "Frontiers of Biological Energetics", Vol. 1; Dutton, P. L.; Leigh, J. S.; Scarpa, A., Eds.; Academic Press: New York, 1978, pp. 91–100.

[220] McCray, J. A.; Kihara, T. *Biochim. Biophys. Acta* **1979,** *548,* 417–426.

[221] Williams, G.; Eley, C. G. S.; Moore, G. R.; Robinson, M. N.; Williams, R. J. P. *FEBS Lett.* **1982,** *150,* 293–299.

[222] Kihara, H. *Biochim. Biophys. Acta* **1981,** *634,* 93–104.

[223] Hopfield, J. J.; Ugurbil, K. *In* "Electron Transport and Oxygen Utilisation"; Chien Ho, Ed.; Elsevier: Amsterdam, 1982, pp. 81–87.

[224] Eley, C. G. S. Ph.D. Thesis, University of Oxford, **1982.**

[225] Takano, T.; Trus, B. L.; Mandel, N.; Mandel, G.; Kallai, O. B.; Swanson, R.; Dickerson, R. E. *J. Biol. Chem.* **1977,** *252,* 776–785.

[226] Robinson, M. N.; Eley, C. G. S.; Moore, G. R.; Williams, G.; Morita, Y.; Weiss, H.; Narita, K. *J. Mol. Biol.* **1984,** submitted for publication.

[227] Dickerson, R. E.; Takano, T.; Eisenberg, D.; Kallai, O. B.; Samson, L.; Cooper, A.; Margoliash, E. *J. Biol. Chem.* **1971,** *246,* 1511–1535.

[228] Koppenol, W. H.; Margoliash, E. *J. Biol. Chem.* **1982,** *257,* 4426–4437.

[229] Perutz, M. F. *Science* **1978,** *201,* 1187–1191.

[230] Rieder, R.; Bosshard, H. R. *J. Biol. Chem.* **1980,** *255,* 4732–4739.

[231] Koppenol, W. H.; Vroonland, C. A. J.; Braams, R. *Biochim. Biophys. Acta* **1978,** *503,* 499–508.

[232] Myer, Y. P.; Pande, A.; Pande, J.; Thallam, K. K.; Saturno, A. F.; Verma, B. C. *Int. J. Quantum Chem.* **1981,** *20,* 513–521.

[233] Marcus, R. A. *Annu. Rev. Phys. Chem.* **1964,** *15,* 155–196.

[234] Marcus, R. A.; Sutin, N. *Inorg. Chem.* **1975,** *14,* 213–216.

[235] Cannon, R. D. "Electron Transfer Reactions"; Butterworth: London, 1980.

[236] Albery, W. J. *Annu. Rev. Phys. Chem.* **1980,** *31,* 227–263.

[237] Hopfield, J. J. *Proc. Natl. Acad. Sci. U.S.A.* **1974,** *71,* 3640–3644.

[238] Chance, B.; DeVault, D. C.; Frauenfelder, H.; Marcus, R. A.; Schrieffer, J. R.; Sutin, N. (Eds.), "Tunneling in Biological Systems"; Academic Press: New York, 1979.

[239] De Vault, D. *Q. Rev. Biophys.* **1980**, *13*, 387–564.

[240] Potasek, M. J.; Hopfield, J. J. *Proc. Matl. Acad. Sci. U.S.A.* **1977**, *74*, 3817–3820.

[241] Austin, R. H.; Hopfield, J. J. *In* "Electron Transport and Oxygen Utilisation"; Chien Ho, Ed.; Elsevier: Amsterdam, 1982, pp. 73–80.

[242] Mauk, A. G.; Scott, R. A.; Gray, H. B. *J. Am. Chem. Soc.* **1980**, *102*, 4360–4363.

[243] Albery, W. J.; Eddowes, M. J.; Hill, H. A. O.; Hillman, A. R. *J. Am. Chem. Soc.* **1981**, *103*, 3904–3910.

[244]. Eddowes, M. K.; Hill, H. A. O. *Faraday Discuss. Chem. Soc.* **1982**, *74*, 331–341.

[245] Wherland, S.; Pecht, I. *Biochemistry* **1978**, *17*, 2585–2591.

[246] Phillips, C. S. G.; Williams, R. J. P. "Inorganic Chemistry"; Oxford Univ. Press (Clarendon): London and New York, 1966.

[247] Taube, H. *Adv. Chem. Ser.* **1977**, *162*, 127–144.

[248] Mailer, C.; Taylor, C. P. S. *Can. J. Biochem.* **1972**, *50*, 1048–1055.

[249] Warshel, A.; Churg, A. K. *J. Mol. Biol.* **1983**, *168*, 693–697.

[250] Moore, G. R.; Huang, Z.-X.; Eley, C. G. S.; Barker, H. A.; Williams, G.; Robinson, M. N.; Williams, R. J. P. *Faraday Discuss. Chem. Soc.* **1982**, *74*, 311–329.

[251] Capaldi, R. A.; Darley-Usmar, V.; Fuller, S.; Millett, F. *FEBS Lett.* **1982**, *138*, 1–7.

[252] Margoliash, E.; Bosshard, H. R. *TIBS* **1983**, *8*, 316–320.

[253] Argos, P.; Mathews, F. S. *J. Biol. Chem.* **1975**, *250*, 74–751.

[254] Poulos, T. L.; Freer, S. T.; Alden, R. A.; Edwards, S. L.; Skogland, U.; Takio, K.; Eriksson, B.; Xuong, N.; Yonetani, T.; Kraut, J. *J. Biol. Chem.* **1980**, *255*, 575–580.

[255] Salemme, F. R. *J. Mol. Biol.* **1976**, *102*, 563–568.

[256] Poulos, T. L.; Kraut, J. *J. Biol. Chem.* **1980**, *255*, 10322–10330; and Poulos, T. L.; Finzel, B. C. *Pept. Protein Rev.* **1984**, in press.

[257] Mauk, M. R.; Reid, L. S.; Mauk, A. G. *Biochemistry* **1982**, *21*, 1843–1846.

[258] Stonehuerner, J.; Williams, J. B.; Millett, F. *Biochemistry* **1979**, *18*, 5422–5427.

[259] Smith, H. T.; Ahmed, A. J.; Millett, F. *J. Biol. Chem.* **1981**, *256*, 4984–4990.

[260] Pettigrew, G. W. *FEBS Lett.* **1978**, *86*, 14–16.

[261] Bisson, R.; Capaldi, R. A. *J. Biol. Chem.* **1981**, *256*, 4362–4367.

[262] Waldmeyer, B.; Bechtold, R.; Bosshard, H. R.; Poulos, T. L. *J. Biol. Chem.* **1982**, *257*, 6073–6076.

[263] Eley, C. G. S.; Moore, G. R. *Biochem. J.* **1983**, *215*, 11–21.

[264] Chothia, C.; Janin, J. *Nature (London)* **1975**, *256*, 705–708.

[265] Ross, P. D.; Subramanian, S. *Biochemistry* **1981**, *20*, 3096–3102.

[266] Waldmeyer, B.; Bechtold, R.; Zurrer, M.; Bosshard, H. R. *FEBS Lett.* **1980**, *119*, 349–351.

[267] Pettigrew, G. W.; Seilman, S. *Biochem. J.* **1982**, *201*, 9–18.

[268] Hill, B. C.; Nicholls, P. *Biochem. J.* **1980**, *187*, 809–818.

[269] Petersen, L. C.; Cox, R. P. *Biochem. J.* **1980**, *192*, 687–693.

[270] Yoshimura, T.; Sogabe, T.; Aki, K. *Biochim. Biophys. Acta* **1981**, *636*, 129–135.

[271] Chapman, S. K.; Davies, D. M.; Vuik, C. P. J.; Sykes, A. G. *J. Chem. Soc., Chem. Commun.* **1983**, 868–869.

Redox Reactions at Solid–Liquid Interfaces

Michael G. Segal and Robin M. Sellers

Central Electricity Generating Board
Technology Planning and Research Division
Berkeley Nuclear Laboratories
Berkeley, Gloucestershire, England

I. INTRODUCTION

A. Scope

Redox reactions at solid–liquid interfaces occur in a wide variety of situations. In the broadest sense, this could include all of electrochemistry, but this article is limited to reactions in which both redox partners undergo chemical change. Many such reactions have received attention in recent years because of their importance in the fields of mineralogy and hydrometallurgy, in oxide scale removal and decontamination of nuclear plants, and also to some extent in the storage of solar energy.

97

Most important in this area of study are reactions in which a species in solution reacts by electron transfer with one (or more) component of a solid, in a process resulting in dissolution of that solid. Among the many examples of reductive dissolution reactions, we can cite that between hematite and V^{2+}, as described by the present authors[1]:

$$Fe_2O_3 + 2V^{2+} + 6H^+ \rightarrow 2Fe^{2+} + 2V^{3+} + 3H_2O \qquad (1)$$

A good example of an oxidative dissolution reaction is that between lead sulphide (galena) and ferric chloride, reaction (2), which is typical of the use

$$PbS + 2Fe^{3+} \rightarrow Pb^{2+} + 2Fe^{2+} + S^0 \qquad (2)$$

of ferric ion to leach metal sulphides, a subject reviewed in great detail by Dutrizac and MacDonald.[2] In many cases the work described in this article has been of an empirical nature. It is, perhaps, inevitable that when a research programme is carried out with the aim of improving an industrial process, detailed studies of kinetics and mechanisms are not always made. The aim of this article is to bring together and interpret this work as far as possible, so as to provide a better understanding of the mechanisms of the fundamental chemical processes involved.

B. Nature of the solid–liquid interface

When a metal oxide or sulphide is immersed in water, charges develop on the surface. Figure 1 shows in diagrammatic form the principal features. According to this simple picture the boundary layer consists of an inner or Helmholtz layer, and an outer or diffuse layer. For the solids under consideration here the particles are generally positively charged at an acidic pH (in the absence of other electrolytes) due to reactions of the following type:

$$>M\text{–}OH + H_2O \rightleftharpoons >M\text{–}OH_2^+ + OH^- \qquad (3)$$

Proton loss from the surface [equilibrium (4)] tends to occur only at much higher pH values:

$$>M\text{–}OH + OH^- \rightleftharpoons >MO^- + H_2O \qquad (4)$$

The important parameters influencing these equilibria are the surface charge density, adsorption densities of other ions, and electrokinetic potential. Equilibria such as (3) are usually characterised in terms of the isoelectric point (IEP), which is the pH at which there are electrically equivalent concentrations of positive and negative complexes. This is usually determined by measuring the electrophoretic mobility as a function of pH. The IEP is often

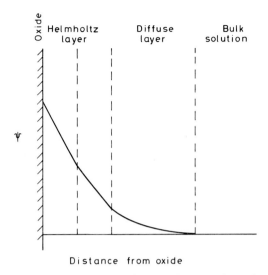

Fig. 1. Schematic representation of the metal oxide–solution interface.

equated with the point of zero charge (PZC), though it is important to recognise that this is not always the case. Some typical values of the IEP for metal oxides are given in Table 1. There is often considerable variation between different measurements, reflecting at least in part the role of the preparative route in determining the surface properties. It is important to

TABLE 1

Isoelectric points for some metal oxides[a]

Oxide	IEP
Cr_2O_3 (hydrous)	7.0
CuO	9.5
Fe_3O_4	6.5
α-Fe_2O_3	5.2, 6.7, 8.6
γ-Fe_2O_3	6.7
α-FeOOH	6.7
γ-FeOOH	7.4
MnO_2	4–4.5
NiO	10.3
PuO_2	8.6–9.0
SnO_2	4.7, 5.5, 7.3
TiO_2	4.7, 6.2
UO_2/U_3O_8	3.5
ZnO	8.7, 9.0, 9.2, 9.3

[a] From Parks.[3]

stress also that several different types of surface site may exist, and these may have different acid–base properties, not reflected in the IEP values.

The development of a surface charge gives rise to the adsorption of ionic or polar species in the Helmholtz layer. This may have a dramatic effect on redox processes at the surface, often causing a reduction in rates, but also accelerating them under certain circumstances. Clearly, the sign and magnitude of the surface charge and those of the redox partner are also likely to be important (cf. comments in following section).

C. General kinetic processes

1. Chemical control

A major consideration in determining the rate laws of heterogeneous reactions is simply that of geometry. Although many physical situations can arise, only two need concern us here: reaction at a plane surface and reaction of a particle in suspension. The reactions of suspensions are most widely studied and most often misinterpreted.

The dissolution of particles is the most common of the processes under consideration. In cases in which the rate is governed by chemical reaction, rather than diffusion control of one kind or another, then it follows that the instantaneous rate must be proportional to the surface area, since that relates directly to the effective concentration of the reacting species in the solid state. The rate law that follows from this can be expressed in a number of ways, which are equivalent.

Segal and Sellers[4] have given a detailed derivation for the case of a uniform distribution of spherical particles, which yields a rate law of the form in Eq. (5):

$$(1 - C_t/C_\infty)^{1/3} = 1 - k_{obs}t \qquad (5)$$

where C_t and C_∞ are the concentrations in solution of the dissolving species at time t and at infinite time, respectively, and k_{obs} is a composite term given by Eq. (6).

$$k_{obs} = k/r_0\rho \qquad (6)$$

where k is the rate constant for reaction (mass/unit area · time), r_0 the initial radius, and ρ the density. These equations apply equally for other particles of regular geometry, e.g., cubic crystals. This form of the equation has the advantage that k can be related directly to the reaction rate constant at a molecular level; this first-order rate constant is then a function of the concentrations of the relevant species in solution—reductant or oxidant, H^+ or OH^-, ligands, etc.

An alternative form of the same cubic rate law is often used:

$$1 - (1 - \alpha)^{1/3} \propto t \tag{7}$$

where α is the fraction reacted at time t. Since $\alpha = C_t/C_\infty$, Eqs. (5) and (7) are clearly identical. For example, Jones and Peters[5] and Dutrizac[6] use plots of $1 - (1 - \alpha)^{1/3}$ versus time as a criterion for chemical, as opposed to diffusion, control in metal sulphide dissolution experiments.

An even simpler version is useful when the products interfere with the reaction, for example by precipitation on the reactive surface. A study of initial rates can be used, and here the consequence of a reaction rate proportional to surface area is the relation (8),

$$k_{initial} \propto 1/r_0 \tag{8}$$

which is merely a special case of Eqs. (5) and (6). This is a useful test in cases in which the rate law indicates diffusion control after the initial stages of reaction.[7]

Derivation of the rate law for reaction at a plane surface is trivial in chemical-controlled cases; since the surface area of the solid, and hence the effective concentration of the active species, is constant, the dissolution is linear with time. This is independent of the reaction order with respect to the reacting species in the solid.

2. Diffusion control

a. Diffusion of reagents through the solution. Rodliffe[8] uses the following definition of a "simple" dissolution reaction, referring to the dissolution of an oxide deposit.

First-order kinetics, dependent only on the concentration of reagent adjacent to the surface and defined by a reaction rate constant.

One molecule of reagent being capable of promoting the dissolution of one molecule of the solid phase.

A negligible back reaction so that the removal of dissolution products can never limit the dissolution rate.

No limitation of dissolution by the precipitation of reaction products either on surfaces or within pores.

This definition can be applied as a criterion to all dissolution reactions, including those dependent on electron transfer. However, a reaction that obeys this definition is not necessarily controlled by the reaction rate constant. As with homogeneous reactions in solution, the rate may be controlled by the transport of reagents through the solution, that is, by diffusion control. In a heterogeneous reaction, the arrival of the reductant or oxidant at the reactive surface can be the rate-determining step; by the criteria above, this is merely a case in which the concentration of reagent adjacent to the surface is small in comparison to the bulk concentration. In the terms used more commonly by chemical engineers, this is depletion in the boundary layer.

We can consider the transfer of reagent to the solid surface, and the subsequent reaction, as described by Rodliffe[8]:

$$J = \frac{kK\beta}{1 + k\beta} \cdot C_b \cdot M_c / M_r \tag{9}$$

where J is the solid dissolution rate (kg m^{-2} s^{-1}), K is the boundary layer mass-transfer coefficient (m s^{-1}), M_c is the molecular weight of the oxide, M_r is the molecular weight of the reagent, k is the first-order rate constant for the dissolution reaction (m s^{-1}), C_b is the reagent concentration in the bulk solution (kg m^{-3}), and β is the surface roughness factor, defined as the ratio of the measured surface area [e.g., using the Brunauer–Emmett–Teller (BET) method] to geometrical surface area.

Although both the notation and the units will be unfamiliar to many inorganic and physical chemists, close inspection reveals that only two of the terms are different from those in general use: k, the first-order rate constant, is the dissolution rate (mass/unit area · time) divided by reagent concentration (mass/volume), and by correcting for the density of the solid, this can be related directly to the reaction rate constant at a molecular level; K, the boundary layer mass-transfer coefficient, is more difficult, as it is a function of the diffusion coefficients of the reacting species and the solution turbulence, etc.

The ratio of dissolution rate to that expected if dissolution kinetics are limiting is then

$$J/J_K = k/(K + k\beta) \tag{10}$$

Values of mass-transfer coefficient can be calculated for different conditions, including flat surfaces and suspended particles. Plots of the ratio J/J_K against k, which show the transition from diffusion control to chemical control, have been given by Rodliffe for various circumstances.[8]

An alternative approach may be used in the case of small particles in suspension. The Smoluchowski treatment, generally used for reactions between molecular species in solution, yields a diffusion-controlled rate constant, assuming neutral reactants:

$$k_{DC} = 8\pi Da \quad \text{ml molecule}^{-1} \text{ s}^{-1} \tag{11}$$

where D is the average diffusion coefficient of the reacting species and a the sum of the radii.[9] Where the reactants are of significantly different size, the sum of the radii is too simple a term; for the motion of a sphere in a liquid, D can also be replaced by more familiar properties, giving Eq. (12):

$$k_{DC} = \left(\frac{2RT}{3000\eta}\right)\left(2 + \frac{r_A}{r_B} + \frac{r_B}{r_A}\right) \quad \text{liter mol}^{-1}\text{ s}^{-1} \tag{12}$$

A good explanation of the derivation of these equations was given by Hague.[9]

It is interesting to compare the results of these two treatments, which are based on quite different assumptions. The very rapid reductive dissolution reaction between Fe_2O_3 and tris(picolinato)vanadium(II) is a suitable case for analysis in this context,[10]: particles of approximately spherical shape, of average radius 20 μm, react with a reductant at 2×10^{-3} M concentration. The oxide density is around 4 g cm^{-3}. Maximum diffusion-controlled rates can be calculated by each method: the mass-transfer coefficient in this system is $\sim 5 \times 10^4$ m s^{-1}, which is equivalent to a second-order rate constant for the reaction of Fe^{3+} in the solid with V(II) in solution of $k_{DC} \simeq 1 \times 10^{15}$ liter mol^{-1} s^{-1}, which compares with a value of $k_{DC} \simeq 7 \times 10^{13}$ liter mol^{-1} s^{-1} from the Smoluchowski treatment.

Experimentally, the second-order rate constant per particle is found to be > 100 liter mol^{-1} s^{-1} at 80°C,[11] which, corrected for the number of reactive sites per unit area of solid, yields $k_{obs} \geq 2.5 \times 10^{15}$ liter mol^{-1} s^{-1}. In view of the approximations and simplifications in these treatments, the discrepancy between the observed rate and the results of the calculation based on boundary layer depletion is not unreasonable. The Smoluchowski treatment predicts a significantly slower reaction, suggesting that it is not readily applicable in cases such as this. However, correction for electrostatic forces would tend to bring the calculated value closer to that observed.

In cases such as this, in which the rate may be governed by the diffusion of reagent through the solution, the actual rate law for reaction remains the same as for the case of rate-determining electron transfer: the reaction rate is proportional only to the instantaneous surface area and the concentrations of relevant reagents. Thus the cubic rate law, Eq. (5), is still obeyed for particles undergoing dissolution. Where the process does not obey the criteria outlined above for "simple" reaction, this may not hold.

b. Diffusion through a solid barrier. In reactions in which one of the products is a solid, the buildup of a product layer on the reactive surface may have a profound effect on the kinetics. Oxidative dissolution of metal sulphides generally produces elemental sulphur, and the reaction of $CuFeS_2$

particles with ferric sulphate provides a good example. A description of the resulting rate law was given by Munoz *et al.*[11] The assumptions are similar to those for other particle dissolution reactions—spherical particles, with rate proportional to instantaneous surface area—but a quasi-steady-state model, in which the oxidant diffuses through the product layer slowly in comparison with the electron transfer reaction, yields the rate law, Eq. (13):

$$1 - \tfrac{2}{3}\alpha - (1 - \alpha)^{2/3} = k_p t \tag{13}$$

where α is the fraction reacted and k_p the rate constant. A detailed derivation of this rate law was given by Habashi[12]; it neglects the change in volume that may occur as the original solid is replaced by the product layer, but it fails only at about 90% reaction.[12] The rate constant k_p is given by Eq. (14).[11]

$$k_p = 8x/\rho(D\varepsilon/\tau)a(1/d_0^2) \quad (\text{time}^{-1}) \tag{14}$$

where x is the stoichiometry factor, moles of solid reacted per mole of reagent in solution; ρ is the density of the solid being dissolved; $D\varepsilon/\tau$ is the effective diffusion coefficient of the incoming reagent through the solid barrier; a is the activity of the reagent in solution; and d_0 is the initial diameter of the particles.

A rather simpler form of Eq. (11) arises when the same processes occur at a flat surface. The derivation given by Habashi yields Eq. (15):

$$1 - (1 - \alpha)^2 = kt \tag{15}$$

This parabolic law is well known in the oxidation of metals, for example, which is a process very similar to those under consideration here.[12]

These parabolic rate laws are often used as a diagnostic test for a surface barrier mechanism, although detailed analysis of k_p values often reveals more complex mechanisms. This is discussed in more detail when we consider the reactions of metal sulphides.

II. REACTIONS OF TRANSITION METAL OXIDES

As noted previously, electron transfer at metal oxide–solution interfaces often (but not always) results in the dissolution of the solid phase. Because of the practical importance of dissolution, much effort has been directed toward technological application, with the result that more fundamental studies either have been lacking, or have become the focus of interest only in recent years. Typical reagents of this type are oxalic acid and permanganate. Many of the processes involved show obvious parallels with the equivalent homogeneous reaction, for example whether they occur by inner- or outer-sphere-type mechanisms, but other factors, particularly those relating to the structure, composition, and morphology of the oxide, may have a pronounced effect on the kinetics, etc. These differences we shall particularly emphasize.

For convenience we consider oxidative and reductive processes separately, and within these broad categories we have further subdivided the section to consider inorganic, organic, and radiation-induced reagents.

A. Reductive dissolution

1. Metal ion reductants

There is a great deal of experience in homogeneous solution chemistry on the kinetics and mechanisms of reduction by the aquo chromium(II) ion Cr^{2+}, and not surprisingly one of the earliest studies of the reductive dissolution of metal oxides employed this reductant.[13] The virtue of using Cr^{2+}, of course, is that the substitution inertness of the Cr(III) product gives important clues as to the mechanism of the electron transfer.[14] A variety of oxides have been studied. With MnO_2, PbO_2, Tl_2O_3, Mn_2O_3, Co_2O_3, and CeO_2, the major product was the hexaaquo ion, $Cr(H_2O)_6^{3+}$, suggesting a simple one-electron transfer, but with Pb_3O_4 and Ca_2PbO_4 a Cr(III) dimer was formed. This was interpreted in terms of a two-electron process, e.g.,

$$Pb(IV) + Cr^{2+} \rightarrow Pb(II) + Cr(IV) \tag{16}$$

$$Cr(IV) + Cr^{2+} \rightarrow (CrOH)_2^{4+} \tag{17}$$

There is an interesting contrast between the reduction of Tl_2O_3 and the equivalent homogeneous solution reaction, the Cr^{2+} reduction of Tl(III), which is a two-electron process. The differences among the various oxides containing lead are also to be noted. Isotopic exchange experiments have indicated that no oxygen is transferred from PbO_2 to chromium. This cannot be attributed to surface exchange being rapid compared to reduction, for extensive transfer has been observed when U^{4+} reacts with PbO_2,[15] even though this reaction is much slower than that of Cr^{2+} with PbO_2. It seemed therefore either that a $Cr^{2+}-O^{2-}$ bond was not formed in this reaction (i.e., it was outer sphere), or that it was formed but that O^{2-} was not released from the lattice when the electron was transferred.

Of course the reduction of a metal ion in homogeneous solution is localized to one centre, while at a solid surface electron transfer may occur at a point remote from the site actually reduced. In the latter case no damage has been done to the lattice at the point of attack, and Zabin and Taube[13] have therefore argued that it may not be possible for the Cr(III) to capture O^{2-} from the oxide, though with other oxides in which the O^{2-} is less tightly held it may be so. This may have been the case with MnO_2, for which Zabin and Taube[13] found approximately one labelled oxygen atom for every two Cr(III) ions formed, corresponding, it was suggested, to complete transfer in one of the steps of the two-step reduction of Mn(IV) to Mn^{2+}.

A potentially clearer demonstration of an inner-sphere pathway occurs in the dissolution of Co_2O_3 by Cr^{2+} in the presence of chloride. Here Zabin and Taube[13] found that the principal chromium-containing product was $CrCl^{2+}$, and that the rate law was of the form

$$-d[Cr^{2+}]/dt = k[Cr^{2+}][H^+][Cl^-]A \qquad (18)$$

where A is the surface area of the oxide. Although these findings are not inconsistent with an inner-sphere pathway, they do not unequivocally demonstrate this to be the mechanism, for it is possible that the chloride ion does not act as a bridging ligand but simply complexes the Cr^{2+}, lowering its net charge and hence the energy for entry into the surface layer. Zabin and Taube[13] also fail to account for the rate law, Eq. (18). The proton dependence in particular is of interest, but it is uncertain whether this arises from protonation at the point of attack or at some more remote site. This system clearly deserves further study. Chloride-mediated reactions have also been observed in the reductive dissolution of α-Fe_2O_3, Fe_3O_4, and $NiFe_2O_4$ by Cr^{2+}.[10,16]

An inner-sphere process is unquestionably involved in the reaction of U^{4+} with PbO_2 or MnO_2, which Gordon and Taube[15] have studied using oxygen tracer experiments. They have found that both oxygen atoms in the product UO_2^{2+} are derived from the oxide, and apparently that this involves transfer of two oxide ions O^{2-}, presumably in some kind of doubly bridged intermediate.

The aquo Fe^{2+}-induced dissolution of MnO_2 illustrates a further complication. Koch[17] investigated the kinetics of this reaction and found that the main factor influencing the rate was the surface area (in this sense the reaction can be said to be normal). The rate was independent of the concentration of products Mn^{2+} and Fe(III) and of Fe^{2+} in the range 0.05–0.07 M, but was first order in Fe^{2+} at ~ 0.01 M. There was also a marked difference between runs in perchlorate media (slow) and sulphate media (fast), and Cu^{2+} had a slight retarding effect. The rate-determining step appears to involve a reaction at the oxide surface that is independent (or nearly so) of the solution composition, and presumably therefore involves adsorption of Fe^{2+} prior to electron transfer. The other factors can also be understood in terms of surface adsorption: Cu^{2+}, by lowering surface site reactivity (or perhaps by retarding surface diffusion), and sulphate, by enhancing reactivity. The local charge was presumably important in these cases.

A number of other metal aquo ion reductions of metal oxides have also been reported.[1,16,18] Valverde,[18] for example, has shown that oxides such as FeO (wustite), Fe_3O_4 (magnetite), or CuO are dissolved more rapidly in the presence of aquo V(III) than in its absence, but mechanistic details are lacking.[1]

As we have described, V^{2+} reduces Fe(III) in hematite at something approaching a diffusion-controlled rate. The reaction clearly has a small activation energy, and in view of the relative substitution inertness of V^{2+} it is virtually certain that this is an outer-sphere electron transfer. This same mechanism almost certainly applies to complexes of V(II) in their reactions with oxides rich in iron(III). These reactions tend to be quite rapid even at pH values of about 4–5, and as a result such reagents are being exploited commercially.[19–23] Part of the development work has been to gain an understanding of how these processes occur. Most of the work to date has been done with the tris(picolinato)vanadium(II) complex and, because of the applications involved (the decontamination of pipework in water-cooled nuclear reactors), is concerned with the dissolution of substituted ferrites (spinel-type oxides). We deal first with the dissolution of $NiFe_2O_4$ by this complex.[4]

The amount dissolved of this oxide varied with time according to the cubic rate law, Eq. (5). The rate constants obtained from the measurements are a function of the reagent concentrations. A linear dependence on $[V(II)]$ has been determined, with a weak inverse dependence on the free picolinate [excess is required to ensure that all the V(II) is present as the tris complex] and a stronger dependence on $[H^+]$. The V(II) dependence clearly establishes the rate-determining step as electron transfer at the oxide surfaces [reaction (19)]. The other dependencies have been interpreted as arising from the effect

---

[1] Valverde's work is concerned with the effect of redox couples on acid-induced dissolutions. In a typical experiment the dissolution rate was measured in the presence of, say, V(III) + V(IV). He noted that the rate varied with [V(IV)] + [V(III)] (at constant [V(IV)/[V(III)]), and interpreted this as a failure to establish equilibrium between ions on the oxide surface and the V(III)/V(IV) couple in solution. While it is unquestionably true that the reaction is under kinetic rather than thermodynamic control, it has not been recognised that V(III) is involved chemically in the reaction or that it is the rate of reaction of this species that determines the overall kinetics. Unfortunately, therefore, few of Valverde's results throw light on the nature of the redox reaction involved.

on reaction (19) of adsorption of picolinate or protons at surface sites. The

$$V(pic)_3^- + >Fe(III) \rightarrow V(pic)_3 + >Fe(II) \tag{19}$$

mechanism can be written as follows:

$$V(pic)_3^- + >s \xrightarrow{k_a} V(pic)_3 + >s^-$$

$$>s + H^+ \xrightleftharpoons{K_b} s-H^+$$

$$V(pic)_3^- + >s-H^+ \xrightarrow{k_b} V(pic)_3 + >s-H$$

$$>s + L \xrightleftharpoons{K_c} >s-L$$

$$V(pic)_3^- + >s-L \xrightarrow{k_c} V(pic)_3 + >s-L^-$$

Scheme 1

where L is picolinate, without regard to its state of protonation, with $>s$ representing surface sites.

Assuming a simple Langmuir adsorption isotherm, this mechanism predicts a rate expression of the form shown in Eq. (20), and this accounts well for the results obtained.

$$k_{obs} = \left(\frac{k_a + k_b K_b[H^+] + k_c K_c[L]}{1 + K_b[H^+] + K_c[L]} \right)[V(II)] \tag{20}$$

The mechanism as written in Eq. (20) is of course something of a simplification. It neglects, for example, the adsorption of picolinate at protonated sites, and treats the anionic and zwitterionic forms of picolinic acid as identical. The latter is certainly not the case in the adsorption of picolinate onto hematite, in which case the binding constant has been shown to reach a maximum at a pH of ~ 4.8.[16]

The surface sites $>s$ are clearly identifiable with iron(III) ions (at the surface), but their precise nature remains elusive. This is because a number of different types of iron(III) sites can be envisaged, but it is unclear whether one of these is more reactive than the others or whether several exist with approximately the same reactivity. Broadly speaking there are three reasons for this:

1. Surface morphology.   Not all iron(III) ions at the surface are equal. In work on the volatilisation of surfaces, it is well established that certain types of sites are much more reactive than others. If we view the surface according to a simple "building block" model, then we can immediately recognise a number of different types of sites, which we can conveniently label as kink sites (k), ledge sites (l), terrace sites (t), etc. (see Fig. 2). The order of reactivity is taken to be k > l > t, since it seems likely that the reorganisation energy increases as the number of bound water molecules decreases.

2. Faults.   Many dissolution processes are known to occur preferentially

at faults in the solid phase, such as screw or edge dislocations or other defects. It therefore seems likely that the proximity to a fault may be important also for redox processes.

3. Oxide structure.   With spinel-type oxides the iron(III) ions may occupy one of two types of environment in the O^{2-} lattice, i.e., tetrahedral holes or octahedral holes.

Examination of oxide powders or, better still, polished crystal surfaces after some dissolution has taken place clearly shows that some localised attack occurs with reducing agents.[4,24] Regular etch pits have been observed, for example, suggesting attack at screw dislocations, while a common feature of attack on magnetite crystals is the formation of deep fissures. On the other hand, there is some evidence for a more general surface attack, e.g., the polished face of a crystal becomes roughened. Some typical results are shown in Fig. 3. It should be apparent from these comments that the surface area increases quite rapidly as the dissolution proceeds, at least in the initial stages. Superficially, this contradicts the assumption underlying either the shrinking-core model of particle dissolution or the linear dissolution kinetics at a planar surface. Although it is customary to develop the theories in terms of surface areas, it is better to think in terms of the numbers of surface sites. Thus, although the surface area of a planar surface increases, we suggest that the number of active sites remains constant.

Distinguishing between octahedral and tetrahedral sites is not straightforward. An attempt to do this was made by Sellers[25] by comparing the reactivity of inverse and normal spinels containing Fe(III). With inverse spinels such as nickel ferrite or cobalt ferrite, the oxide ions exist in a cubic

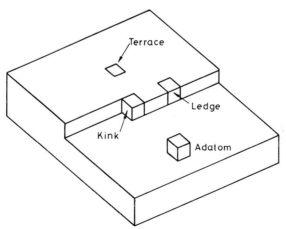

Fig. 2. Schematic representation of a solid surface, showing adatom, kink, ledge, and terrace sites.

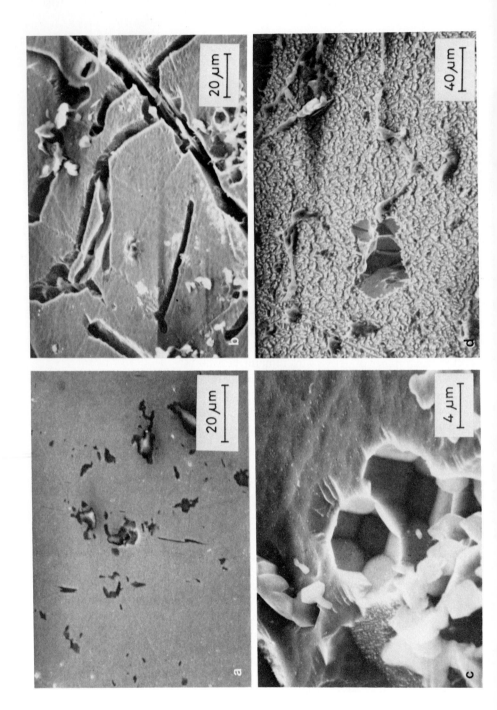

close-packed array, with half the iron(III) ions occupying tetrahedral holes and half occupying octahedral holes, while with normal spinels such as manganese ferrite all the iron(III) ions occupy octahedral holes. If, therefore, there is no difference between these sites, the reactivity should be independent of the nature of the divalent cation, whose ligand field stabilisation energy determines the structure. If, however, only iron(III) ions in octahedral sites react, then the rate should double in going from inverse to normal, while if only iron(III) in tetrahedral environments reacts, the normal spinel should not dissolve. In practice it has been found that manganese ferrite dissolves about twice as rapidly as nickel or cobalt ferrite (under otherwise identical conditions), so it is tempting to conclude that only iron(III) in octahedral holes reacts. However, some caution is required in making a comparison of this sort, for although every attempt was made to prepare and treat the oxides in the same way, it is possible that some small change in the density of faults arose during their synthesis. For the present, then, it is perhaps safest to conclude only that the evidence suggests a preference for iron(III) in octahedral holes, but that reaction with iron(III) in tetrahedral holes cannot be entirely excluded.

Ideally, of course, this could be tested with an oxide in which the iron(III) ions occupied only tetrahedral holes. This is approximately the case with iron(III) titanate Fe_2TiO_5, which has a pseudo-brookite structure, with the iron(III) ions in what may be described as highly deformed octahedral coordination or deformed tetrahedral coordination.[26] Preliminary results suggest that this material does not dissolve at an appreciable rate in vanadium(II) picolinate at 80°C, a result consistent with the observation described previously. However, the material was in the form of a crystal prepared from a high-temperature melt, and this may have affected its reactivity adversely. Furthermore, X-ray photoelectron spectroscopy (XPS) analysis revealed that the surface had become very enriched in titanium, and it may be that the apparent lack of reactivity is associated more with the slow rate of dissolution of the titanium than with that of the iron.[27]

Some difference in reactivity between iron(III) in octahedral and in tetrahedral holes is not unexpected on theoretical grounds. Assuming rigid coordination geometry in the lattice, the change in crystal field stabilisation energy in going from Fe(III) to Fe(II) is more favourable in an octahedral site than in a tetrahedral one. There may also be an effect of the relative sizes of

---

Fig. 3. Changes in the surface morphology of magnetite crystals following treatment in tris(picolinato)vanadium(II). (a) Polished surface before treatment, (b) formation of deep fissures on the $\langle 100 \rangle$ plane, (c) formation of regular pits on the $\langle 100 \rangle$ plane, and (d) general surface roughening on the $\langle 100 \rangle$ plane. Reagent conditions in (b), (c), and (d) were 6.5×10^{-3} M V(pic)$_3^-$, pH ~ 3.6, 80°C; crystals treated for 1 h. (From Allen *et al*.[27])

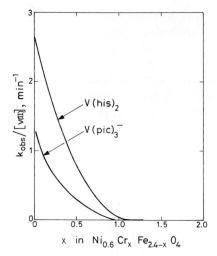

Fig. 4. The effect of chromium content on the dissolution of nickel chromium ferrites of general formula $Ni_{0.6}Cr_xFe_{2.4-x}O_4$ by tris(picolinato)vanadium(II) and bis(histidinato)-vanadium(II). (From Sellers.[25])

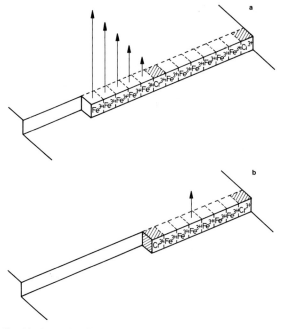

Fig. 5. Building block model of dissolution inhibition by Cr(III) ions. (a) Rapid sequential dissolution of kink sites; (b) slow dissolution of ledge site.

the different holes in the oxide ion array. If, on the other hand, the iron(III) ion is partly coordinated by water or hydroxide ions at the oxide surface, then the ligand reorganisation energy also favours reaction of octahedrally coordinated Fe(III).

In these mixed oxides the metal ions do not necessarily dissolve at the same rate. For instance the iron:nickel ratio in solution in the tris(picolinato)-vanadium(II) + $NiFe_2O_4$ reaction was much greater than the expected value of 2 in the initial stages.[4] Similar observations have been made in the dissolution of magnetite induced by Fe(II)–edta (see later).[28] Changes in the surface composition of the solid have also been detected. For example, reductive dissolution of magnetite crystals led to a surface enriched in Fe^{2+} ions as detected by XPS.[27] It is unlikely that these Fe^{2+} ions result from reduced iron(III) ions, for no Fe^{2+} ions are observed on hematite crystals treated in the same way. The surface enrichment is thought to arise because the divalent cations in these spinel oxides pass into solution only after removal of most or all of the neighbouring iron(III) ions.[4] Indeed, some Fe(III) ions may pass into solution in this way, but in the presence of powerful reductants such as $V(pic)_3^-$ they would be immediately reduced, making such a pathway indistinguishable from that involving reduction prior to dissolution.

Some very interesting results have been obtained when the iron(III) in these spinel oxides is substituted by Cr(III). Complexes based on V(II) are not sufficiently powerful reductants to reduce the Cr(III) ion, and thus can dissolve these mixed oxides only by reductive attack on the Fe(III) ions. The rate of this dissolution is markedly affected by the presence of the chromium, as illustrated in Fig. 4 for both the picolinate and histidine complexes. The effect is not simply a question of dilution, but appears to involve a real inhibition of Fe(III) reduction and dissolution. It is suggested that this occurs through a surface blocking mechanism, with the dissolution of terrace sites becoming progressively more important as the chromium content increases. Figure 5 summarises the processes involved according to a simple "building block" model.

Reduction of Cr(III) ion in oxides can be achieved using complexes of Cr(II). In principle, such processes are catalytic dissolutions and require only small amounts of reductant to effect complete dissolution if sufficient free ligand is present. The mechanism is as follows:

$$Cr(II)L_n + >Cr^{3+} \rightarrow Cr(III)L_n + Cr^{2+} \quad (21)$$

$$Cr^{2+} + nL \rightarrow Cr(II)L_n \quad (22)$$

Dissolution has been obtained with a variety of complexants, mainly amino-carboxylate compounds such as nitrilotriacetic acid (NTA).[29] In practice the reactions are not very efficient because of the prevalence of side reactions of the chromium(II) complex with water or acid:

$$Cr(II)L_n + H_2O \rightarrow Cr(III)L_n + \tfrac{1}{2}H_2 + OH^- \tag{23}$$

$$Cr(II)L_n + H^+ \rightarrow Cr(III)L_n + \tfrac{1}{2}H_2 \tag{24}$$

The highest turnover achieved was with the NTA complex at 140°C, but even here $\Delta[>Cr(III)]/\Delta[Cr(II)L_n]$ was only ~ 2.5.[25] As in the reductive dissolution of other oxides, the rate increased with decreasing pH (at 80°C and using Cr_2O_3).[29,30]

2. Other inorganic reductants

Quite a number of other inorganic reductants are known that bring about reductive dissolution, especially in cases in which the oxide is rich in iron(III). These include hydrazine, hydroxylamine, dithionite, metabisulphate, and hypophosphite, but in no case are any details available (see, e.g., Refs. [31 and 32]).

3. Organic reductants

A number of organic reductants can bring about the dissolution of metal oxides, but only two have been investigated in any detail.

Historically, one of the first practical uses of reductive dissolution was in the analysis of iron(III) oxide particulate matter, for example in power plants, and for this purpose thioglycolic acid has been widely used.[33-36] The effectiveness of this reagent shows a marked pH dependence; it has long been recognised that the pH optimum is ~ 4–5 and that this is associated with the acid–base behaviour of the sulphydryl group. Bradbury[31] studied the kinetics of the dissolution of α-Fe$_2$O$_3$ by a proton replacement method using an autotitrator. This reaction obeyed a cubic rate law and was practically independent of pH in the range 1.5–4.5, but fell rapidly above this. The mechanistic implications of this pH dependence were not commented upon, but it seems likely that acid–base equilibria on both the oxide and the reductant are involved.

Baumgartner *et al.*[37] have measured the kinetics of magnetite dissolution by this reagent by a periodic sampling method. The main concentration dependencies obtained were (1) a rate constant increasing to a plateau as the [thioglycolic acid] increased and (2) a rate increasing with increasing pH to a maximum at a pH of ~ 4.5, then falling sharply up to at least pH 6. Surface adsorption by thioglycolic acid is involved here, while the proton dependence is associated with both the oxide surface and the reductant. The reaction scheme can be written as follows:

$$> FeOH_2^+ \rightleftharpoons > FeOH + H^+ \tag{25}$$

$$HL \rightleftharpoons L^- + H^+ \tag{26}$$

$$> FeOH_2^+ + L^- \rightleftharpoons > FeOH_2^+ L^- \quad (or \quad > FeL + H_2O) \tag{27}$$

$$> FeOH_2^+ L \quad or \quad > FeL \rightarrow products \quad (rate\text{-}determining\ step) \tag{28}$$

where HL is $HSCH_2CO_2H$. This scheme accounts in a general way (but not in detail) for the experimental findings. We consider that the adsorbed complex is likely to be similar to the complex formed in the reduction of aquo Fe(III),[38] but Baumgartner et al.[37] write this in a form implying a "contact-pair" type complex.

Oxalic acid behaves in a manner similar to thioglycolic acid toward both magnetite and hematite, at least in solutions being bubbled with oxygen or containing some other oxidant.[25,39] The mechanism is as shown in reactions (25)–(28) but with HL representing $H_2C_2O_4$. In deaerated solutions, however, there is a marked acceleration in rate following an initial slow phase. Some typical results are shown in Fig. 6 for the dissolution of hematite. The cause of this difference is the formation of ferrous oxalate, which can

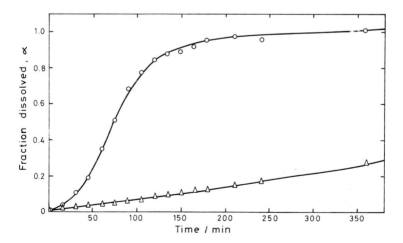

Fig. 6. Dissolution of hematite by oxalic acid. Measurements made in solutions containing 8.0×10^{-3} M oxalic acid, pH 3.7, 80°C. (○) Solution bubbled with argon; (△) solution bubbled with O_2. (From Sellers.[25])

bring about the reductive dissolution of the iron(III) in these oxides. The mechanism of the autocatalysis can be written as follows:

$$> Fe(III) + Ox \rightarrow Fe(II) + CO_2 \qquad (29)$$

$$Fe(II) + 2Ox \rightarrow Fe(II)(Ox)_2 \qquad (30)$$

$$> Fe(III) + Fe(II)(Ox)_2 \rightarrow Fe(II) + Fe(III)(Ox)_2 \qquad (31)$$

The effect of oxidants is to scavenge the Fe(II) intermediates, thus preventing reactions (30) and (31) from taking place. Baumgartner et al.[39] have confirmed this mechanism with magnetite by the addition of Fe(II), which shortened or, at high enough concentrations, eliminated the induction period. Sellers,[25] working with hematite, has developed rate laws for the auto-catalytic sequence represented by reactions (29) and (30) and has shown that these account well for the experimental results.

As with reductants such tris(picolinato)vanadium(II), dissolutions induced by oxalic acid are markedly affected by the substitution of chromium(III) for iron(III) in the magnetite structure.[40] Figure 7 shows some typical results obtained at 140°C. As explained earlier, this is thought to be due to a surface blocking mechanism.

Practical experience with oxalic acid suggests that it is sensitive to the nature of the divalent cation in the spinel structure, for in the decontami-nation of nuclear reactor components superior results are obtained with

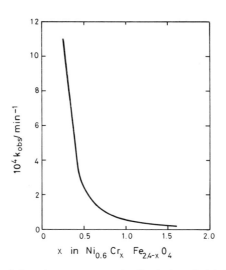

Fig. 7. The effect of chromium content on the dissolution of nickel chromium ferrites of general formula $Ni_{0.6}Cr_xFe_{2.4-x}O_4$ by 5.3×10^{-3} M oxalic acid at 140°C, pH 2.4. (From Sellers and Williams.[40])

magnetite as opposed to, for example, nickel ferrite. It is also suggested that zinc ferrite may be difficult to dissolve. Preliminary work on this using crystals of franklinite [zinc ferrite containing some manganese(II)] has shown that iron is released slowly, but that zinc is not dissolved.[27] It is far from clear why there should be this dependence, and further work is required.

Other organic reductants are known to be able to bring about the reductive dissolution of oxides, especially those containing iron(III), and these include ascorbic acid,[41] erythorbic acid,[41] catechol,[16] etc., but in no case are any kinetic details available.

4. Free-radical reductants

When a colloidal suspension of a metal oxide is irradiated, for example with ^{60}Co γ rays, changes may be induced in the colloid. In general, these originate from free-radical or stable species produced in the liquid phase, not from the direct deposition of energy at the solid–solution interface. For water the effect of radiation can be summarised as follows:

$$H_2O \rightsquigarrow e_{aq}^-, H, OH, H_2, H_2O_2, H^+ \tag{32}$$

It is the reactions of some or all of these species with the colloidal particles that cause the observed changes. Much of the work published on these systems was done before the processes involved in water radiolysis had been completely worked out, and so some caution is required. A general review of this early work was given by Sellers.[42] Two processes appeared to predominate, redox (chiefly reductive) dissolution and charge storage, and the recent resurgence of interest in the radiation and photochemistry of colloidal systems has emphasised both these processes.

The radiolysis of water, reaction (32), produces both oxidising and reducing intermediates, and if all of these react, complex changes can occur that may be difficult to interpret. It is therefore often convenient to add scavengers to control the chemistry, and in the work of Buxton et al.[28,43] this has been done by adding 2-propanol and saturating with nitrous oxide. The reactions which take place are as follows:

$$e_{aq}^- + N_2O \xrightarrow{H_2O} N_2 + OH + OH^- \tag{33}$$

$$H/OH + (CH_3)_2CHOH \longrightarrow H_2/H_2O + (CH_3)_2COH \tag{34}$$

In these systems all (or practically all) of the unstable intermediates are converted into the $(CH_3)_2\dot{C}OH$ radical, which is a powerful one-electron reductant. In the presence of monodisperse hematite colloids the radical has three fates, summarised by the following reactions [where $\dot{R}OH$ is $(CH_3)_2\dot{C}OH$]:

$$\dot{R}OH + Fe(III) \rightarrow Fe(II) + RO + H^+ \qquad (35)$$

$$\dot{R}OH + {>}Fe(III)_{colloid} \rightarrow Fe(II) + RO + H^+ \qquad (36)$$

$$\dot{R}OH + \dot{R}OH \rightarrow RO + RHOH \qquad (37)$$

Reaction (35) represents reaction with the small amount of iron(III) ion present in solution as a result of the colloid preparation; all this material is rapidly reduced in the initial stages of the radiolysis. Reaction (36) results ultimately in the dissolution of the colloids, but the rate at which this occurred was limited by the disproportionation of reaction (37). This could be overcome by the addition of edta, and even with only 1×10^{-3} M edta the dissolution rate was greatly enhanced. The explanation for this is that the edta complexes dissolved iron(II) ion to form Fe(II)–edta, which is a sufficiently powerful reductant to reduce and dissolve iron(III) ions in the hematite, reaction (38). The oxidised complex Fe(III)–edta can be reduced back to the iron(II) form by reaction with the 2-propanol radical, reaction (39), a process that competes favourably with reaction (34). As the dose

$$Fe(II)\text{–}edta + {>}Fe(III)_{colloid} \rightarrow Fe(III)\text{–}edta + Fe^{2+} \qquad (38)$$

$$\dot{R}OH + Fe(III)\text{–}edta \rightarrow RO + H^+ + Fe(II)\text{–}edta \qquad (39)$$

accumulates, therefore, the thermal dissolution pathway becomes increasingly important. This catalytic dissolution continues even after the sample is removed from the radiation field and, if sufficient edta is present, results ultimately in the complete dissolution of the colloid. No dissolution was observed under comparable conditions at pH 9.2, in which the particles were negatively charged. The reasons for this are not understood.

Magnetite colloids at pH ~ 2 paralleled the behaviour of hematite both in the presence and in the absence of edta. The iron(II) ions in the lattice were released at a slower rate than were reduced iron(III) ions, however, whether dissolution was induced by 2-propanol radicals or Fe(II)–edta. This presumably led to the surfaces on the particles becoming enriched in Fe(II), but this had no obvious effect on the kinetics of the dissolution process. These results mirror closely the findings in the tris(picolinato)vanadium(II) + $NiFe_2O_4$ system (see earlier).

Dissolution was also observed when TiO_2 colloids were irradiated in the presence of 2-propanol.[44] The overall process was again dissolution, but the mechanism was more complex. Two different colloidal preparations were used. That yielding the larger sized particles ($\sim$150-nm diameter) changed from white to blue on irradiation, the intensity of the colouration reaching a maximum after quite small doses and declining at higher doses. The blue colour disappeared on standing, on admission of oxygen or tetranitromethane after irradiation, or on photobleaching. Its rate of formation was halved when ethanol was used as a hydroxyl radical scavenger and was eliminated with either methanol or *t*-butanol, suggesting that the nature of the free-radical reductant was important in determining whether reaction took place or not. The main product from the 2-propanol was acetone, but this was formed in only half the yield expected if all the 2-propanol radicals reacted with the colloid by electron transfer. With a second TiO_2 colloid having a mean particle diameter of only ~ 10 nm no colouration was seen, but the dissolution rate was enhanced by a factor of ~ 20.

Henglein[44] has attributed the blue colouration to the formation of titanium(III) hydroxide, though other work[45] suggests that it is better described as arising from mixed valence species such as Ti(III)/Ti(IV). He suggests that these are predominantly associated with the surface, and speculates that they may pass into solution (or disappear by some other pathway) more rapidly from the smaller than from the larger sized particles. If true, this would explain why the blue colouration of the larger particles became less intense as the fraction dissolved increased. The major route for the destruction of the colour centres was suggested to be by reaction with the 2-propanol radical, regenerating 2-propanol (hence the low acetone yields), with the concomitant dissolution of the titanium site to give TiO^{2+} in solution, though why dissolution should occur is not obvious. Further, it is not clear why the initial electron transfer does not bring about dissolution in the way reduction of iron(III)-containing oxides does. Indeed, it seems possible that this mode of reaction (with perhaps a small percentage of the electron transfer events representing addition of an electron to the conduction band or to buried sites to give the blue colouration), followed by oxidation to TiO^{2+} in the bulk of the solution, can equally well account for the results.

Charge storage to give what Henglein has dubbed an "electron pool" is the predominant mode in the reduction of a variety of metal colloids such as cadmium,[46] gold,[47,48] silver,[49,50] and thallium.[51] This is somewhat beyond the scope of this article, but there are interesting lessons from these experiments that may have a bearing on redox processes at metal oxide– or metal sulphide–solution interfaces. It is perhaps worth mentioning also that with thallium metal colloids the 2-propanol radical may either transfer an electron to the particle or remove one from it,[51] in the way suggested by Henglein for TiO_2 colloids. With thallium, however, no dissolution is involved.

B.   Oxidative dissolution

Oxidative dissolution has been much less well studied than has reductive dissolution, and in no instance have we been able to find any detailed investigations of the kinetics or mechanism of the processes involved. This is all the more surprising given the commercial importance of such reagents, particularly formulations utilising permanganate. The commercial applications are concerned primarily with the treatments of chromium-rich oxides and their conversion to a form more readily attacked by reducing agents or complexing acids. The major reaction taking place is as follows:

$$>Cr(III) + MnO_4^- + 2H_2O \rightarrow CrO_4^{2-} + MnO_2 + 4H^+ \tag{40}$$

Pick[52] has shown that there are significant differences in the interaction of permanganate with oxide films formed on stainless steel and on inconel, although the oxides formed are quite similar. In general, permanganate in nitric acid media was more effective at releasing chromium from the stainless-steel oxides than from the inconel oxides, while the reverse was true at alkaline pH. Under the latter condition, the rate of release was very rapid from the inconel oxides, but in the presence of nitric acid the rate of release decreased. This appeared to be related to the deposition of manganese dioxide, which appeared to be more prevalent in the acidic media, especially on the inconel specimens. Pick has suggested that this may be associated with differences in the surface charge, but the underlying metal may also be implicated.[52]

Preliminary results in our laboratory[53] using chromia (Cr_2O_3) powders suggest that the mechanism of permanganate oxidation initially involves adsorption at the oxide surface, reaction (41), since above a certain concentration the rate becomes independent of permanganate concentration.

$$>S + MnO_4^- \rightleftharpoons >S-MnO_4^- \xrightarrow[\text{transfer}]{\text{electron}} \text{products} \tag{41}$$
$$\text{(rds)}$$

It is not widely appreciated that the function of permanganate in conditioning these chromium-rich oxides is not solely to leach out the chromium, but also to oxidise ferrous ions to ferric. That this can occur has been demonstrated by Allen *et al.*[24] using magnetite crystals. Prior to treatment both Fe(II) and Fe(III) ions could be detected at the surface by XPS. After treatment [5×10^{-3} *M* KMnO$_4$, pH 1.4 (HNO$_3$) at 80°C for 3.7 h] practically no Fe(II) ions were detectable. The photoelectrons are usually considered to emerge from the uppermost 5 nm of the surface, so it seems that the reagent has oxidised practically all Fe(II) ions in the outer 12 molecular diameters or so, a considerable penetration. No material in solution could be detected following these treatments. Further experiments to investigate how the depth of penetration varies with reagent concentration and time of exposure are planned.

A large number of other oxidants have been used, including such compounds as H_2O_2, I_2, O_3, Ce(IV) (see, for example, Ref. [18]), but so little is known of the chemistry involved that further comment is not warranted at this juncture. We note, however, an interesting use of Ce(IV) to enhance the dissolution of PuO_2, apparently via an oxidative dissolution. Some of the early work on this reaction has been roundly criticised by Ryan and Bray,[54] but there seems no doubt that such a process can operate. We note in passing also that a patent has been issued for the reductive dissolution of PuO_2 in nuclear fuel reprocessing.[55] Although this is not a field which lends itself to laboratory work, its practical importance will, we hope, ensure further detailed investigations.

C. Other comments

Most of the work described in the previous two sections has dealt with redox reactions with metal oxide powders. In practical applications, of course, the oxide is present on a metal substrate. Experience with permanganate (cf. Section II,B) suggests that the underlying metal can exert an influence on both the kinetics and the mechanism of the reaction, though Pick's work suggests that this may be more associated with the behaviour of products than of reactants. This may be true, too, with oxalic acid, for it is known that corrosive processes can become important (see, for example, Ref. [56]), but appears not to be the case with tris(picolinato)vanadium(II).[19–23] Some effect might be expected with conducting or semiconducting oxides, since the metal may impose a potential on the oxide that is different from that of the rest potential of the free oxide. Unfortunately, the critical experiments seem not to have been performed.

III. REACTIONS OF METAL SULPHIDES

A. Thermodynamics

The redox reactions of metal sulphides, especially those of the transition metals, are often more complicated than the equivalent reactions of the oxides. Either or both the metal cation and the sulphide anion can react with an oxidant, and a variety of products including elemental sulphur and/or sulphate can be formed. The importance of sulphide minerals as a source of metals has led to a large number of studies of these reactions, since oxidation processes are used both in the froth flotation used to separate ores and in the subsequent extraction.[57–60]

The complexity of these systems is illustrated by the E_h/pH diagram for the Fe–S–H_2O system (Fig. 8); this is a phase diagram showing the stable species as a function of electrical potential and pH. Kelsall has used electrochemical methods to identify the redox steps that can occur on oxidation or reduction of such materials. For example, in acid solution (pH 0), pyrrhotite ($Fe_{1-x}S$) is oxidised to Fe(II), S^0, and SO_4^{2-} at potentials below 0.7 V, while the iron is dissolved as Fe(III) at higher potentials. At higher pH values, $Fe(OH)_3$ can be formed on the surface. Chalcopyrite, $CuFeS_2$, is even more complex, and processes can include selective dissolution of iron as Fe(II), with formation of S^0 and Cu_2S and/or CuS.[58]

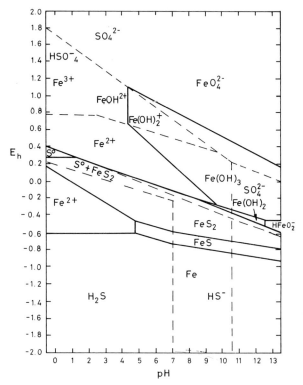

Fig. 8. E_h–pH diagram for the iron–sulphur–water system. (From Kelsall and Page.[57])

B. Kinetics

1. Dissolution by oxidation with Fe(III)

Reactions of various sulphides with oxidising agents have been of great interest as a possible route to the extraction of valuable metals from natural ores. Sulphides of (among others) Zn, Pb, Fe, and especially Cu have been studied, and of the latter the most attention has been paid to chalcopyrite, $CuFeS_2$, because of its relative abundance and because it is generally the most difficult of the copper sulphides to leach.[61] Although a number of oxidants have been used, detailed kinetic studies have been limited almost exclusively to Fe(III). The main reaction with chalcopyrite is

$$CuFeS_2 + 4Fe(III) \rightarrow Cu^{2+} + 5Fe^{2+} + 2S^0 \qquad (42)$$

although a minor pathway involving further oxidation of S to SO_4^{2-} has been observed under some conditions. The possible oxidation of the metal ion components of the solid seems to be of little significance in most cases. Most of the early work in this field was highly empirical, aimed at development of industrial processes. Some elementary studies on the effects of temperature, reagent concentrations, and other factors include the work of Ermilov[62] and Ablamov *et al.*[63,64] The addition of xylene to a well-mixed reaction, to remove the sulphur formed on the surface of the particles, was found to have a significant effect on the reaction rate under some circumstances.[62,63] Sullivan[65] studied reactions of several ores with iron(III) solutions, without any determination of reaction mechanism, but observed that iron(III) chloride was more rapid than the sulphate. Other points of interest worth noting are the observations of Ichikuni and Kamiya,[66] who found that Cu^{2+} inhibits the reaction with O_2 but catalyses the reaction of ZnS with Fe(III), and those of Kuzminkh and Yakhontova, who found that the latter reaction is much less than first order in $[Fe(III)]$.[67] This is consistent with Sullivan's earlier observations that the reaction of Cu_2S with Fe^{2+} in sulphuric acid media was virtually independent of $[Fe^{2+}]$[65]; however, in this case the reaction is reported to be independent of $[H^+]$ and the particle size as well. A two-stage reaction was observed in which the rate of the first step

$$Cu_2S + 2Fe(III) \rightarrow Cu^{2+} + 2Fe^{2+} + CuS \qquad (43)$$

is controlled by some diffusion process in the solid state.

There is little doubt that such processes, in which the reaction rate is controlled by diffusion of some reacting species through either the initial solid reactant or some product layer, are of fundamental importance in many, if not all, of the reactions under consideration here. However, as Dutrizac and MacDonald pointed out in a detailed review of the topic in 1974, "several aspects of the dissolution of chalcopyrite are not fully resolved, and certainly warrant additional detailed study."[68] Equally pertinent is the same authors' observation that "there is little unanimity among the various authors concerning the dissolution kinetics in Fe(III) solution."[68] In spite of a considerable amount of "additional detailed study" since that time, both comments remain equally valid today.

Jones and Peters[5] found that the reaction of $CuFeS_2$ with $FeCl_3$ obeyed a cubic rate law very well, with no inhibition by the product layer. The rate increased by a factor of about 2 for a 10-fold increase in iron(III) ion concentration (0.1–1 M). The results of these and a variety of electrochemical measurements were interpreted on the basis of the rate-determining step being electron transfer from the solid to Cu^{2+} in solution; the rate effect of the iron(III) ion is said to be to raise the Cu^{2+}/Cu^+ rates by rapid oxidation of copper(II). The behaviour of iron(III) sulphate as oxidant is different, and no clear mechanism is given.

Dutrizac also found that natural chalcopyrite particles obeyed the cubic rate law with $FeCl_3$, and that sintered disks of synthetic material obeyed linear kinetics.[6] A systematic study of the iron(III) ion concentration dependence yielded

$$k \propto \left[Fe(III)\right]^{0.3} \tag{44}$$

Addition of $CuCl_2$ greatly enhanced the reaction rate, but there was no evidence for significant catalysis by the product under the conditions of the simpler experiments. The kinetic results preclude diffusion control, and the reaction is described as chemically controlled, although the rate-determining step is not defined. The addition of sulphate retards the reaction, and at high concentrations there is significant loss of linearity in the reaction, which approaches the "parabolic" form expected for a reaction controlled by diffusion through an insoluble product.

Although this last paper[6] seems to present one of the most detailed and careful studies of the reaction kinetics, it cannot be said that the conclusions are generally accepted. Several other groups of workers have found good evidence for diffusion control in the $CuFeS_2/FeCl_3$ system, and their interpretations are not always in agreement either. Three long and closely argued papers by Ammou-Chokroum et al.[58,69,70] are based on the observation of "paralinear" kinetics in the reaction of compressed disks of chalcopyrite with $FeCl_3$. The initial stages of reaction are said to be the formation of a layer of copper-rich sulphide, which is compact, and a porous layer of sulphur. The reaction rate is governed by the diffusion of metal ions through the intermediate product, according to their model. This interpretation is based largely on electrochemical measurements. The observed "paralinear" kinetics have been attributed by Dutrizac to the wide range of particle sizes present in the unsintered disks studied, a criticism which, if true, renders the rest of their interpretation irrelevant.[61]

Linge used ferric nitrate and also found a rapid initial reaction, followed in this case by parabolic rather than linear reaction.[59] A diffusion-controlled reaction model was derived, and, on the basis of the absolute rate and its temperature dependence, he also suggested a solid-state diffusion step within the lattice of the starting material or intermediate product. The data are not consistent with rate being controlled by the inward diffusion of Fe(III) through a layer of elemental sulphur at the surface, a mechanism proposed by Rath *et al.* for the reaction of ZnS with $FeCl_3$[711] and by Bauer *et al.* for the reaction of $CuFeS_2$ with Fe(III) and other oxidants in H_2SO_4.[61] In the latter case it is suggested that basic iron sulphates are present in the sulphur layer. It is possible that such species are responsible for the differences in behaviour of the chalcopyrite/ferric system in different media and for the effect of added sulphate in $FeCl_3$,[6] but no attempts have yet been made to rationalise the observed results.

A very detailed study of the chalcopyrite/ferric sulphate system has been carried out by Munoz *et al.*[111] Using monosized particles, they observed good parabolic kinetics and a rate proportional to surface area. The activation energy is 83.7 kJ mol^{-1}, and the reaction is apparently zero order in iron(III) ions. Unambiguous evidence is given for the formation of a layer of sulphur on the particle surface during the reaction. The experimental rate data and a theoretical analysis using the Wagner theory of oxidation suggest that the rate-determining step is the transport of electrons through the sulphur layer to the solid–liquid interface, where rapid reduction of Fe(III) occurs.

More recent work by Dutrizac on natural chalcopyrite in chloride and sulphate media is in broad agreement with the results obtained on synthetic samples, but shows clearly why it is impossible to compare directly the results of studies carried out with different samples of the material.[7] For example, the reaction with $FeCl_3$ shows a dependence on $[Fe]^{0.3}$, compared with the 0.8 power dependence with synthetic material.[6] A cubic rate law is obeyed, confirming chemical reaction control in the chloride system, but the activation energy of 63 kJ mol^{-1} is about 20 kJ mol^{-1} higher than found previously for other natural material and the synthetic sulphide.[6] The high activation energy is attributed in this case to the presence of impurities in the natural solid, but it indicates how unreliable such measurements can be as a guide to mechanisms in systems of this kind. Very significant effects have been found when impurities were deliberately added to synthetic chalcopyrite disks.[72]

2. Other oxidants

Various other oxidants have been considered as possible leaching agents for sulphide ores and their reactions have been studied under controlled conditions. None has received the attention paid to Fe(III).

The most common alternative to iron(III), for obvious reasons, is O_2. Sullivan[65] found that the reaction of chalcocite with air was extremely slow:

$$Cu_2S + \tfrac{5}{2}O_2 + H_2O \rightleftharpoons CuSO_4 + Cu(OH)_2 \qquad (45)$$

Because the experiment was carried out in neutral pure water, the result may be misleading, but it is consistent with a number of other studies. Ichikuni and Kamiya also found no oxidation in the reaction of ZnS with sulphuric acid in aerated solutions. A mixed zinc/lead sulphide was found to react with O_2 in HCl, but only slowly; the reaction is catalysed by copper ions.[73]

A few detailed kinetic studies have been carried out. Munoz *et al.*[11] found that $CuFeS_2$ reacts with O_2 in sulphuric acid by a complex mechanism, involving both an adsorption/desorption equilibrium of O_2 on the sulphide and a diffusion-controlled step when the product layer of sulphur and iron sulphates builds up. McKay and Halpern[74] found that the rate of reaction of FeS_2 was first order in O_2 partial pressure, but was independent of other species in solution including H^+, Fe^{2+}, Fe(III), Cu^{2+}, and SO_4^{2-}. The reaction was studied in the range 100–130°C, and the products included both S^0 and SO_4^{2-}; the ratio was a function of pH and temperature. In alkaline solution MoS_2 reacts with oxygen at relatively moderate temperature and pressure,[75] yielding molybdate and sulphate if the reaction proceeds to completion:

$$MoS_2 + \tfrac{9}{2}O_2 + 6OH^- \rightleftharpoons MoO_4^{2-} + 2SO_4^{2-} + 3H_2O \qquad (46)$$

However, thiosulphate is found as the intermediate sulphur product. A mechanism involving an activated complex with two O_2 molecules at a single molybdenum site is proposed, with subsequent rapid electron transfer steps requiring OH^- participation.[75]

Chlorine is more effective than oxygen on zinc–lead sulphates, reacting rapidly with or without copper ions as catalyst.[73] In sulphuric acid, both dichromate and hydrogen peroxide react with $CuFeS_2$ by the same mechanism as do iron(III) ions, governed by the diffusion of oxidant through the product sulphur/iron sulphate layer.[11]

IV. CONCLUDING REMARKS

Apart from their intrinsic interest, redox reactions at metal oxide– or metal sulphide–solution interfaces are of immense practical importance because they offer a way of bringing about the dissolution of the solid phase that is more rapid than alternative methods, such as attack by acid or complexants. We have reviewed here some of the factors that influence the kinetics and mechanisms of such reactions, but much remains to be elucidated. There are parallels with the equivalent processes in homogeneous solution (where these occur), but a host of other factors may impinge. The electrical double layer obviously plays an important "modifying" role, but other aspects of the solid surface may be important: kink versus ledge or terrace sites, for example, or the nature and density of faults in the solid phase. These require detailed study. For oxidative processes even elementary studies will provide much needed information.

Acknowledgments

It is a pleasure to be able to thank the many people with whom we have discussed the subject covered in this article. In particular, we wish to acknowledge the help and encouragement of our colleagues at Berkeley Nuclear Laboratories, especially Drs. T. Swan, D. Bradbury, W. J. Williams, M. E. Pick, R. S. Rodliffe, G. C. Allen, P. M. Tucker, C. Kirby, and J. A. Crofts. Our thanks go also to Mr. G. Kelsall and Dr. M. A. Blesa for making available copies of their work prior to publication.

This article is published by permission of the Central Electricity Generating Board.

References

[1] Segal, M. G.; Sellers, R. M. *J. Chem. Soc., Chem. Commun.* **1980,** 991.

[2] Dutrizac, J. E.; MacDonald, R. J. C. *Miner. Sci. Eng.* **1974,** *6*, 59.

[3] Parks, G. A. *Chem. Rev.* **1965,** *65*, 177.

[4] Segal, M. G.; Sellers, R. M. *J. Chem. Soc., Faraday Trans. I* **1982,** *78*, 1149.

[5] Jones, D. L.; Peters, E. *In* "Extractive Metallurgy of Copper", Vol. 2; Yannopoulos, J. C.; Agarwal, J. C., Eds., AIME: New York, 1976.

[6] Dutrizac, J. E. *Metall. Trans. B* **1978,** *9B*, 431.

[7] Dutrizac, J. E. *Metall. Trans. B* **1981,** *12B*, 371.

[8] Rodliffe, R. S. "Water Chemistry of Nuclear Reactor Systems 2"; British Nuclear Energy Society: London, 1981, p. 383.

[9] Hague, D. N. "Fast Reactions", Wiley-Interscience: New York, 1971.

[10] Segal, M. G. Unpublished.

[11] Munoz, P. B.; Miller, J. D.; Wadsworth, M. E., *Metall. Trans. B* **1979,** *10B*, 149.

[12] Habashi, F. "Extractive Metallurgy", Vol. 1; Gordon & Breach: New York, 1969.

[13] Zabin, B. A.; Taube, H. *Inorg. Chem.* **1964,** *3*, 963.

[14] Basolo, F.; Pearson, R. G. "Mechanisms of Inorganic Reactions"; Wiley: New York, 1967.

[15] Gordon, G.; Taube, H. *Inorg. Chem.* **1962,** *1,* 69.

[16] Bradbury, D.; Segal, M. G.; Sellers, R. M.; Swan, T.; Wood, C. J. *Rep. Electr. Power Res. Inst.* **1983,** *EPRI NP-3177.*

[17] Koch, D. F. A. *Aust. J. Chem.* **1957,** *10,* 150.

[18] Valverde, N. *Ber. Bunsenges. Phys. Chem.* **1976,** *80,* 333.

[19] Bradbury, D.; Segal, M. G.; Sellers, R. M.; Swan, T.; Wood, C. J. "Water Chemistry of Nuclear Reactor Systems 2"; British Nuclear Energy Society: London, 1981, p. 403.

[20] Swan, T.; Bradbury, D.; Segal, M. G.; Sellers, R. M.; Wood, C. J. *CEGB Res.* **1982,** *13,* 3.

[21] Bradbury, D.; Segal, M. G.; Swan, T.; Comley, G. C. W.; Ferrett, D. J. *Nucl. Energy (Br. Nucl. Energy Soc.)* **1981,** *20,* 403.

[22] Bradbury, D.; Segal, M. G.; Sellers, R. M.; Swan, T.; Wood, C. J. "Decontamination of Nuclear Facilities"; American Nuclear Society: 1982, pp. 3–21.

[23] Bradbury, D.; Pick, M. E.; Segal, M. G.; Sellers, R. M.; Swan, T.; Large, N. R.; Monahan, J. "Water Chemistry of Nuclear Reactor Systems 3"; British Nuclear Energy Society: London, 1983, p. 203.

[24] Allen, G. C.; Sellers, R. M.; Tucker, P. M. *Philos. Mag. [Part] B* **1983,** *48,* L5.

[25] Sellers, R. M. Unpublished.

[26] Wells, A. F. "Structural Inorganic Chemistry"; 4th ed., Oxford Univ. Press (Clarendon): London and New York, 1975

[27] Allen, G. C., Kirby, C.; Sellers, R. M.; Tucker, P. M. Unpublished.

[28] Buxton, G. V.; Rhodes, T.; Sellers, R. M. *J. Chem. Soc., Faraday Trans. I,* **1983,** *79,* 2961.

[29] Bennett, S.; Bradbury, D.; Daniel, B.; Sellers, R. M.; Segal, M. G.; Swan, T. "Water Chemistry of Nuclear Reactor Systems 3"; British Nuclear Energy Society: London, 1983, p. 361.

[30] Bennett, S.; Daniel, B.; Sellers, R. M. Unpublished.

[31] Bradbury, D. "Water Chemistry of Nuclear Reactor Systems 1"; British Nuclear Energy Society: London, 1978, p. 373.

[32] McKeague, J. A.; Day, J. H. *Can. J. Soil Sci.* **1966,** *46,* 13.

[33] Swank, H. W.; Mellon, M. G. *Ind. Eng. Chem., Anal. Ed.* **1938,** 10.

[34] Klump, W.; Busch, H. *Mitt. Ver. Grosskesselbetr.* **1962,** *81,* 433.

[35] Wilson, A. L. *Analyst* **1964,** *89,* 402.

[36] Tetlow, J. A.; Wilson, A. L. *Analyst* **1964,** *89,* 442.

[37] Baumgartner, E.; Blesa, M. A.; Maroto, A. J. G. *J. Chem. Soc., Dalton Trans.* **1982,** 1649.

[38] Leussing, D. L.; Kolthoff, I. M. *J. Am. Chem. Soc.* **1953,** *75,* 390.

[39] Baumgartner, E.; Blesa, M. A.; Marinovich, H. A.; Maroto, A. J. G. *Inorg. Chem.* **1983,** *22,* 2226.

[40] Sellers, R. M.; Williams, W. J. *Discuss. Faraday Soc.* **1984,** *77* (Pap. 21).

[41] Miyazaki, M.; Amemiya, M.; Sato, Y.; Takamura, T. Japan-Kokai **1982,** 72/25,073.

[42] Sellers, R. M. *Cent. Electr. Gener. Board Rep.* **1976,** No. RD/B/N3707.

[43] Buxton, G. V.; Rhodes, T.; Sellers, R. M. *Nature (London)* **1982,** *295,* 583.

[44] Henglein, A. *Ber. Bunsenges. Phys. Chem.* **1982,** *86,* 241.

[45] Allen, G. C.; Wood, M. B.; Dyke, J. M. *J. Inorg. Nucl. Chem.* **1973,** *35,* 2311.

[46] Henglein, A.; Lilie, J. *J. Phys. Chem.* **1981,** *85,* 1246.

[47] Meisel, D. *J. Am. Chem. Soc.* **1979,** *101,* 6133.

[48] Westerhausen, J.; Henglein, A.; Lilie, J. *Ber. Bunsenges. Phys. Chem.* **1981,** *85,* 182.

[49] Henglein, A. *J. Phys. Chem.* **1979,** *83,* 2209.

[50] Henglein, A.; Lilie, J. *J. Am. Chem. Soc.* **1981,** *103,* 1059.

[51] Buxton, G. V.; Rhodes, T.; Sellers, R. M. *J. Chem. Soc. Faraday Trans. I* **1982,** *78,* 3341.

[52] Pick, M. E. "Water Chemistry of Nuclear Reactor Systems 3"; British Nuclear Energy Society: London, 1983, p. 61.

[53] Williams, W. J. Unpublished.

[54] Ryan, J. L.; Bray, L. A. *In* "Actinide Separations"; Navratil, J. D.; Schulz, W. W. (Eds.); Am. Chem. Soc.; Washington, D.C., 1980, p. 499.

[55] Crofts, J. A.; Mills, A. L.; Weatherley, L. R. UK Patent Appl. **1979**, No. 7,914,978.

[56] Shoesmith, D. W.; Lee, W.; Owen, D. G. *Power Ind. Res.* **1981**, *1*, 253.

[57] Kelsall, G. H.; Page, P. *Poster Int. Soc. Electrochem. Meet., 1982* **1982**.

[58] Ammou-Chokroum, M.; Steinmetz, D.; Malve, A. *Bull. Mineral.* **1978**, *101*, 26.

[59] Linge, H. G. *Hydrometallurgy* **1976**, *2*, 51.

[60] Agracheva, R. A.; Volskiya, A. N.; Egorov, A. M. *Izv. Akad Nauk SSSR* **1959**, *3*, 37.

[61] Bauer, J. P.; Gibbs, H. L.; Wadsworth, M. E. *Rep. Invest.—U.S., Bur. Mines* **1974**, *RI-7823*.

[62] Ermilov, V. V. *Tr. Inst. Metall. Obogashch., Akad. Nauk Kaz. SSR* **1960**, *3*, 163.

[63] Ablamov, A. B. *Tr. Inst. Metall. Obogashch., Akad. Nauk Kaz. SSR* **1960**, *3*, 90.

[64] Tseft, A. L.; Ablamov, A. A.; Tkachenko, O. B.; Batyrbekova, S. A.; Tulenkov, L. N.; Kartasheva, L. A. *Tr. Inst. Metall. Obogashch., Akad. Nauk Kaz. SSR* **1963**, *14*, 41.

[65] Sullivan, J. D. *Trans. Am. Inst. Min., Metall. Pet. Eng.* **1935**, *106*, 515.

[66] Ichikuni, M.; Kamiya, H. *Bull. Chem. Soc. Jpn.* **1961**, *34*, 1780.

[67] Kuzminkh, I. N.; Yakhontova, E. L. *Zh. Prikl. Khim.* **1950**, *23*, 1197.

[68] Dutrizac, J. E.; MacDonald, R. J. C. *Miner. Sci. Eng.* **1974**, *6*, 59.

[69] Ammou-Chokroum, M.; Cambazoglu, M.; Steinmetz, D. *Bull. Soc. Fr. Mineral. Cristallogr.* **1977**, *100*, 149.

[70] Ammou-Chokroum, M.; Cambazoglu, M.; Steinmetz, D. *Bull. Soc. Fr. Mineral. Cristallogr.* **1977**, *100*, 161.

[71] Rath, P. C.; Paramgru, R. K.; Jena, P. W. *Hydrometallurgy* **1981**, *6*, 219.

[72] Dutrizac, J. E.; MacDonald, R. J. C. *Can. Met. Q.* **1973**, *12*, 409.

[73] Muir, D. M.; Gale, D. C.; Parker, A. J.; Giles, D. E. *Proc. Australas. Inst. Min. Metall.* **1976**, *259*, 23.

[74] McKay, D. R.; Halpern, J. *Trans. Metall. Soc. AIME* **1958**, *212*, 301.

[75] Dresher, W. H.; Wadsworth, M. E.; Fassell, W. M. *J. Met.* **1956**, 794.

Pulse Radiolysis Studies on Metalloproteins and Metalloporphyrins

G. V. Buxton

Cookridge Radiation Research Centre
University of Leeds
Cookridge Hospital
Leeds, England

I. INTRODUCTION

The variable valency of transition metals plays a key role in the catalysis of redox processes in biological and chemical systems. In many cases reactions in which there is multiple electron transfer are known to occur by a sequence of one-electron steps involving oxidation states of the metal that are inherently unstable and/or have only a transient existence during the reaction sequence. A convenient method of generating and studying these transient intermediates is that of pulse radiolysis,[1] in which a short pulse of ionising radiation of 10^{-6} s or shorter duration is used to generate free radicals from the aqueous solution. Under appropriate conditions either oxidising or reducing radicals can be generated and their reactions with metal complexes and other substrates can be followed by optical spectroscopy or conductivity.

This article is concerned with the redox properties of metalloproteins and metalloporphyrins, for which, in aqueous solution, the pulse radiolysis technique has been used to advantage, either as the only way of generating the species of interest or because of its superior time resolution over that of other fast reaction techniques, such as stopped-flow and small-perturbation methods. Attention is focused mainly on work published during the period 1979–1983. After a brief description of how water radiolysis may be used to generate free radicals having a wide range of redox properties, there follow sections on metalloproteins and metalloporphyrins. Metalloproteins are grouped under headings of superoxide dismutases, heme proteins, nonheme iron proteins, and blue copper proteins. The section on metalloporphyrins also mentions cobalamins and other cobalt macrocyclic complexes.

II. GENERATION OF FREE RADICALS BY THE RADIOLYSIS OF WATER

A. The primary radicals e_{aq}^-, H, and OH

The interaction of ionising radiation with matter is nonspecific and the amount of energy absorbed by any one component of a chemical system is directly related to the abundance of that component.[2] In a dilute aqueous solution, for example, virtually all the energy is absorbed by the water.

The radiation chemistry of water is well understood and, in the context of the work to be reviewed here, can be summarised by reaction (1) at 10^{-12} s after the ionisation event:

$$4.6H_2O \rightsquigarrow 4.6e_{aq}^- + 4.6OH + 4.6H^+ \tag{1}$$

The figures are G values, defined as the number of molecules formed (or destroyed) per 100 eV of absorbed energy, and refer to radiation of low linear energy transfer (LET)[2] such as ^{60}Co γ rays or energetic electrons (≥ 1 MeV). By 10^{-7} s after the ionisation event, the yields of radicals have decreased and molecular products have been formed through radical–radical reactions, and the situation in neutral water is then described by reaction (2):

$$4.1H_2O \rightsquigarrow 2.7e_{aq}^- + 0.6H + 2.7OH + 2.7H^+ + 0.4H_2 + 0.7H_2O_2 \tag{2}$$

In water, 10^{-7} s is the lifetime of a radical reacting at the diffusion-controlled rate with a solute S, whose concentration is 10^{-3} mol dm^{-3}, i.e., $k[S] = 10^7$ s^{-1}, where k is the bimolecular rate constant. Thus, under these conditions the numbers of primary radicals available are as shown in reaction (2). When $k[S] < 10^7$ s^{-1} the number of radicals scavenged scarcely changes, provided $[S] \gg [\text{radicals}]$, but when $k[S] > 10^7$ s^{-1} the yield of available radicals can increase by 0.3–0.5 for each 10-fold increase in $k[S]$, depending upon the particular circumstances.[3] Under most practical conditions of dilute ($\leq 10^{-2}$ mol dm^{-3}), near-neutral solutions, use of the G values given in reaction (2) provides an accurate enough estimate of the number of free radicals available, but it should be remembered that in more concentrated solutions significantly higher values apply.

B. Conversion of the primary radicals into secondary radicals

Of the primary reactants, e_{aq}^- and H are powerful reductants and OH is a powerful oxidant (see Table 1). It is usually desirable, however, to have totally oxidising or totally reducing conditions and this is achieved by adding appropriate solutes,[3] e.g., for oxidising conditions N_2O is invariably used.

$$e_{aq}^- + N_2O \longrightarrow N_2 + O^{\cdot-} (\xrightarrow{H_2O} OH + OH^-) \tag{3}$$

For reducing conditions an organic scavenger RH for OH (and H) is used

$$OH(H) + RH \rightarrow \dot{R} + H_2O(H_2) \tag{4}$$

where $\dot{R}$ is a reducing radical or a relatively unreactive radical (see Table 1). A combination of reactions (3) and (4) provides a way of converting all the primary radicals into a single kind of secondary radical, e.g., reactions (5) and (6)

$$OH(H) + HCO_2^- \rightarrow CO_2^{\cdot-} + H_2O(H_2) \tag{5}$$

$$OH + X^- \rightarrow X + OH^- \tag{6}$$

where X^- is an inorganic anion. In this case H remains, but as it comprises only 10% of the yield of primary radicals it can often be neglected without seriously affecting the interpretation of the chemistry. Further radical conversion can also be achieved

$$CO_2^- + O_2 \rightarrow O_2^- + CO_2 \tag{7}$$

$$CO_2^- + ArNO_2 \rightarrow ArNO_2^- + CO_2 \tag{8}$$

where $ArNO_2$ is a nitroaromatic compound. These few examples serve to illustrate that the radiolysis of water provides, under well-controlled conditions, a clean method of generating a single kind of radical having the desired redox properties. Compilations of the rate constants of aliphatic carbon-centred radicals,[4] inorganic radicals,[5] metal ions in unusual valency states,[6] as well as those of e_{aq}^-,[7] H,[8] OH/O^-,[9] and HO_2/O_2^- [9] are available.

TABLE 1

Some free radicals commonly utilised as redox agents in aqueous radiation chemistry

Radical	E° (V)	Comments
Primary radicals		
e_{aq}^-	-2.9 (1)[a]	Very reactive; nucleophile
H	-2.3 (1)	Abstracts H from C—H; adds to C=C
OH	2.8 (2)	Electrophile; abstracts H from C—H; adds to C=C; $pK = 11.9$ (1)
	1.8 (3)	Refers to neutral solution
Secondary radicals		
O^-	1.6 (3)	Nucleophile; abstracts H from C—H; does not add to C=C
HO_2	1.5 (4)	$pK = 4.7$ (4)
O_2^-	1.0 $\rbrace$ (4)	Standard state: O_2 at 1 atm
O_2^-	-0.33	
CO_2^-	-2.0 (5)	$pK_{CO_2H} = 1.4$ (6)
$(CH_3)_2\dot{C}OH$	-1.5 (5)	$pK = 12.2$ (7)
$(CH_3)_2\dot{C}O^-$	-2.2 (5)	
$\dot{C}H_2(CH_3)_2COH$		Relatively unreactive generally, but easily reduced to yield $(CH_3)_2C=CH_2 + OH^-$
$\dot{C}H_3$		Formed from $e_{aq}^- + CH_3X$
Br_2^-		Reacts with ions by electron transfer; does not abstract H from C—H

[a] Number in parentheses is source of data: (1) Swallow, A. J. "Radiation Chemistry"; Longmans, Green: New York, 1973. (2) Baxendale, J. H. *Radiat. Res., Suppl.* **1964**, *4*, 114. (3) Koppenol, W. H. Unpublished. (4) Allen, A. O.; Bielski, B. H. J. *In* "Superoxide Dismutase", Vol. 1; Oberley, L. W., Ed.; CRC Press: Boca Raton, Florida, 1982, pp. 125–141. (5) Breitenkamp, M.; Henglein, A.; Lilie, J. *Ber. Bunsenges. Phys. Chem.* **1977**, *81*, 556. (6) Buxton, G. V.; Sellers, R. M. *J. Chem. Soc., Faraday Trans. 1* **1973**, *69*, 555–559. (7) Asmus, K.-D.; Henglein, A.; Wigger, A.; Beck, G. *Ber. Bunsenges. Phys. Chem.* **1966**, *70*, 756.

III.   METALLOPROTEINS

A.   Superoxide dismutases

Native superoxide dismutases (SODs) contain either Cu and Zn or Mn or Fe and they catalyse reaction (9):

$$2O_2^- + 2H^+ \rightarrow H_2O_2 + O_2 \tag{9}$$

1.   (Cu,Zn)-superoxide dismutase

The most widely studied dismutase has been bovine SOD, which contains two atoms each of Cu and Zn of which only the Cu atoms are catalytically active. The catalytic mechanism has been shown[10,11] to involve alternate reduction and oxidation of copper by O_2^-

$$O_2^- + Cu(II) \longrightarrow Cu(I) + O_2 \tag{10}$$

$$O_2^- + Cu(I) \xrightarrow{2H^+} H_2O_2 + Cu(II) \tag{11}$$

with k_{10} and k_{11} in the range $2.0–3.7 \times 10^9$ dm^3 mol^{-1} s^{-1}. Under catalytic, or so-called turnover, conditions the observed rate constant k_{cat} for the dismutation of O_2^- is equal to $2k_{10}k_{11}/(k_{10} + k_{11})$, so that in this case $k_{cat} \simeq k_{10} \simeq k_{11}$. Although the activity of the enzyme is constant between pH 6 and 9, it decreases[12] when pH < 5.3, and the question arose as to whether this pH dependence is due to a lower reactivity of $H\dot{O}_2$ (pK_a = 4.7) or to a reversible transition involving Cu^{2+} that is known to occur[13] in bovine SOD between pH 3 and 5. The data were shown[12] to be best described by a kinetic scheme in which it is assumed that $H\dot{O}_2$ and O_2^- both reduce and oxidise copper as in reactions (10) and (11), with a common rate constant of 3.0×10^9 dm^3 mol^{-1} s^{-1}, and that acid inactivation of SOD occurs with $K_{12} = 4.0 \times 10^{-5}$ mol dm^{-3}.

$$SOD_{active} + H^+ \rightleftharpoons SOD_{inactive} \tag{12}$$

This kinetic treatment gave a good fit to the data and it was also shown that acid does not inactivate SOD irreversibly.

A further question concerns the reaction of SOD with organic peroxy radicals, e.g., from lipids. To gain information on the likelihood of such a reaction, the peroxy radical from 2-methyl-2-propanol [$\dot{O}_2CH_2C(Me)_2OH$; see Table 1] was generated in the presence of SOD. This radical was chosen because, unlike α-hydroxyperoxy species, it does not eliminate O_2^-. No reaction was observed with SOD, indicating a rate constant $< 10^8$ dm^3 mol^{-1} s^{-1}, but reactions of other organic peroxy radicals with SOD cannot yet be ruled out.[12]

Zinc is not essential to the enzyme activity of bovine SOD and is believed to have a structure function. Indeed, in a recent pulse radiolysis study[14] it was shown that removal of the Zn atoms does not affect significantly either the turnover rate constant ($[SOD] \ll [O_2^-]$) or the rate constants when $[SOD] > [O_2^-]$ for oxidation and reduction of SOD. It was concluded, in agreement with previous findings, that zinc is not essential for catalytic activity but may contribute stability to the active centre, and that all the Cu atoms are involved in the catalytic cycles both for the native and for the Zn-free enzyme.

Both Cu and Zn may be substituted by Co in the native SOD to yield (Cu,Co)-, (Co,Zn)-, and (Co,Co)-proteins. The (Cu,Co)-derivative has been shown to be fully active[14] whereas the (Co,Zn)- and (Co,Co)-proteins have activities that are some three orders of magnitude less than that of the native (Cu,Zn)-protein.[15] The rate constants under turnover conditions have been found to be 4.8×10^6 and 3.1×10^6 dm^3 mol^{-1} s^{-1}, respectively, and were independent of pH in the range 7.4–9.4. It was also shown[15] that the Co-substituted protein has no activity by comparing its effect on the decay of O_2^- with that of the apoprotein.

Under stoichiometric conditions ($[O_2^-] \simeq [Co]$), rate constants for the reaction of O_2^- with the Co-proteins are 1.5–1.6×10^9 dm^3 mol^{-1} s^{-1}, but are an order of magnitude lower (1.9–2.3×10^8 dm^3 mol^{-1} s^{-1}) in the presence of phosphate or pyrophosphate (7 mmol dm^{-3}).[15] This partial inhibition by phosphate supports the idea[16] that a net positive charge in the vicinity of the active site is important for the substrate–metal interaction.

The rate constants for the reaction of O_2^- with the Co-proteins are similar to those for O_2^- reacting with Co(II) macromolecules for which the initial product has been assigned as a Co(III)–peroxo or a Co(II)–superoxo complex.[17] Superoxide has also been shown[18] to react with Co(III)–superoxo complexes to form O_2 and the corresponding peroxo complex. By analogy, it was suggested[15] that a similar mechanism may operate in the case of the cobalt-substituted SOD (p is the SOD protein):

$$p\text{--}Co(II) + O_2^- \longrightarrow (p\text{--}Co\text{--}O_2)^{\ddagger}$$
$$\downarrow O_2^- + 2H^+$$
$$p\text{--}Co(II) + O_2 + H_2O_2$$

However, it was pointed out[15] that an alternative mechanism involving oxidation and reduction of the metal centre is kinetically indistinguishable from the scheme given above. This work provides the first demonstration that substitution of copper by another metal in (Cu,Zn)-protein results in a protein that retains superoxide dismutase activity.

2.   Iron-containing superoxide dismutases

Iron-containing SOD generally consists of two identical subunits, each binding a single Fe(III). In a study of the Fe-SOD from *Photobacterium leiognathi* it was shown[19] that the catalytic mechanism is described by the redox cycle:

$$\text{Fe(III)} + O_2^- \longrightarrow \text{Fe(II)} + O_2 \tag{13}$$

$$\text{Fe(II)} + O_2^- \xrightarrow{2H^+} \text{Fe(III)} + H_2O_2 \tag{14}$$

with a rate constant under turnover conditions of 5.5×10^8 dm^3 mol^{-1} s^{-1}, which is constant for $6.2 < \text{pH} < 9.0$. This behaviour is closely similar to that of bovine (Cu,Zn)-SOD and there is no evidence of saturation of the catalytic activity. On the other hand, Fe-SOD from *Escherichia coli* B does exhibit saturation kinetics that can be described by the Michaelis–Menten-type rate law[20]

$$-d[O_2^-]/dt = 2k_s[O_2^-]^2 + k_{cat}[\text{Fe}][O_2^-]/(K_m + [O_2^-]) \tag{15}$$

where k_s is the rate constant for spontaneous dismutation and the parameters k_{cat} and K_m are treated as composite constants. Over the pH range 7.2–10.4 $k_{cat} \simeq 3 \times 10^4$ s^{-1} and $K_m \simeq 10^{-4}$ [1 + exp(pH − 8.8)] M. This dependence of K_m on pH is interpreted in terms of an ionisation of the protein with $pK_a = 8.8$, with only the low-pH form able to bind O_2^-:

$$\text{Fe–p–H} \rightleftharpoons \text{Fe–p} + H^+ \tag{16}$$

Some similarities with azide binding to Fe(III) in this protein, which also shows saturation behaviour, have been noted.[20]

3.   Other systems showing superoxide dismuting activity

Before concluding this section mention should be made of the fact that SOD activity is not limited to enzyme systems. Certain metalloporphyrins and low-molecular-weight transition metal complexes are also efficient catalysts of reaction (9). The aquo ion Cu(H$_2$O)$_6^{2+}$ has a catalytic efficiency about twice that of the native (Cu,Zn)-enzyme, whereas the Cu(II)–edta complex shows no activity, and it has been suggested[21] that the much lower activity of some other copper complexes is due to the presence of a small amount of free Cu(II) in equilibrium with the complex. A number of Cu(II)–histidine complexes have been studied[22] as model systems for SOD. Of the six known complexes, only $(Cuhis_2H)^{3+}$ shows catalytic activity, with k_{cat} = 3.4 $\pm$ 0.9 $\times$ 10^8 dm^3 mol^{-1} s^{-1} for 2 < pH < 7. From the kinetic data it is not possible to distinguish between electron transfer (17) and adduct formation (18)

$$(Cuhis_2H)^{3+} + O_2^- \rightarrow (Cuhis_2H)^{2+} + O_2 \qquad (17)$$

$$(Cuhis_2H)^{3+} + O_2^- \rightarrow [(Cuhis_2H)\cdots O_2]^{2+} \qquad (18)$$

but theoretical considerations favour the latter mechanism.[22] Of the Cu(II)–histidine complexes, the active $(Cuhis_2H)^{3+}$ is unique in having one imidazole nitrogen protonated, leaving just one octahedral site on the copper accessible to O_2^- or $H\dot{O}_2$. Similar conclusions have been drawn concerning the active site in the native (Cu,Zn)-protein.[14]

By contrast, hydroxo and aquo complexes of iron show no activity while edta complexes are weakly active (see later). SOD activity is also exhibited[21] by iron(III) and cobalt(III) derivatives of tetrakis(4-N-methylpyridyl)porphine, as well as by tetraphenylporphinesulphonatoferrate(III), as shown by assay methods. The catalytic efficiency of a number of metalloporphyrins has been investigated by pulse radiolysis.[23,24]

In one study[23] only tetrakis(4-N-methylpyridyl)porphineiron(III) [Fe-(III)TMpyP] was investigated. Experiments were carried out in 1 mmol dm^{-3} phosphate at pH 7.8 and ionic strength 5 $\times$ 10^{-2} mol dm^{-3}. At low Fe(III)-TMpyP concentration, in which the metalloporphyrin is in the monomeric form, the decay of O_2^- obeyed the rate law

$$-d[O_2^-]/dt = k_{cat}[Fe(III)TMpyP] \qquad (19)$$

with $k_{cat} \sim 3 \times 10^7$ dm^3 mol^{-1} s^{-1}. Rate constants for the reaction of e_{aq}^-, CO_2^-, and O_2^- were measured at higher [Fe(III)TMpyP], in which μ-oxo-and/or μ-hydroxo-bridged dimers are present. Under these conditions only apparent rate constants were measured, i.e., no distinction was made between reaction with monomer and dimer, and were as follows: $k(e_{aq}^-)$ = 2.4 $\pm$ 0.5 $\times$ 10^{10} dm^3 mol^{-1} s^{-1}, $k(CO_2^-)$ = 7.1 $\pm$ 1 $\times$ 10^9 dm^3 mol^{-1} s^{-1}, and $k(O_2^-) \simeq 7 \pm 2 \times 10^8$ dm^3 mol^{-1} s^{-1}. For e_{aq}^- and CO_2^-, absorbance changes at 580 (growth) and 350 nm (bleaching) occurred at the same rate and resulted in the same spectral changes. For O_2^- the spectral changes at these wavelengths were different, indicating that O_2^- reacts differently from e_{aq}^- and CO_2^-. One possibility suggested[23] was that e_{aq}^- and CO_2^- reduce Fe(III) to Fe(II) by electron transfer, while O_2^- forms an adduct that reacts with a second $O_2^{\cdot-}$.

$$O_2^{\cdot-} + \text{Fe(III)TMpyP} \rightarrow \text{Fe(III)TMpyP-O}_2^{\cdot-} \qquad (20)$$

Such a scheme is analogous to that for Co(III)–peroxo complexes and $[(\text{Cuhis}_2\text{H})-\text{O}_2]^{2+}$ mentioned earlier. The value of $k_{cat} \sim 3 \times 10^7$ dm^3 mol^{-1} s^{-1} indicates that Fe(III)TMpyP is about 1% as active as the native (Cu,Zn)-enzyme. This compares with 3% as determined by the standard superoxide assay.[21]

In a more extensive study[24] of Fe(III)TMpyP, $k_{20} = 2 \times 10^9$ dm^3 mol^{-1} s^{-1} and $k_{21} = 2.3 \times 10^9$ dm^3 mol^{-1} s^{-1}

$$\text{Fe(III)TMpyP-O}_2^{\cdot-} + O_2^{\cdot-} \xrightarrow{2\text{H}^+} \text{Fe(III)TMpyP} + \text{H}_2\text{O}_2 \qquad (21)$$

at pH 8.1 and zero ionic strength were reported. Thus, this iron porphyrin appears to have the same SOD activity as the native (Cu,Zn)-protein, in contradiction to the result[23] presented above. Both k_{20} and k_{21} decreased with increasing ionic strength, indicating charges on the pophyrin of 5 ± 0.2 at pH 5.6 and 4 ± 0.2 at pH 8.0 and showing that both axial ligands at pH 5.6 are water molecules. At an ionic strength of 5×10^{-2} mol dm^{-3}, k_{cat} is $\sim 3 \times 10^8$ dm^3 mol^{-1} s^{-1}, which is an order of magnitude greater than that reported by Ilan *et al.*[23] The reason for this discrepancy is unclear.

When the axial water molecules were replaced by ligands to form the dicyano, dihistidyl, and bis(imidazole) complexes, $O_2^{\cdot-}$ was found[24] to react 10^3 times more slowly, whereas the rates of reduction of the complexes by e_{aq}^-, $\text{CO}_2^{\cdot-}$, and $\dot{\text{C}}\text{H}_2\text{OH}$ were much less affected. These findings are in accord with the idea[23,24] that $O_2^{\cdot-}$ reacts by an inner-sphere mechanism while the other reducing radicals react by an outer-sphere process.

Several other water-soluble metal porphyrins have been investigated for SOD activity,[25] but none is as active as Fe(III)TMpyP. The order of catalytic efficiency found was Fe(III)TMpyP $\gg$ Mn(III)TMpyP > Co(III)TMpyP $\simeq$ Mn(III)TAP > Fe(III)TPPS$_4$, where TAP is tetra(4-*N,N,N*-trimethylanilinium)porphyrin and TPPS$_4$ is tetra(4-sulphonatophenyl)porphyrin. Co-(III)TPPS$_4$ and Mn(III)TPPS$_4$ showed no activity, nor did any of these porphyrins having Cu(II), Ni(II), or Zn(II) as the metal centre. The activity of the less reactive metalloporphyrins decreases with decreasing reduction potential, suggesting that an outer-sphere mechanism operates in these cases. An exception is Co(III)TMpyP, which has a high reduction potential, but this is compensated for by a low self-exchange rate constant of 20 dm^3 mol^{-1} s^{-1}.[26]

That direct methods of observing SOD activity are to be preferred over indirect assays is illustrated by the conflicting reports on the activity shown by Fe–edta. Although direct observation of $O_2^{\cdot-}$ has shown[27] that Fe–edta is weakly active, this activity could not be demonstrated using the cytochrome c/xanthine oxidase system.[28] A recent pulse radiolysis study[29] has confirmed the weak activity of Fe–edta as a SOD and also resolved the apparent conflict noted above. When cytochrome c was added to the O_2^-/Fe–edta system it was shown that Fe(II)–edta, formed in the reduction half of the catalytic cycle, reacted sufficiently rapidly with ferricytochrome c under the conditions of the assay to mask the catalytic activity.

Reduction of ferricytochrome c [cyt(III)] by O_2^- as a method of detecting the formation of this radical by enzyme systems and of assaying for SOD activity has prompted several pulse radiolysis studies of the kinetics of the reaction. Generally, the results show that the reaction rate is pH dependent, but no good agreement has been established on rate constants at any one pH.[30] Some of the discrepancies have been attributed[31] to the presence of copper impurities that are present in trace amounts in cytochrome c preparations and that, as noted earlier, can act as efficient catalysts.

In a recent study[30] the influence of Cu^{2+} on the apparent rate of reaction of O_2^- with cyt(III) was clearly demonstrated. Addition of only 2×10^{-8} mol dm^{-3} Cu^{2+} to a solution containing 10^{-5} mol dm^{-3} cyt(III) at pH 7 increased the rate of decay of O_2^- threefold. The Cu^{2+} has no effect on the rate at pH > 9 and its enhancing effect at lower pH is eliminated by 10^{-6} mol dm^{-3} edta. At pH 7.8, $k[O_2^- + \text{cyt(III)}] = 2.6 \pm 0.1 \times 10^5$ dm^3 mol^{-1} s^{-1}; using this value it has been calculated[30] that one unit of (Cu,Zn)-SOD activity is equivalent to 3.6 ± 0.3 pmol of the enzyme, as compared with 9 pmol estimated earlier.[32] Other aspects of the reactions of free radicals with cytochrome c are presented in Section III,B,1.

4. Related reactions of O_2^-

While it is generally assumed that the role of SOD is to protect the biological system from harmful effects of the superoxide radical, it has been pointed out[33] that O_2^- is relatively unreactive, reducing and oxidising the heme groups of methemoglobin and oxyhemoglobin, respectively, with rate constants of only $\sim 4 \times 10^3$ dm^3 mol^{-1} s^{-1} at pH 7.[34] Because of this low reactivity, it has been suggested[33] that O_2^- may be the precursor of more reactive species, including semiquinones formed from naturally occurring quinones, Q, through reaction (22)

$$O_2^- + Q \rightleftharpoons Q^- + O_2 \tag{22}$$

and that the role of SOD might be to prevent the formation of such reactive secondary species.[35] Pulse radiolysis studies have shown[33] that semiquinone radicals from 9,10-anthraquinone-2-sodium sulphonate, menadione, duroquinine, and 2,5-dimethylbenzoquinone all reduce methemoglobin and cytochrome c at much faster rates than does O_2^-. Moreover, it was shown[33] that equilibrium (22) is maintained under steady-state conditions in which O_2^- is generated by γ radiolysis or by enzymes, and that SOD suppresses the reduction of the heme proteins by Q^-.

Another way in which O_2^- may initiate biological damage is through "superoxide-driven Fenton chemistry"[36] whereby hydroxyl radicals are generated from reduced metal ions and hydrogen peroxide.

$$M^{(n-1)+} + H_2O_2 \rightarrow M^{n+} + \dot{O}H + OH^- \tag{23}$$

In a study of the radiation-induced damage of penicillinase it was shown[37] that O_2^- is effective only when Cu^{2+} is present and that added H_2O_2 further enhances the damage. The effect is eliminated completely by edta. The suggested mechanism is shown in reactions (24) and (25)[37]

$$\text{protein–Cu(II)} + O_2^- \rightarrow \text{protein–Cu(I)} + O_2 \tag{24}$$

$$\text{protein–Cu(I)} + H_2O_2 \rightarrow \text{protein–Cu(II)} + \dot{O}H + OH^- \tag{25}$$

with the $\dot{O}H$ being formed in the immediate vicinity of the biological target. Because of the general unreactivity of O_2^- the radical could be expected to migrate long distances and to react selectively with protein–metal complexes as in reaction (24).

B.   Heme proteins

1.   Cytochrome c

Cytochrome c is involved in biological electron transport, with the heme ion shuttling between (II) and (III) oxidation states. Pulse radiolysis studies have been aimed at discovering the mechanism of electron transfer between simple reducing and oxidising free radicals in the expectation that this may help to reveal the mechanisms of electron transfer in the enzymatic systems. Both the reduction of ferricytochrome c by radicals such as e_{aq}^-,[38–40] H,[40,41] O_2^-,[30,31,42–46] CO_2^-,[40,41] and a number of organic radicals[40,41,47] and the oxidation[48] of ferrocytochrome c by Br_2^-, $(SCN)_2^-$, N_3^-, and $\dot{O}H$ have been investigated. The hydroxyl radical is also known[49–51] to reduce ferricytochrome c.

The general feature that has emerged from these studies is that free-radical attack may lead to indirect reduction via the protein moiety or more directly through interaction with the exposed edge of the heme group. In the case of reduction by H and $\dot{O}H$, there is good evidence that free radicals are formed on the protein surface by addition to unsaturated bonds or by hydrogen atom abstraction reactions followed by intramolecular electron transfer to the buried ferriheme. The extent of reduction is of the order of 50% and it is not clear whether the intramolecular transfer of the electron occurs by a tunnelling mechanism[51] or by a pathway involving a series of fast steps. Some of the most recent work on cytochrome c is reviewed in the following section.

  a. *Oxidation of ferrocytochrome c.* The oxidation of ferrocytochrome [cyt(II)] has received much less attention than the reduction of cyt(III). A recent study[48] has shown that cyt(II) is oxidised with 100% efficiency by Br_2^-, $(SCN)_2^-$, or N_3, whereas $\dot{O}H$ is only 5% efficient. The corresponding rate constants in neutral solution at ionic strength $I = 0.073$ mol dm^{-3} are $> 10^{10}$ dm^3 mol^{-1} s^{-1} for $\dot{O}H$ and $\sim 10^9$ dm^3 mol^{-1} s^{-1} for the others. The low yield of oxidation by $\dot{O}H$ is consistent with nonspecific attack of this radical on the amino acid residues of the protein. On the other hand, the reactions of Br_2^-, $(SCN)_2^-$, and N_3 are simple second-order processes and there is no evidence of any intermediate species during the oxidation. Since the observed rate constants are ~ 10-fold greater than those calculated for attack at the amino acid residues, it was concluded[48] that the radicals attack cyt(II) at a specific site, which is most probably the partially exposed edge of the porphyrin ring. Ionic strength effects were observed for oxidation of cyt(II) by Br_2^- that correspond well with the overall charge of $+8$ for the heme protein, which also supports the idea that Br_2^- oxidises cyt(II) by an outer-sphere-type process at the exposed edge of the heme group.

  Oxidation of cyt(II) by $Fe(CN)_6^{3-}$ has been studied[52] by pulse radiolysis by reducing cyt(III) rapidly with e_{aq}^-. At low $[Fe(CN)_6^{3-}]$ (<1 mmol dm^{-3}) the observed pseudo-first-order rate constant was proportional to $[Fe(CN)_6^{3-}]$, but became independent at concentrations >2 mmol dm^{-3}. These observations are consistent with the occurrence of the intermolecular and intramolecular electron transfer reactions (26) and (27)

$$\text{cyt(II)} + Fe(CN)_6^{3-} \rightarrow \text{cyt(III)} + Fe(CN)_6^{4-} \qquad (26)$$

$$\text{cyt(II)}Fe(CN)_6^{3-} \rightarrow \text{cyt(III)}Fe(CN)_6^{4-} \qquad (27)$$

with $k_{26} = 4 \times 10^7$ dm^3 mol^{-1} s^{-1} and $k_{27} = 4.6 \times 10^4$ s^{-1} at $I = 0.02$ mol dm^{-3}. From the data it was calculated[52] that the binding constants of $Fe(CN)_6^{4-}$ or $Fe(CN)_6^{3-}$ to either cyt(II) or cyt(III) are almost indistinguishable (0.87–2.1×10^3 dm^3 mol^{-1}), which shows that the binding is insensitive to the conformational differences between cyt(II) and cyt(III).

  On the other hand, experiments[53] using the stopped-flow method showed no evidence for limiting kinetics in the oxidation of cyt(II) by $Fe(CN)_6^{3-}$ or in the reduction of cyt(III) by $Fe(CN)_6^{4-}$ at $I = 0.1$ mol dm^{-3}, and it was concluded[53] that the binding constants are < 200 dm^3 mol^{-1} at this ionic strength. These binding constants, being derived from kinetic measurements, relate to specific association of the iron ions at the site of electron transfer on the surface of the protein molecule. Assuming that this specific association is an electrostatic interaction and is therefore dependent on ionic strength, the two sets of binding constant data[52,53] would be compatible if the local effective charge on the protein is $\sim +3$.

b.   Modified cytochrome c.   Factors that govern the redox reactions of cytochrome *c* include electrostatic charge and conformation. The contributions of these two factors in determining the rate of electron transfer have been separated to some extent by comparing the redox reactions of modified cytochromes *c* with those of the native protein.[46] Three derivatives were chosen for study: (1) acylated (ac), in which the ε-amino groups of the lysines are transformed to neutral groups; (2) succinylated (suc), in which these groups become negatively charged and the conformation of the heme region is disrupted; and (3) carboxymethylmethionine 65,80 cytochrome *c* (cxm), in which the conformation of the heme region is disrupted. In the first two cases the overall charge of the protein is negative; in the last case it is positive and similar to that of the native protein. Second-order rate constants for the reduction and oxidation of these proteins are shown in Table 2.

TABLE 2

Second-order rate constants for the reduction and oxidation of cytochrome *c*
and its derivatives at pH $\simeq 7^a$

		Cytochrome *c* derivative			
		Native	Ac	Cxm	Suc
Redox potential (mV)		260	165	-120	-120
Charge on protein		9/8	$-10/-11$	$-29/-30$	9/8
	I (mol dm^{-3})	$k \times 10^{-8}$ (dm^3 mol^{-1} s^{-1})			
Reductant					
e_{aq}^-	0.10	300	200	330	400
CO_2^-	0.10	1.0	15	1.3	40
O_2^-	0.10	0.008	0.0035	NR^b	NR
$(CH_3)_2\dot{C}OH$	0.005	1.8	2.5	3.5	18
Oxidant					
$Fe(CN)_6^{3-}$	0.10	0.069	0.002	2.7	0.38

^{a} From Ref. [46].
^{b} NR, No reduction.

The reactivity of $CH_3\dot{C}HOH$, being neutral, reflects the effect of conformational modification in the heme region. A more exposed heme edge lowers the reduction potential of the cyt(III), so that the observed order of reactivity, suc $\gg$ cxm $>$ ac $\simeq$ native cytochrome c, suggests that exposure of the heme edge outweighs any decrease in reactivity caused by a lowering of the reduction potential. Ferricyanide also oxidises the reduced derivatives in the order cxm $\gg$ suc $\gg$ native $\gg$ ac, which follows the expected pattern if conformational and electrostatic factors both contribute to the redox reactivity of charged redox agents. However, this pattern does not hold at all for CO_2^-, for which the order of reactivity is suc $>$ ac $\gg$ cxm $\simeq$ native.

Electrostatic effects were examined in more detail by measuring kinetic salt effects,[46] from which rate constants at zero (k_0) and infinite (k_∞) ionic strength were calculated such that electrostatic effects should be maximal and zero, respectively. Thus, k_∞ reflects the effect of conformational changes and k_0 emphasises electrostatic effects. The rate constants are listed in Table 3 and illustrate that these two factors are separable and can be quite large. The data generally support the idea that electron transfer occurs in the region of the exposed heme edge.

The reduction of cytochrome c by O_2^- has been studied in detail to obtain a better understanding of the mechanism and site of reaction.[30] The activation energy for this reaction was found to be 31 ± 5 kJ mol^{-1}, compared with 14 ± 3 kJ mol^{-1} for reduction of cyt(III) by CO_2^- under the same conditions (pH 7.0, $I - 3 \times 10^{-3}$ mol dm^{-3}). This difference is interpreted[30] in terms of the orbitals that overlap between the donor radical and the acceptor iron–porphyrin system. For CO_2^-, overlap can occur between the half-filled π_u^* orbital of CO_2^- and the half-filled orbital of the low-spin heme group to produce singlet-state CO_2. In the case of O_2^-, formation of singlet O_2 is most unlikely on energetic grounds, and to form triplet-state O_2 requires overlap of the filled π_g^* orbital of O_2^- and a higher lying vacant orbital of the heme group. Consequently, a larger activation energy is required for reduction by O_2^-. The pH dependence of the reduction of cyt(III) by O_2^- was shown[30] to fit a single pK_a of 9.1. In these experiments edta was present to complex any free Cu(II) ions that, as already mentioned in Section III,A,4, enhance the apparent rate of reaction of O_2^- with cyt(III) at pH $<$ 8.5. The Cu(II) ion has no effect above pH 9 due, presumably, to the formation of unreactive hydroxo complexes.

To assess the reactive site for O_2^- a series of modified cytochromes c have been studied in which different single lysine residues were substituted[30] by 4-carboxy-2,4-dinitrophenyllysine (CDNP-lysine). This substitution neutralises the positive charge at the ε-nitrogen atom of the modified lysine and introduces a negative charge via the 4-carboxy group of the aromatic ring. The rates of reaction of O_2^- with these modified proteins are in the order

TABLE 3

The effect of ionic strength on rate constants for reactions of native and
modified cytochrome c at pH $\simeq 7^a$

Reaction	Z^b	Z^c	$k_0{}^b$ $(dm^3\ mol^{-1}\ s^{-1})$	$k_\infty{}^c$
Native (red) + Fe(CN)$_6{}^{3-d}$	7.7	4.6	1.1×10^9	3.0×10^6
Native (red) + Fe(CN)$_6{}^{3-e}$	4.4	3.1	3.2×10^8	4.4×10^6
Cxm (red) + Fe(CN)$_6{}^{3-}$	3.1	1.0	1.9×10^9	1.7×10^8
Ac (red) + Fe(CN)$_6{}^{3-}$	-5.8	-9.2	4.3×10^2	5.9×10^5
Suc (red) + Fe(CN)$_6{}^{3-}$	-5.0	-8.2	1.5×10^3	7.9×10^5
Native (ox) + CO$_2{}^{-f}$	6.9	6.3	6.5×10^9	6.4×10^8
Ac (ox) + CO$_2{}^{-}$	-8.0	-11.1	1.3×10^7	2.6×10^8

[a] From Ref. [46].
[b] Charge Z calculated from

$$\ln k_{obs} = \ln k_0 - \frac{A(Z_A + Z_B)I^{1/2}}{1 + \kappa R^{\ddagger}_{AB}} + \frac{AZ_A^2 I^{1/2}}{1 + \kappa R_A} + \frac{AZ_B^2 I^{1/2}}{1 + \kappa R_B}$$

[c] Charge Z calculated from

$$\ln k_{obs} = \ln k_\infty - 3.576 \left(\frac{e^{-\kappa R_A}}{1 + \kappa R_B} + \frac{e^{-\kappa R_B}}{1 + \kappa R_A} \right) \left(\frac{Z_A Z_B}{R_A + R_B} \right)$$

where Z_i and R_i are the charge and radius of reactant i; $A = 1.174$; $\kappa = 0.329 I^{1/2}$.
[d] Data from Peterman, B. F.; Morton, R. A. *Can. J. Chem.* **1977**, *55*, 796–803.
[e] Data from Ref. [53].
[f] Data from Ref. [45].

native > CDNP-lysine 60 > -lysine 13 > -lysine 87 > -lysine 27 > -lysine
86 > -lysine 72. Taking into account electrostatic effects, the low reactivities
of cyt(III) modified at positions 87, 86, and 72 point to $O_2{}^-$ reacting at the
left-hand side of the heme edge.

The reaction of $O_2{}^-$ with cytochrome c has also been used to probe the
interaction of the heme protein with sodium dodecyl sulphate (SDS) as a
model for the electrostatic and hydrophobic interactions of cytochrome c
with components of the mitochondrial membrane *in vivo*.[54] The kinetic
data are consistent with a rapidly attained two-state equilibrium between
cyt(III) (native) and cyt(III)*, an SDS-modified conformation of the native
protein

$$\text{cyt(III)} + \text{SDS} \rightleftharpoons \text{cyt(III)}^* \tag{28}$$

with $O_2{}^-$ reacting at different rates with each form.

$$O_2{}^- + \text{cyt(III)} \rightarrow \text{cyt(II)} + O_2 \tag{29}$$

$$O_2{}^- + \text{cyt(III)}^* \rightarrow \text{cyt(II)} + O_2 \tag{30}$$

The observed rate constant is given by

$$k_{obs} = (k_{29} + k_{30}K_{28})/(1 + K_{28})$$

and since

$$\ln K_{28} = -\Delta G_{28}/RT$$

ΔG_{28} can be found from the kinetic data. ΔG_{28} was expressed[54] as $\Delta G_{H_2O} + \Delta G_{SDS}[SDS]$ and the data yielded the following values: $k_{29} = 8 \times 10^5$ dm³ mol⁻¹ s⁻¹, $k_{30} = 5 \times 10^4$ dm³ mol⁻¹ s⁻¹, $\Delta G_{H_2O} = 7.9 \pm 0.8$ kJ mol⁻¹, and $\Delta G_{SDS} = 2.0 \pm 0.2 \times 10^4$ kJ mol⁻².

The effect of SDS on the rate of reduction by O_2^- is essentially independent of the cyt(III)/SDS ratio, indicating that the lowering of k_{obs} by SDS is not solely due to a lowering of the net positive charge on cytochrome *c* by the binding of SDS anions. It was concluded,[54] therefore, that the interaction is not purely of an electrostatic nature. Spectral changes of the protein solution on adding SDS are consistent with displacement of methionine 80, which is associated with the opening of the heme crevice, and ΔG_{H_2O} is identified with the free energy change accompanying this process.

Intramolecular transfer from the surface region of cytochrome *c* to the heme group has been measured[55] in a neat experiment in which the electron donor is the $(NH_3)_5Ru(II)$ moiety bound to histidine 33. The modified cyt(III) was obtained with the Ru(III) complex bound in this position and this was then reduced by CO_2^- generated pulse radiolytically at pH 6.7. Optical measurements indicated no detectable difference between the modified and native proteins; the modified protein also showed activity in the biological electron transfer assay with cytochrome *c* oxidase.

From the kinetic measurements two processes were identified: a rapid one corresponding to direct reduction of cyt(III) by CO_2^- and a slow one due to intramolecular transfer [reaction (31)]

$$Ru(II)–cyt(III) \rightarrow Ru(III)–cyt(II) \tag{31}$$

with $k_{31} = 82 \pm 20$ s⁻¹. Since the reduction potential of cyt(III) is 0.26 V at pH 7 and 25°C and that of $[(NH_3)_5Ruhis]^{2+}$ is 0.07 V, the driving force for reaction (31) is ~ 0.2 eV. The distance between His 33 and the heme group is estimated to be 12–16 Å for tuna cytochrome *c*, so that these experiments (with horse heart cytochrome *c*) demonstrate that, given quite a small driving force of ~ 0.1 eV, electron transfer can take place quite rapidly at long distances within proteins.

Cytochrome *c* interacts with cytochrome aa_3, heme proteins that are the two terminal components of the mitochondrial respiratory chain. This interaction has been investigated[56] by observing the conditions under which electron transfer occurs from reduced (by e_{aq}^-) iron-free (porphyrin) cytochrome and native cytochrome c to cytochrome aa_3. Both reduced proteins transfer an electron to cyt aa_3 with $k \sim 2 \times 10^7$ dm^3 mol^{-1} s^{-1}. When cyt c is associated with cyt aa_3, neither heme c nor heme a is easily accessible to e_{aq}^-, indicating that the exposed heme edge of cyt c is masked by the cyt c binding site of cyt aa_3. However, cyt(II) is able to transfer an electron to the cyt c–cyt aa_3 adduct, and similarly the (porphyrin–cyt c)$^-$ transfers its electron to the porphyrin–cyt c–cyt aa_3 adduct. Since the iron-free protein maintains the native structure, these results indicate the existence of two catalytically active sites for cytochrome c on cytochrome aa_3, and also that the electron is transferred via the exposed edge. It is considered unlikely that that the electron leaves the porphyrin ring of iron-free protein via residues that function as axial ligands in the native protein.

The effect of ionic strength on some of these reactions has also been investigated.[54] The dependence of the reaction of cyt c with cyt aa_3 confirmed that electrostatic forces are involved in the interaction of the two proteins. However, the ionic strength dependence of the reaction of cyt c with the cyt c–cyt aa_3 complex is much smaller and no firm conclusions were drawn from this behaviour. Two possible reasons offered[56] are that the negative charge of cyt aa_3 is partially neutralised by the positive charge of cyt c, and that the complex dissociates at high ionic strength to release more reactive cyt aa_3.

2. Cytochrome c_3

Cytochrome c_3 is a multiheme molecule having four closely packed heme c groups with midpoint potentials between -350 and -250 mV. Pulse radiolysis studies have been carried out to determine whether the redox properties of cyt c_3 correspond to the action of four sites of fixed potential, or whether the midpoint potentials vary as the overall degree of reduction varies. In one study,[57] the reduction of the protein by CO_2^- followed biphasic kinetics and it was concluded that the four heme groups are equivalent to two pairs, a result that was also deduced from the reduction by dithionite in stopped-flow experiments.[57]

More recent experiments on cyt c_3 from *Desulfovibrio vulgaris* have been carried out using the methyl viologen radical (MV$^+$) as reductant.[58] The advantage in using this radical is that it is stable under O_2-free conditions so that equilibrium positions between reductant and oxidant can be measured. Thus, under conditions in which $[\text{MV}^+] \ll [c_3[4\text{Fe(III)}]]$, the reduction is described by

$$c_3[4\text{Fe(III)}] + \text{MV}^+ \rightleftharpoons c_3[3\text{Fe(III)}\cdot\text{Fe(II)}] + \text{MV}^{2+} \qquad (32)$$

and $k_{obs} = k_{32}[c_3[4Fe(III)]] + k_{-32}[MV^{2+}]$. The measured rate constants are $k_{32} = 4.5 \pm 0.2 \times 10^8$ dm^3 mol^{-1} s^{-1} and $k_{-32} = 7.8 \pm 1.5 \times 10^4$ dm^3 mol^{-1} s^{-1}. From the value of K_{32}, expressed as $4k_{-32}/k_{32}$, the midpoint potential for the reduction of one heme was found to be -262 ± 5 mV. No spectral changes occurred after reduction of cyt c_3, indicating no conformational changes, but NMR and ESR measurements on the equilibrium system show that the electron is delocalised over all four hemes. Thus delocalisation does not appear to cause optical changes in the α and β absorption bands.

By measuring the $[MV^+]$ at equilibrium, an empirical expression, Eq. (33), was obtained[58] for the redox potential

$$E = -0.250 - 0.088\alpha + (RT/F)\ln[(1 - \alpha)/\alpha] \qquad (33)$$

where α is the degree of reduction of cyt c_3, so that $E^\circ = -0.250 - 0.88\alpha$ (pH 8). The usefulness of Eq. (33) is that it describes the redox properties of cyt c_3 without requiring specification of the separate midpoint potentials of the hemes or the nature of any heme–heme interactions. For $0 < \alpha < 0.1$, $E^\circ = -250$ to -260 mV, which is in good agreement with the kinetic value of -262 ± 5 mV.

From the effect of ionic strength on k_{32}, and assuming reaction (32) is diffusion controlled, the effective charge on the protein was estimated[58] as $+4.7 \pm 0.7$, in fair agreement with $+6$ calculated from the amino acid composition. Assuming that cyt c_3 is a sphere of 17 Å radius and of charge $+5$, it was calculated that the energy required to put the fourth electron on the protein is 32 mV more positive than is required for the first electron. This represents the difference in midpoint potentials of the first and last reduced

TABLE 4

Rate constants for the reaction of MV$^+$ and CH$_3$ĊHOH with cytochrome c_3, cytochrome c, and metmyoglobin[a]

		$k \times 10^{-8}$ (dm^3 mol^{-1} s^{-1})	
Heme protein	I (mol dm^{-3})	MV$^+$	CH$_3$ĊHOH
Cytochrome c_3	5×10^{-3}	4.5 ± 0.5	6.0 ± 1.0
	3.05×10^{-1}	9.7 ± 1	
Cytochrome c	4×10^{-3}	0.72 ± 0.14	1.4 ± 0.2^b
	2×10^{-1}	2.3 ± 0.2	
Metmyoglobin	4×10^{-3}	0.23 ± 0.02	0.55 ± 0.05
	2×10^{-1}	0.54 ± 0.07	

[a] From Ref. [58]; pH 8 (5 mmol dm^{-3} Tris).
[b] pH 7 (3.3 mmol dm^{-3} phosphate).

hemes due to electrostatic interaction alone, and it will become smaller as the ionic strength is increased. For $0 < I < 0.1$ mol dm^{-3} the difference in the midpoint potentials of MV^{2+} and cyt c_3 was estimated to vary between -12 and $+6$ mV for reduction of the first heme, and between 0 and $+26$ mV for reduction of the fourth heme.

The rates of reduction of cytochrome c_3, cytochrome c, and metmyoglobin by MV^+ and $CH_3\dot{C}HOH$ are compared in Table 4. Interesting features to note in the table are that the rate constants for cyt c_3 are approximately fourfold greater than those for cyt c, which means the average rate constant per heme is the same for both proteins, although the reduction potentials are quite different. Reduction of the heme in metmyoglobin by these radicals is more difficult, which reflects the configurational change of the heme iron from low spin to high spin that occurs on reduction of this protein.

3. Myoglobin

Myoglobin contains a single heme group; its radiation chemistry is of particular interest from the standpoint of the use of ionising radiation to preserve meat, as well as being of more general interest in terms of the reactions of free radicals with metalloproteins. The redox forms of myoglobin are ferrimyoglobin (ferriMb), ferro- or deoxymyoglobin (Mb), oxymyoglobin (MbO_2), and an Fe(III)–peroxo form $[(Mb(IV)]$ that is generated by the reaction of H_2O_2 with ferriMb.

Rdox transformations of ferriMb by the free radicals $\dot{O}H$ and e_{aq}^- generated radiolytically from water have been investigated in detail.[59] In deaerated solution, in which both radicals are present, ferriMb is converted to a mixture of the deoxy and Fe(III)–peroxo forms. Exposure of the irradiated solution to oxygen quantitatively converts Mb to MbO_2. Similarly, in N_2O-saturated solutions, in which e_{aq}^- is converted to $\dot{O}H$ through reaction (3), Mb and Mb(IV) are again formed, but when O_2 is present during the irradiation only Mb(IV) is formed. These results clearly demonstrate (1) that reaction of ferriMb with OH results in its reduction and (2) that the presence of O_2 prevents the intermediate products from undergoing reductive conversion to Mb. Under conditions of continuous radiolysis, a steady state is set up between Mb, ferriMb, and Mb(IV) in deaerated solutions under conditions in which only OH or e_{aq}^- are effective reactants. The steady-state compositions are quite different, however, reflecting the oxidative nature of OH and the reductive nature of e_{aq}^-:

	Mb $\rightleftharpoons$ ferriMb $\rightleftharpoons$ Mb(IV)		
OH	18%	44%	38%
e_{aq}^-	64%	24%	12%

The reduction of ferriMb by OH is reasonably explained by the formation of a carbon- or nitrogen-centred radical on the protein moiety followed by intramolecular transfer from this reducing radical to the metal centre. The mechanism of oxidation of Mb by e_{aq}^- is more obscure. Without giving mechanistic detail it was suggested[59] that e_{aq}^- adds to the heme group and then an intramolecular two-electron transfer occurs to form a modified ferriMb molecule.

In addition to these simple redox transformations, dimeric products were observed[59] whose yield was independent of dose rate over an enormous range. This is a very interesting result for it shows that e_{aq}^-- and OH-generated globin radicals either reduce the heme group or dimerise in noncompeting reactions, which in turn points to the globin radicals being of two distinctly different types. It was suggested[59] that this duality may be explained by radicals formed by addition and abstraction, respectively, or by the structural disposition of the radical on the protein moiety. The OH-induced formation of the Fe(III)–peroxo entity in the presence of O_2 may involve the formation[59] of HO_2 from the globin peroxy radical followed by its reaction with ferriMb.

Reduction of MbO_2 by e_{aq}^- ($k = 4 \times 10^{10}$ dm^3 mol^{-1} s^{-1}) also results in the formation of Mb(IV).[60] Here the initial step is thought to be

$$e_{aq}^- + \text{\textasciitilde}\text{Fe(II)–O}_2 \longrightarrow \text{\textasciitilde}\text{Fe(II)–O}_2^- \tag{34}$$

followed by intramolecular electron transfer from the heme iron to the bound molecular oxygen. No evidence was obtained for the +5 oxidation state of Mb by pulse radiolysis, indicating that this state, which is similar to Compound I of peroxidase, has a lifetime shorter than 70 ns if it exists at all.[60] It was suggested[60] that the intramolecular electron transfer gives rise to OH as the reduction product of the bound dioxygen:

$$\text{\textasciitilde}\text{Fe(II)–O}_2^- \xrightarrow{\text{2H}^+} \text{\textasciitilde}\text{Fe(IV)} \cdot \text{OH}^- + \text{OH} \tag{35}$$

On the other hand, pulse radiolysis experiments with the oxy form of diacetyl-deuteroperoxidase[60] show that e_{aq}^- reduces it to Compound I of this enzyme ($k = 4 \times 10^{10}$ dm^3 mol^{-1} s^{-1} at pH 7.4). The one-electron reduction of oxyperoxidase to Compound I is not observable by other methods because the product reacts further with reducing agents and oxyperoxidase.[61]

4. Hemoglobin

Reduction of the heme iron in methemoglobin from the Fe(III) to Fe(II) state is accompanied by a spin-state transition from low spin to high spin. By use of pulse radiolysis it has been shown[62] that the spin-state transition is well separated in time from the reduction of the heme Fe(III) ion by e_{aq}^-: the effects of external parameters on this relaxation process have been examined. Reduction was carried out at pH 9 in 1 mol dm^{-3} methanol solution and 60% of e_{aq}^- reacted as follows:

$$e_{aq}^- + \text{Fe(III)} \xrightarrow{k_a} \text{Fe(II) (low spin)} \xrightarrow{k_b} \text{Fe(II) (high spin)} \tag{36}$$

with $k_a = 3.0 \pm 0.3 \times 10^{10}$ dm^3 mol^{-1} s^{-1} and $k_b \sim 1.2 \times 10^3$ s^{-1}. A more detailed analysis of absorbance changes in the Soret band indicated a strong possibility of two rates for the reduction and spin-state transitions, suggesting inequivalence of the α and β subunits in methemoglobin and in the non-equilibrium hemochrome intermediate. The rate of the spin-state transition increased by 100-fold between pH 9 and 6 and was also increased by approximately 20-fold at pH 8 by addition of 330 mmol dm^{-3} phosphate or pyrophosphate anions.

Spin-state transitions are also observed for partially reduced methemoglobin complexed with F$^-$, N$_3$$^-$, and CN$^-$. The results for F$^-$ and N$_3$$^-$ indicate that these anions are no longer bound to the reduced iron in the unrelaxed Fe(II) heme state. On the other hand, the half-life of the partially reduced cyanide complex is ~ 30 s, and it was concluded[62] that CN$^-$ remains bound to iron in the hemochrome form for this time. Since the iron atom of the fluoromethemoglobin is in a high-spin state, while the azo- and cyanomethemoglobin complexes have iron in the low-spin state, the similar relaxation rates of the hemochromes derived from aquo-, fluoro-, and azomethemoglobin suggest that the relaxation rate is insensitive to the initial spin state of the iron.[62] A mechanism was proposed[62] to account for these reduction–relaxation processes in which e_{aq}^- is accepted by the distal histidine that is hydrogen bonded to the OH$^-$ ligand of the aquomethemoglobin. The imidazole anion then substitutes for OH$^-$ to form the nonequilibrium hemochrome state that relaxes through the iron atom moving out of the porphyrin plane to achieve the high-spin state.

In another study[63] of hemochrome relaxation it was reported that the unrelaxed hemochrome state can bind oxygen. This means that the distal histidine cannot be bound to the iron of the hemochrome as was proposed earlier.[62] For example, the rate of oxygen binding at pH 8.4 and 260 μmol dm^{-3} O$_2$ is 25-fold faster than the decay of the hemochrome state.[63] Instead, it was proposed that the initial hemochrome state comprises a hexacoordinated structure in which the OH$^-$ ligand interacts weakly with the iron, its release being retarded by the neighbouring protonated distal histidine. This proposal invokes the novel idea that the presence of OH$^-$ raises the pK of the imidazole proton so that the distal histidine remains protonated above pH 8. It also explains why deprotonation of the water ligand in methemoglobin is accompanied by protonation of another group in the protein. Oxygen binding follows biphasic kinetics between pH 8.4 and 9.8, which is ascribed[63] to deprotonation of the distal histidine with a pK of 8.8 and the concomitant release of the OH^- ligand. When OH^- is weakly bound, the iron is held in the low-spin state and oxygen binding is faster; release of OH^- allows the iron to relax to the high-spin state and oxygen binding becomes slower. This interpretation of biphasic kinetics of oxygen binding to partially reduced methemoglobin differs from that put forward earlier by Ilan et al.,[64] who ascribed the kinetics to changes in quaternary structure from high (R) to low (T) affinity states.

The decay of the hemochrome state is interpreted[63] as a change from the weakly hexacoordinated to a pentacoordinated structure, with simultaneous change of spin state from low to high. The enthalpy of activation for this process is ~ 53 kJ mol^{-1}, but it increases to 90 kJ mol^{-1} in 1 mol dm^{-3} methanol solution, indicating that methanol strengthens the weak interaction between the OH^- ligand and the heme iron. There is also a large entropy of activation in the methanolic solution, indicating release of one or more methanol molecules in forming the transition state; these features point to a strong interaction between methanol and hemoglobin, as has been found by other investigators.[65]

5.   Subunits of heme proteins

The relative reactivities of the heme and protein moieties of heme proteins toward hydrated electrons can be determined by looking at the reactivities of the separate constituents. Myoglobin is a suitable system for this kind of study because the heme can be removed reversibly to leave the apoprotein

TABLE 5

Reaction of e_{aq}^- with myoglobin and its derivatives[a]

Reactant	$k \times 10^{-10}$ $(dm^3\ mol^{-1}\ s^{-1})$	$k_D \times 10^{-10}$	Heme binding-site area (Å^2)	Percentage reaction at heme[b]
FerriMb	2.5 ± 0.2	8.6	1000	67
Protoporphyrin IX-Mb	2.0 ± 0.2	8.6	680	58
FerroMb	1.6 ± 0.7	8.6	450	47
ApoMb	0.8 ± 0.04	8.6	500[c]	0

[a] From Ref. [66]; pH 7, 20°C; $I = 10^{-3}$ mol dm^{-3}.
[b] Defined as 100 $(k_{Mb^+} - k_{ApoMb})/k_{Mb^+}$.
[c] Effective reactive area of the protein.

(apoMb), which retains the basic structural features of ferroMb. Such a study was made[66] in which the kinetics of the reaction of e_{aq}^- with ferriMb, ferroMb, apoMb, and apoMb combined with iron-free protophorphyrin IX were measured. The results are summarised in Table 5. These data show two features: first, apoMb reacts at a significant rate with e_{aq}^-; second, the presence of the heme, either as the iron-free or the Fe(II) form, raises the rate to nearly that of ferriMb. Thus it is the porphyrin that acts as the most effective electrophilic sink for e_{aq}^-.[66]

The sperm whale ferroMb used in this work contains no cysteine or cystine,[66] which are the most reactive amino acids with e_{aq}^-. Protonated histidine is known to react more than 15-fold faster than the other amino acid residues, so it is likely that this is the reactive site of the apoprotein. Calculation shows[66] that at pH 7 there are 3.1 protonated histidine residues out of a total of 9 titratable residues in apoMb, so that the rate constant per protonated residue is 2.5×10^9 dm^3 mol^{-1} s^{-1} (see Table 5), which compares well with the value of 2.8×10^9 dm^3 mol^{-1} s^{-1} reported for protonated histidine at pH 4.8.[67]

Table 5 also shows the calculated diffusion-controlled rate constant k_D for reaction of e_{aq}^- with Mb, assuming the Smoluchowski equation is valid, together with heme binding-site areas calculated[66] on the basis that the fraction of the molecular surface that is reactive is given by k_{obs}/k_D. It is argued that these areas are much larger than the area of the exposed heme edge, which is estimated to be 12 Å,[66] so that intramolecular migration of the electron from its original site of attachment in the heme group region to the heme seems to occur. A complete reaction scheme has been written as[66]

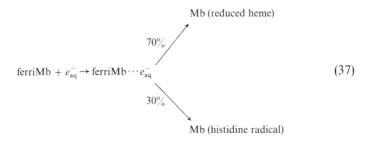

$$\text{ferriMb} + e_{aq}^- \rightarrow \text{ferriMb} \cdots e_{aq}^- \qquad (37)$$

Mb (reduced heme)

70%

30%

Mb (histidine radical)

The mechanism of electron migration to the heme is still an open question. Proposals which have been put forward include electron tunnelling,[51] migration via peptide hydrogen bonds, and sequential stepping between aromatic amino acid residues.[67,68] Whatever the mechanism in the case of ferriMb, intramolecular electron transfer to the heme from the protein moiety occurs too rapidly to be observed and must have a rate constant in excess of 10^6 s^{-1}.[59,66]

The spectral changes that occur when e_{aq}^- reacts with protoporphyrin IX–Mb are consistent with the formation of the porphyrin radical anion.[66] For ferroMb the spectral changes are qualitatively similar to those reported for Zn(II)–porphyrin complexes[69] and, since neither metal is easily reduced, the product is probably the corresponding porphyrin radical anion in each case.

The iron porphyrin ferrideuteroporphyrin IX, DPFe(III), has been used as a model system in the investigation of the mechanism by which cytochrome P-450 reacts with chlorinated methyl radicals, which are implicated in the metabolism of carbon tetrachloride, a hepatoxic molecule, and other polyhalogenated compounds.[70]

DPFe(III) is reduced by $(CH_3)_2\dot{C}OH$ with $k_{38} = 3.7 \times 10^8$ dm^3 mol^{-1} s^{-1} in aqueous 2-propanol (6.5 mol dm^{-3}) at pH 7.2.[70]

$$(CH_3)_2\dot{C}OH + DPFe(III) \rightarrow DPFe(II) + (CH_3)_2CO + H^+ \tag{38}$$

Reaction (38) was observed to occur in a single step, although, under the conditions used, the counterion $[OH^-$ or $(CH_3)_2CHO^-]$ coordinated to Fe(III) must also be released or protonated and equilibrium (39) must be established

$$DPFe(II) + L \rightleftharpoons DPFe(II)-L \tag{39}$$

where L = H_2O or $(CH_3)_2CHOH$. Thus these processes must occur in less than 2 μs, but protonation of the OH^- counterion is observed when the hemin is intercalated in micelles,[71] presumably because here the proton donor water has restricted access to the inner coordination sphere of the iron.

The radical $\dot{C}Cl_3$ formed in reaction (40)

$$e_{aq}^- \text{ or } (CH_3)_2\dot{C}OH + CCl_4 \rightarrow \dot{C}Cl_3 + Cl^- + (CH_3)_2CO + H^+ \tag{40}$$

does not react at a measurable rate with DPFe(III) $(k \leq 10^6$ dm^3 mol^{-1} s$^{-1})$ under pulse radiolysis conditions, but does react rapidly with DPFe(II) to form a transient product that undergoes further change. A mechanism which conforms to these observations is

$$DPFe(II) + \dot{C}Cl_3 \rightarrow DPFe(II)-\dot{C}Cl_3 \tag{41}$$

$$DPFe(II)-\dot{C}Cl_3 \rightarrow products \tag{42}$$

where $k_{41} = 2 \pm 1 \times 10^9$ dm^3 mol^{-1} s^{-1} and $k_{42} = 70$ s^{-1}.[70] Reaction (42) is thought to be a structural rearrangement of $[DPFe(II)-CCl_3]$ or of its coordination shell. Similar results were obtained with $\dot{C}HCl_2$ and $\dot{C}H_2Cl$ radicals, except that structural rearrangement was observed to be 10-fold faster. The product $[DPFe(II)-CH_{3-n}Cl_n]$ is suggested[70] to contain an iron–carbon bond.

These results indicate that $\dot{C}Cl_3$ radicals will be captured rapidly by

Fe(II) cytochrome P-450 and that the lifetime of the iron–carbon complex may be sufficiently long for it to be reduced to the dichlorocarbene complex.[70] The formation of the latter is thought[72] to be the step by which CCl_4 is metabolised to CO_2. Other metalloporphyrins are reported in Section IV.

C. Nonheme iron proteins

1. Methemerythrin

Hemerythrin is an oxygen carrier but contains no heme group. It occurs in certain marine worms as an octomer in the coelomic fluids and as a monomer in the retractor muscle, each monomeric unit containing two linked iron atoms. In the Fe(II) form it reversibly binds oxygen and is readily oxidised to the Fe(III) met form.

The reduction of methemerythrin by e_{aq}^-, CO_2^-, and the methyl viologen radical MV^+ has been reported and compared with the similar reductions of metmyoglobin and cytochrome c_3.[73] The relevant data are summarised in Table 6, which also includes data for reduction by SO_2^- from dithionite. The main feature of the data is that methemerythrin is less reactive than the heme proteins, although the orders of reactivity of the reducing radicals are qualitatively similar, with $e_{aq}^- > CO_2^- > MV^+ > SO_2^-$. The efficiency of reduction of the iron(III) centre is also lower for the methemerythrin; estimates are 20–30% for e_{aq}^- and $\sim 10\%$ for CO_2^-,[73] in contrast to 50–70% for the respiratory globins. Another interesting and significant difference is that the rate of reduction of methemerythrin by e_{aq}^- appears to be little

TABLE 6

Rate constants for reaction of radicals with heme proteins and methemerythrins[a]

Free radical	k (dm^3 mol^{-1} s^{-1})			
	Metmyoglobin	Cytochrome c_3[b]	Metmyohemerythrin	Methemerythrin
e_{aq}^-	6.0×10^{10}	2×10^{10}	4.5×10^9	3.9×10^9
CO_2^-	2.9×10^9	2.1×10^8		6.8×10^9
MV^+	2.3×10^8			2.5×10^6
SO_2^-	2×10^6	6.8×10^6	1.1×10^6	4.7×10^5

[a] From Ref. [73]; pH 8.2, 25°C; $I = 3 \times 10^{-2}$ mol dm^{-3}; except data for cytochrome c_3.
[b] From Ref. [57]; pH 8–9, 20°C.

influenced by pH between 6.4 and 8.2,[73] whereas there is a 10-fold decrease in the rate for methemoglobin between pH 6 and pH 10.[57,74] It has been suggested that this difference[73] is due to the presence of an Fe–OH bond in the basic form of the heme protein, rendering it more difficult to reduce, whereas one of the pair of iron atoms in the subunit of methemerythrin may still remain coordinated to a water molecule.

2. Iron–Sulphur proteins

Iron–sulphur proteins have emerged in recent years as an important category of redox metalloproteins, functioning in such diverse processes as photosynthesis, respiration, N_2 fixation, biosynthesis, and degradation mechanisms.[75,76] The redox-active centre comprises clusters of two or four iron atoms and two or four atoms of inorganic sulphur, the clusters being held in the protein moiety by Fe–S bonds to the sulphur atoms of cysteine residues:

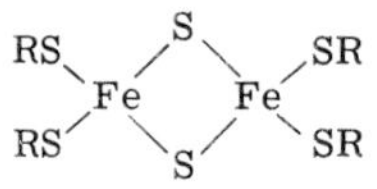

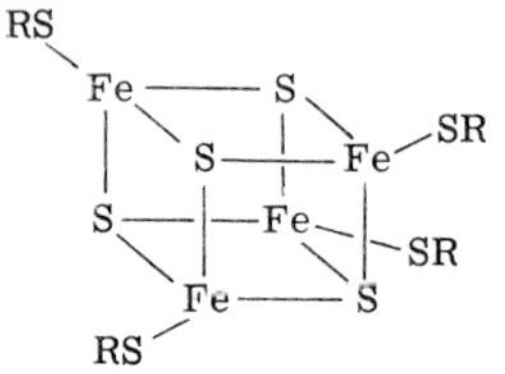

Iron–sulphur proteins that have been studied by pulse radiolysis include [2Fe–2S] ferredoxins from spinach[77] and parsley,[78] 2[4Fe–4S] ferredoxins from *Clostridium pasteurianum*,[78,79] and high-potential iron–sulphur protein (Hipip) from *Chromatium vinosum* D.[78,80] The 2[4Fe–4S] ferredoxin contains two [4Fe–4S] clusters, both with a reduction potential of -400 mV, whereas Hipip, with a single [4Fe–4S] centre, has a reduction potential of 350 mV. All these iron–sulphur clusters are normally simply one-electron active and the redox states may be represented as $Fe_2S_2^{2+,1+}$, $Fe_4S_4^{2+,1+}$, and $Fe_4S_4^{3+,2+}$ for [2Fe–2S] ferredoxins, 2[4Fe–4S] ferredoxins, and Hipip, respectively. As with other metalloproteins, pulse radiolysis studies have been aimed largely at obtaining kinetic and mechanistic information for the one-electron changes.

a. [2Fe–2S] ferrodoxins. Reduction of spinach ferredoxin by CO_2^- and e_{aq}^- is essentially quantitative in the pH range 5.05–9.67.[77] A single rate process is observed for CO_2^- ($k = 6.2 \pm 0.6 \times 10^7$ dm^3 mol^{-1} s^{-1}), but reduction of the cluster by e_{aq}^- has been observed to occur in two steps with $k = 9.4 \pm 0.3 \times 10^9$ dm^3 mol^{-1} s^{-1} and $8.3 \pm 1.7 \times 10^2$ s^{-1}. These biphasic kinetics were interpreted[77] in terms of direct reduction of the cluster and reaction with the protein moiety followed by intramolecular electron transfer to the cluster. However, no intermediates corresponding to radical anions of amino acid residues could be detected optically.[77]

Reduction of parsley ferredoxin by e_{aq}^- is also quantitative at pH 7–8, but in this case the [2Fe–2S] cluster is reduced in a single process in step with loss of absorption due to e_{aq}^-.[78] The rate constant for this reaction is 9.7×10^9 dm^3 mol^{-1} s^{-1}, which is the same as that for the direct reduction of spinach ferredoxin noted previously. It is possible that the second stages observed in some protein studies are due to the presence of contaminants (see later).

b.   2[4Fe–4S] ferredoxin.   Under appropriate conditions, i.e., $[e_{aq}^-] \ll$ [2[4Fe–4S](oo)] (o, oxidized), the half-reduced ferredoxin 2[4Fe–4S](or) (r, reduced) can be obtained by reaction of e_{aq}^- with 2[4Fe–4S](oo). Reoxidation of 2[4Fe–4S](or) by Co(NH$_3$)$_5$Cl^{2+} has been shown[78,79] to occur at the same rate ($k \sim 5 \times 10^5$ dm^3 mol^{-1} s^{-1}) as the overall oxidation of 2[4Fe–4S](rr),[79] indicating that there is no redox cooperativity between the two [4Fe–4S] clusters.

The 2[4Fe–4S] ferredoxin is reduced by e_{aq}^-, CO$_2^{\cdot -}$, and (CH$_3$)$_2\dot{C}$OH. In one study[79] biphasic kinetics and low efficiency (30–40%) were observed for the reduction by e_{aq}^-, while in another,[78] single-step reduction was observed with efficiencies ranging from 35 to 96% from one protein preparation to another. The latter behaviour is probably the more meaningful. Variations in protein treatment, including not purifying it to the correct A_{390}/A_{285} absorbance peak ratio,[81] exposing the protein to O$_2$ for 1 h prior to use, and varying the amount of buffer (1–10 mmol dm^{-3} phosphate) or $\dot{O}$H scavenger (0.1–0.2 mol dm^{-3} *t*-butanol), provided no clearly identifiable explanation for the conflicting results reported. However, the tendency of [4Fe–4S] clusters to denature to [3Fe–3S] clusters has been noted (e.g., see Armstrong, Vol. 1 of this series, pp. 65–120), and the "free" Fe present is possibly attached to the protein surface where it can act as an electron sink. If this is true then the second stage could correspond to electron transfer from this adventitious Fe to the cluster. The magnitude of such rate constants (although spurious) is of interest.

Reduction of 2[4Fe–4S] in the oxidised form by CO$_2^{\cdot -}$ and (CH$_3$)$_2\dot{C}$OH is slower than by e_{aq}^-. Butler *et al.*[79] observed a slow reaction corresponding to the slow stage of the reduction by e_{aq}^- ($k = 1.1 \pm 0.1 \times 10^3$ s^{-1}), implying that these radicals do not reduce the cluster directly. On the other hand, Adzamli *et al.*[78] (unpublished data) found that (CH$_3$)$_2\dot{C}$OH reacts at a 20-fold faster rate ($k = 2 \times 10^4$ s^{-1}), equivalent to having a bimolecular rate constant of 1.2×10^9 dm^3 mol^{-1} s^{-1}. This result suggests that (CH$_3$)$_2\dot{C}$OH does react directly with the iron cluster.

c. Hipip. Both the oxidised form, Hipip(o), and the reduced form, Hipip(r), are reduced by e_{aq}^- ($\sim 30\%$ efficiency) in a single-stage process with $k = 2 \times 10^{10}$ dm^3 mol^{-1} s^{-1} (see Table 7).[78,80] The absorbance change that occurs in the reduction of Hipip(r) shows that superreduced Hipip is produced, having a spectrum that closely resembles that of 2[4Fe–4S](rr). The Hipip(r) form is generally inert toward strong reducing agents such as Cr(II)–edta,[82] although it can be reduced by dithionite in an unfolded state in 80% dimethyl sulphoxide.[83] Pulse radiolysis experiments show that CO_2^- is also without effect[80] despite its low reduction potential (-2.0 V; see Table 1), so that rapid reduction by e_{aq}^- may reflect its ability to tunnel through an energy barrier.

Superreduced Hipip is stable for at least 2 s, the maximum time of observation in the pulse radiolysis experiment,[80] and is reoxidised by O_2 with $k = 4.8 \times 10^6$ dm^3 mol^{-1} s^{-1} at pH 7 and 25°C. The reoxidation appears to be quantitative as judged by absorbance changes, but there is no confirmation that the product Hipip(r) retains its high-potential properties.

The kinetic data for the reactions of e_{aq}^- with iron–sulphur proteins are summarised in Table 7, where the rate constants k_e and k_p are obtained by monitoring absorbance changes of e_{aq}^- and the protein, respectively. Values of $k(e_{aq}^- + \text{protein})$ are generally in quite good agreement, but k_p and the effi-

TABLE 7

Kinetic data for reactions of e_{aq}^- with iron–sulphur proteins[a]

Protein	Concentration (μmol dm^{-3})	$k_e \times 10^{-5}$ (s^{-1})	$k_p \times 10^{-5}$ (s^{-1})	$k_2 \times 10^{-10}$ (dm^3 mol^{-1} s^{-1})
[2Fe–2S](o)	12	2.4	2.9	0.93
	29	4.3	—	0.96
	47	6.5	5.6	0.98
2[4Fe–4S](oo)	19	7.4	6.0	3.1
	19	10.6	0.3	4.5
	19	12.5	7.5	5.1
	20	8.6	5.6	3.2
	22	9.3	6.7	3.2
	(6–20)	Range[b]	0.01	3.4[c]
Hipip(o)	23	5.8	4.8	1.7
	(10–40)	Range[b]	Range[b]	1.7[d]
Hipip(r)	33	9.1	5.0	2.1
	(10–40)	Range[b]	Range[c]	1.8[d]

[a] With permission, from Ref. [78] and Elsevier Science Publishing Co., Inc.
[b] Dependent on protein concentration.
[c] From Ref. [79].
[d] From Ref. [80].

ciency of reduction show considerable discord. Because values of k_e and k_p that are in step, or nearly so, have been observed, it was suggested[78] that a single rate-determining step may represent the correct behaviour of the proteins. Out-of-step ($k_p < k_e$) reduction may reflect the presence of contaminants that lead to minor denaturation of the protein, thereby providing a temporary resting place or sink for the electron on its way to the cluster. It is interesting to note that the clusters in [2Fe–2S] and 2[4Fe–4S], for which close to 100% efficiency of reduction has been observed,[78] are known to be close to the surface and have exposed cysteinyl S atoms, whereas in Hipip(o) and Hipip(r) the clusters are buried ~ 5 Å and reduction efficiencies greater than $\sim 50\%$ have yet to be observed. This difference may reflect permanent electron trapping by amino acids in the latter case, as has been argued in the case of heme proteins.

D. Copper proteins

The redox properties of a number of copper proteins have been studied using pulse radiolysis. These include azurin[84,85] and stellacyanin,[86] which contain only one copper atom, Type 1 Cu(II), per molecule, and the oxidases ceruloplasmin,[87] laccase,[86,88] and ascorbate oxidase,[85] which have several different Cu redox centres and catalyse the reduction of O_2 to water. The copper atoms in these oxidases are grouped into three classes according to their spectroscopic features: Type 1 Cu(II), absorbing at 610 nm; Type 2 Cu(II), absorbing at 330 nm; and Type 3 Cu(II), also absorbing at 330 nm. The Type 1 Cu(II), which is responsible for the intense blue colour of these copper proteins, is often suggested to be the site of initial reduction by substrates, which is followed by intramolecular electron transfer to form eventually a pair of Cu(I) ions at the Type 3 site, which binds O_2. Pulse radiolysis experiments using free radicals as the reductants have been carried out with the aim of discovering mechanistic details of these processes.

1. Stellacyanin and azurin

Reduction of Type 1 Cu(II), the only copper in these proteins, by e_{aq}^- and CO_2^- appears to be mainly the result of direct reaction of the radicals with the Cu(II),[84–86] although reported extents of reduction differ. With stellacyanin about 60% of the Cu(II) was reduced in the direct reaction with e_{aq}^- ($k = 2.3 \times 10^{10}$ dm^3 mol^{-1} s^{-1}) and the remainder was reduced in a slower step attributed to intramolecular transfer of electrons from primary trapping sites on the protein moiety ($k = 5 \pm 1 \times 10^2$ s^{-1}).[86] Direct reduction of azurin was reported to be only 20% efficient.[84] More recent measurements[85] confirm that the Type 1 Cu(II) in azurin is reduced by direct reaction with e_{aq}^- and CO_2^-, and by using a train of pulses to generate the reducing radicals it was shown that there is no saturation of the primary acceptor sites for the electron (see later).

2. Multicopper proteins

In contrast to the single-copper proteins stellacyanin and azurin, the Type 1 Cu(II) in the multicopper proteins ceruloplasmin, laccase, and ascorbate oxidase appears to be reduced by e_{aq}^- and CO_2^- almost entirely by the indirect route involving nonspecific reduction of the protein, followed by a rate-determining intramolecular step or steps. Type 1 Cu(II) reduction has been shown to be brought about by $\dot{O}H$, O_2^-, and $\dot{C}H_2(CH_3)_2COH$ in fungal laccase,[88] by O_2^- in ceruloplasmin,[87] and by nitroaromatic radical anions (RNO_2^-) in ascorbate oxidase.[85] Some of the most prominent findings are summarised here.

 a. Laccase.[88] The rates and extent of reduction of Type 1 Cu(II) in fungal laccase by a number of free radicals are listed in Table 8. Spectral

TABLE 8

Rate constants for the intramolecular reduction of Type I Cu(II)
in fungal laccase by free radicals at pH 6[a]

Adduct radical	k (s^{-1})	Fraction of Type 1 Cu(II) reduced	
		Zero dose[b]	Infinite dose[c]
$\dot{O}H$	$5.5 \pm 1.4 \times 10^3$	0.43	0.32
$\dot{O}H$ (with 1 mmol dm^{-3} KF)	5×10^3	0.31	0.30
$\dot{C}H_2(CH_3)_2COH$	5×10^3	0.026	0.10
$\dot{C}H_2(CH_3)_2COH + e_{aq}^-$	5×10^3	0.065	0.14
CO_2^-	$1.0 \pm 0.2 \times 10^4$	0.28	0.32
O_2^-	$1.3 \pm 0.5 \times 10^2$	0.24	0.60
O_2^- (with 1 mmol dm^{-3} KF)	$2-3 \times 10^3$	0.48	0.61
O_2^- (with 4 μmol dm^{-3} SOD)	$8.2 \pm 0.9 \times 10^3$	0.043	0.31

[a] From Ref. [8].
[b] [Reduced laccase, Type 1 Cu(II)]/[$\dot{R}$] extrapolated to zero dose.
[c] Total reducible fraction of Type 1 Cu(II). These results show that the radical acceptor sites on the protein become saturated.

changes show that each radical interacts with the protein, probably at a limited number of sites on the peptide backbone and with such components as disulphide bridges (R–S–S–R), histidine, and aromatic residues. The transient protein species all absorb in the region of 310 nm and thus mask any changes involving Type 2 and/or Type 3 Cu(II). However, the lack of an effect of fluoride, which is known to bind to the Type 2 Cu(II) site, on reduction with O_2^- is thought to argue against involvement of this site.

In the reduction of Type 1 Cu(II) in tree laccase, evidence has been presented[86,89] for electron trapping by RSSR to form $(RSSR)^-$, followed by transfer to the Type 1 Cu(II) site. Disappearance of $(RSSR)^-$, which absorbs at ~ 400 nm,[90] occurs synchronously with loss of the Type 1 Cu(II) absorption at 600 nm. Similar findings have been reported for ceruloplasmin.[87]

b. Ascorbate oxidase. This protein has eight copper atoms per molecule that are reported[91] to comprise three Type 1, a Type 2, and possibly two pairs of Type 3 Cu(II). However, it exhibits much greater catalytic activity than either ceruloplasmin or laccase. Reaction of the enzyme with e_{aq}^- and CO_2^- results in reduction of Type 1 Cu(II) by an intramolecular process as observed for laccase and ceruloplasmin, but reduction initiated by e_{aq}^- is an order of magnitude faster.[85] Although an absorbing species was observed at 410 nm, and was assumed to be $(RSSR)^-$, its rate of decay was some 20-fold slower than the rate of reduction of Type 1 Cu(II), so that $(RSSR)^-$ is not involved in this reduction, in contrast to the behaviour of ceruloplasmin and tree laccase already noted. On the other hand, transient absorbance changes at 330 nm occurred at a rate similar to the rate of loss of the Type 1 Cu(II) colour. This, coupled with the fact that there is more than one component involved in the rate, is in accord with the idea that a number of different primary acceptor sites for the electron are involved in the reduction of Type 1 Cu(II).[88]

With e_{aq}^- and CO_2^- as reductants, the maximum bleaching of Type 1 Cu(II) was never more than 50% of that observed upon full reduction of the protein, even when $[CO_2]^- \gg [\text{Type 1 Cu(II)}]$. Furthermore, when CO_2^- was produced in a train of pulses at 400 pulses s^{-1}, the extent of colour change decreased with each successive pulse, whereas in a similar experiment with azurin the extent of the colour change was approximately the same for each of the first six pulses. It was concluded[85] from these observations that the primary acceptor sites on ascorbate oxidase can become saturated, so that further reducing equivalents cannot be passed to the Type 1 Cu(II) site, while with azurin such saturation does not occur. It was inferred[85] that e_{aq}^- and CO_2^- react directly with Type 1 Cu(II) in azurin, as was concluded previously for this protein[84] and stellacyanin.[86] However, in view of the apparent nonspecific attack of e_{aq}^- and CO_2^- on the protein moiety, followed by intramolecular reduction of Type 1 Cu(II) in the multicopper proteins, it would be surprising if such reactions did not also occur with single-copper proteins. An alternative explanation may be that the intramolecular step is not rate determining in the latter instances, as seems to be the case with at least some of the iron proteins discussed earlier. Possibly the rate-determining intramolecular electron transfers to Type 1 Cu(II) that occur in the multicentre proteins are a manifestation of the involvement of Type 2 Cu(II) and/or Type 3 Cu(II) sites.

A much clearer and simpler picture of the reduction of ascorbate oxidase is obtained when less powerful reducing agents in the form of nitroaromatic radical anions are used.[85] Reduction potentials at pH 7 for the RNO_2^- radicals studied are metronidazole, -486 mV; misonidazole, -389 mV; and nitrofurazone, -257 mV. These radicals react with ascorbate oxidase with rate constants in the range $1–3 \times 10^7$ dm^3 mol^{-1} s^{-1} and effect almost complete reduction of Type 1 Cu(II). However, under approximately stoichiometric conditions, i.e., $[RNO_2^-] \simeq [\text{Type 1 Cu(II)}]$, only 40% of Type 1 Cu(II) was reduced even though all RNO_2^- was removed on the same time scale. There was no evidence of radicals being formed on the protein, but the absorbance change at 330 nm occurred at a rate similar to that at 610 nm, indicating reduction of Type 2 and/or Type 3 Cu(II) by RNO_2^-. From the number of RNO_2^- that reacted with ascorbate oxidase in a train of pulses it was concluded[85] that the protein can accept six to seven reducing equivalents. This is almost equal to the number of copper atoms per molecule and, since no protein radicals were observed, indicates that Type 2 and Type 3 Cu(II) are involved in the reduction. It was further concluded[85] that the oxidoreductive states of the different types of copper are all in equilibrium. Since RNO_2^- reacts rather slowly with ascorbate oxidase, the observation that this bimolecular reaction is the rate-determining step does not rule out the occurrence of subsequent intramolecular electron transfer processes in the reduction of the coppers.

IV. METALLOPORPHYRINS

Iron porphyrins have already been mentioned both in connection with superoxide dismutase activity[24] (Section III,A,3) and as a model for cytochrome P-450 in its reaction with carbon tetrachloride[70] (Section III,B,5). These are just two examples of the use of metalloporphyrins as models for redox reactions of metalloproteins; further examples that illustrate the various kinds of reactions that metalloporphyrin species can undergo are described here.

A. Electron transfer

One-electron oxidation and reduction of porphyrins to form the corresponding radical cations and anions and the study of their electron transfer reactions with electron donors and acceptors can be conveniently carried out using pulse radiolysis. Water-soluble porphyrins, for example, can be reduced[69,70] by e_{aq}^-, $CO_2^{\cdot -}$, or $(CH_3)_2\dot{C}OH$ and oxidised by Br_2^-,[92] $\dot{C}H_2CHO$,[93] or $\dot{N}_3$.[94] Water-insoluble porphyrins can be reduced to the radical anion in alcohols[95] and oxidised to the radical cation in dichloroethane.[96] When the central metal atom of the metalloporphyrin can be reduced or oxidised, the corresponding overall redox change occurs at the metal atom; otherwise, the reaction occurs at the ligand. However, the redox change at the metal atom may involve either direct reaction with the redox agent or indirect reaction, e.g., through reduction of the ligand followed by intramolecular transfer to the metal centre. The two alternatives may be distinguished using pulse radiolysis because the porphyrin radical anions and cations have characteristic absorption bands at 600–700 nm.

As mentioned above, neutral metallopophyrins can be oxidised to their radical cations in dichloroethane. Radiolysis of the solvent generates solvent radical cations ($S^{\ddot +}$) that oxidise the metalloporphyrins (MP) at either the ligand or the metal

$$S^{\ddot +} + MP \rightarrow MP^{\ddot +} \tag{43a}$$

$$S^{\ddot +} + M^{n+}-P \rightarrow M^{(n+1)+}-P \tag{43b}$$

Tetraphenylporphyrin (TPP) complexes of several metals have been investigated in this way.[96] The Co(II)TPP complex is oxidised to Co(III)TPP, whereas the Cd(II), Cu(II), Pb(II), Zn(II), Mg(II), and VO^{2+} complexes yield the radical cations. The Ni(II)TPP complex is oxidised at the metal centre in the presence of pyridine, but at the ligand in the absence of pyridine. The presence of pyridine is generally required to prevent demetallation of the porphyrins by the HCl, which is a radiolysis product of the solvent.[97]

The subsequent one-electron oxidation of Co(II)TPP and chlorophyll *a* by the radical cations was found[96] to occur generally with rate constants of $3-5 \times 10^8$ dm^3 mol^{-1} s^{-1}. These values show little variation with the oxidation potentials of the electron acceptors (0.5–1.0 V) and are about 100-fold slower than the diffusion-controlled rate, even though the difference in oxidation potentials of the donors and acceptors exceeds 0.3 V. It was suggested[96] that pyridine, which binds to the metal centre, has to be at least partially removed or reorientated before electron transfer can occur. This suggestion is supported by the fact that rate constants for electron transfer from metal-free porphyrin radical anions to several quinones in aqueous solution[95] vary with reduction potential from 10^7 to 10^9 dm^3 mol^{-1} s^{-1}, in accord with the Marcus theory.[69]

One-electron transfer reactions involving zinc and cobalt porphyrins have been studied in aqueous solution.[69] Water-soluble porphyrins used were tetra(4-sulphonatophenyl)porphyrin (TPPS), tetra[4-(N,N,N-trimethylamino)phenyl]porphyrin (TAPP), and tetra(4-N-methylpyridyl)-porphyrin (TMPyP). Reduction of the Zn complexes in neutral and alkaline solution yields the ZnP$^{\bar{}}$ radical anion, the rate of reduction increasing in the order ZnTPPS < ZnTAPP < ZnTMPyP, which is the order of the reduction potentials. On the other hand, the Co(III) complexes appear to be reduced directly to Co(II), i.e., reduction occurs at the metal centre. The only exception noted was Co(III)TMPyP, for which the radical anion was observed as a long-lived intermediate. The formation of the radical anion was attributed[69] to the presence of the electron-attracting pyridinium ring in TMPyP. The final product is Co(II)TMPyP.

The three zinc porphyrins are rapidly oxidised by Br$_2{}^-$ (neutral solution) to yield the radical cation ZnP$^+$, whereas Co(II)TPPS is oxidised to Co(III)TPPS. Electron transfer rates between ZnP$^{\bar{}}$ and Co(III)P, and between Co(II)TPPS and ZnP$^+$, have been measured and are shown in Table 9. These rate constants could be expected to be affected by the reduction

TABLE 9

Rate constants for electron transfer between porphyrins[a]

Donor	E°_{Donor} (V)[b]	Z_{Donor}	Acceptor	$E^\circ_{Acceptor}$ (V)[b]	$Z_{Acceptor}$	$k \times 10^{-7}$ (dm^3 mol^{-1} s^{-1})
(ZnTPPS)$^{\bar{}}$	−1.1	−5	Co(III)TPPS	−0.4	−5	2
(ZnTPPS)$^{\bar{}}$	−1.1	−5	Co(III)TPPS	0.1	+3	55
(ZnTPPS)$^{\bar{}}$	−1.1	−5	Co(III)TMPyP	0.2	+3	140
(ZnTAPP)$^{\bar{}}$	−0.7	+3	Co(III)TPPS	−0.4	−5	38
(ZnTAPP)$^{\bar{}}$	−0.7	+3	Co(III)TAPP	0.1	+3	54
(ZnTAPP)$^{\bar{}}$	−0.7	+3	Co(III)TMPyP	0.2	+3	49
(ZnTMPyP)$^{\bar{}}$	−0.6	+3	Co(III)TPPS	−0.4	−5	110
(ZnTMPyP)$^{\bar{}}$	−0.6	+3	Co(III)TAPP	0.1	+3	16
(ZnTMPyP)$^{\bar{}}$	−0.6	+3	Co(III)TMPyP	0.2	+3	40
Co(II)TPPS		−6	(ZnTPPS)$^+$		−4	11
Co(II)TPPS		−6	(ZnTMPyP)$^+$		+4	280

[a] With permission, from *J. Phys. Chem.* **1981,** *85,* 3678–3684 and American Chemical Society. Rates of reduction of Co(II)TPPS were determined at pH 13 in the presence of 0.1 mol dm^{-3} 2-propanol. Rates of oxidation of Co(II)TPPS were determined at pH 7 in the presence of 0.1 mol dm^{-3} Br$^-$.

[b] Estimated values.

potentials of the donors and acceptors and by their charge. As was pointed out,[69] the reaction between ZnTPPS⁻ and Co(III)TMPyP is most favoured, since $\Delta E \simeq 1.3$ V and the charges on the reactants are -5 and $+3$, respectively, yet it is probably slower than diffusion control. The general conclusion drawn from the data in Table 9 is that the effect of the multiple charge on the rate of electron transfer is quite small, i.e., only the charges close to the site of interaction are kinetically significant.

The reduction of Co(III)TMPyP by ZnTPPS⁻ results predominantly in direct reduction of the Co(III), in contrast to radical-anion formation when the reductant is $(CH_3)_2\dot{C}O^-$.[69] This result indicates that reduction by the milder agent ZnTPPS⁻ is more selective and directs the electron to the metal centre. Thus there appears to be a parallel with the reduction of the copper centres in ascorbate oxidase by RNO_2^-, described in Section III,D.

B. Alkylation

1. Cobalt

The reaction of Co(II) porphyrins with alkyl and hydroxyalkyl radicals yields Co alkyl derivatives[98] containing a cobalt–carbon bond. Formation and homolytic cleavage of these bonds is believed to be involved in the catalytic mechanism of enzymes containing the coenzyme derivative of vitamin B_{12}, which has a porphyrin-like corrin ring chelated to tervalent cobalt.

Alkylation of Co(II)TPPS has been studied using pulse radiolysis.[99] Co(II)TPPS is reduced to Co(I)TPPS by CO_2^- and $(CH_3)_2\dot{C}O^-$ by an outer-sphere electron transfer, but hydroxyalkyl radicals such as $\dot{C}H_2OH$ and $(CH_3)_2\dot{C}OH$, and alkyl radicals such as $\dot{C}H_3$ and $(CH_3)_2\dot{C}H$, form products containing a cobalt-to-carbon σ bond. Rate constants for these reactions are all $\sim 10^9$ dm³ mol⁻¹ s⁻¹; these are comparable with those observed for reactions of methyl radicals with cobalamin (B_{12r}),[100] but are an order of magnitude faster than the rate for $\dot{C}H_3$ reacting with the macrocyclic complex Co(III)(4,11-diene N_4).[101]

$$\dot{R} + Co(II)TPPS \rightarrow R{-}Co(III)TPPS \qquad (44)$$

When $\dot{R} = \dot{C}H_3$, the alkylation product is stable,[99] but for several other radicals the product is unstable. Hydroxylalkyl radicals are oxidised to the corresponding carbonyl compound through heterolysis of the cobalt–carbon bond.

$$HOR{-}Co(III)TPPS \rightarrow Co(I)TPPS + RO + H^+ \qquad (45)$$

Secondary alkyl radicals yield olefins through heterolytic cleavage of the cobalt–carbon bond and elimination of a β proton,[102] e.g.,

$$(CH_3)_2CH{-}Co(III)TPPS \rightarrow Co(I)TPPS + (CH_3)_2{\rightleftharpoons}CH_2 + H^+ \qquad (46)$$

Similar reactions are observed with cob(II)alamin B_{12r}[103] and Co(II)(4,11-diene N_4),[104] in which cobalt–carbon bonds are formed rather than the metal being oxidised or reduced by outer-sphere electron transfer as previously suggested.[105,106] However, unlike k_{44}, the rate constants for these reactions vary considerably in the case of Co(II)(4,11-diene N_4). For example, no reaction is observed between the macrocyclic complex and $(CH_3)_2\dot{C}OH$ ($k < 10^7$ dm^3 mol^{-1} s^{-1}),[104] whereas the corresponding reaction with B_{12r} and Co(II)TPPS is rapid. On the other hand, the rate of heterolysis of the cobalt–carbon bond formed by secondary alkyl adducts to Co(II)TPPS is several orders of magnitude faster[99] than those for the organocobalamins.[102] Thus neither Co(II)TPPS nor Co(II)(4,11-diene N_4) serve as good models for vitamin B_{12r} in all respects.

The complex Co(I)TPPS formed in reaction (45) or by direct reduction of Co(II)TPPS is strongly nucleophilic. It reacts with CH_3I to yield $CH_3Co(III)TPPS$ ($k = 3 \times 10^5$ dm^3 mol^{-1} s^{-1}) and with N_2O to form Co(II)TPPS ($k \le 3 \times 10^2$ dm^3 mol^{-1} s^{-1}).[99] Comparison with the reactivities of cob(I)alamin[107] and Co(I)-macrocyclic complexes[108] indicates that the nucleophilic character of Co(I)TPPS is intermediate between the nucleophilicities of these two types of complex. It remains uncertain in some cases whether these reactions occur via simple two-electron transfer or by consecutive one-electron steps involving free-radical intermediates. The classical S_N2 mechanism is found for cob(I)alamin,[107] where $k(B_{12s} + CH_3I) = 3.4 \times 10^4$ dm^3 mol^{-1} s^{-1}, but the free-radical mechanism is considered more likely for the corresponding reaction of Co(I)(4,11-diene N_4), for which $k = 4.7 \times 10^8$ dm^3 mol^{-1} s^{-1}.[108] In the case of N_2O, however, the free-radical intermediate $\dot{O}H$ would have to react directly with the cobalt(I) centre to account for the results,

$$Co(I)P + N_2O \xrightarrow{H_2O} Co(II)P + \dot{O}H + OH^- \qquad (47a)$$

$$\dot{O}H + Co(I)P \longrightarrow Co(II)P + OH^- \qquad (47b)$$

which is unlikely in view of the reactivity of $\dot{O}H$ toward the organic moiety. The alternative two-electron oxidation has been proposed to occur by an inner-sphere mechanism,[109]

$$Co(I)P + N_2O \xrightarrow{slow} {}^-N_2O{-}Co(II)P \xrightarrow[fast]{H^+} HO{-}Co(III)P + N_2 \qquad (48)$$

$$Co(III)P + Co(I)P \xrightarrow{fast} 2Co(II)P \qquad (49)$$

in which the Co(II)P state does not build up to significant concentrations.

2. Iron

As mentioned earlier, ferrideuteroporphyrin IX has been studied as a model system for possible cytochrome P-450-induced toxicity of carbon tetra-chloride and other polyhalogenated compounds (Section III,A,5). It was proposed[70] that Fe(II) porphyrins form iron–carbon σ bonds with $\dot{C}Cl_3$, $\dot{C}HCl_2$, and $\dot{C}H_2Cl$; this has prompted the study of the reactions of other carbon-centred radicals with iron porphyrins by means of pulse radiolysis.

Methyl radicals have been shown[110] to react rapidly with ferri- ($k = 2.3 \times 10^9$ dm^3 mol^{-1} s^{-1}) and ferrodeuteroporphyrin ($k = 3.9 \times 10^9$ dm^3 mol^{-1} s^{-1}) and slowly with the iron-free porphyrin ($k = 1$–2×10^7 dm^3 mol^{-1} s^{-1}) in 1:1 v/v 2-propanol–water mixtures. This is a clear demonstration that the radicals react with iron porphyrin at the iron atom. The reaction of $\dot{C}H_3$ with DPFe(III) generates an unstable species absorbing at 525 nm. This absorption peak is assigned[110] to DPFe(IV)CH$_3$, which may decay by reaction with CH$_3$I [the source of $\dot{C}H_3$ (Table 1)] or by alkaline hydrolysis

$$\text{DPFe(IV)CH}_3 + \text{OH}^- \rightarrow \text{DPFe(II)} + \text{CH}_3\text{OH} \tag{50}$$

DPFe(IV)CH$_3$ also reacts with DPFe(II) ($k = 4 \times 10^8$ dm^3 mol^{-1} s^{-1}), which may involve electron transfer or exchange of the methyl group

$$\text{DPFe(IV)CH}_3 + \text{DPFe(II)} \rightarrow \text{DPFe(III)CH}_3 + \text{DPFe(III)} \tag{51}$$

The first alternative is considered the more likely[110] in view of the fact that electron transfer between porphyrins is rapid (Section IV,A).

Reaction of $\dot{C}H_3$ with DPFe(II) in neutral 2-propanol–water generates a stable species absorbing at 540 nm,[110] assigned to DPFe(III)CH$_3$. Although this product is stable under anaerobic conditions, in the presence of air it is oxidised to DPFe(III). It was also shown[110] that DPFe(III)CH$_3$ can be prepared by normal chemical means by reducing DPFe(III) with sodium dithionite in the presence of methyl iodide, but the low solubility of the porphyrin in the alcohol–water solvent prevented characterisation of the DPFe(III)CH$_3$ using NMR spectroscopy. ESR and proton NMR measurements of alkyliron porphyrins electrogenerated in dimethyl formamide[111] show that low-spin complexes are formed. It has been pointed out[110] that mesomeric forms are likely to be important:

$$\text{DPFe(III)CH}_3 \leftrightarrow \text{DPFe(II)CH}_3$$

$$\text{DPFe(IV)CH}_3 \leftrightarrow \text{DPFe(III)CH}_3 \leftrightarrow \text{DPFe(II)CH}_3{}^+$$

Thus, the hydrolysis of DPFe(IV)CH$_3$ to DPFe(II) suggests that DPFe(II)-CH$_3{}^+$ represents the structure of this complex to some extent. Similarly,

DPFe(III)CH$_3$ behaves more like an Fe(II) porphyrin with regard to its oxidation by O$_2$, and less like an Fe(III) porphyrin in that it is stable in alkaline solution, i.e., it does not undergo anionic exchange.

A similar study has been made of the reaction of the halothane-derived radical CF$_3$ĊHCl (from CF$_3$CHClBr) with Fe(III) and Fe(II) porphyrins in acidic 2-propanol and acidic and alkaline 2-propanol–water mixtures.[112] The radical CF$_3$ĊHCl reacts rapidly with DPFe(II) in alkaline 2-propanol–water ($k = 7 \times 10^9$ dm^3 mol^{-1} s^{-1}) and with the dimethyl ester DPDMeFe(II) in acidic 2-propanol and acidic 2-propanol–water ($k = 1.5 \times 10^{10}$ dm^3 mol^{-1} s^{-1}). The products exhibit absorption bands at 540–550 nm, which are characteristic of σ-bonded alkyliron porphyrins[111] and can be assigned to DPFe(III)CF$_3$CHCl and DPDMeFe(III)CF$_3$CHCl, respectively.[112] No reaction was observed between the halothane radical and the ferrideutero-porphyrins; thus CF$_3$ĊHCl resembles ĊCl$_3$ in its reactivity toward ferric porphyrins and, so far, only the methyl radical has been reported to react rapidly. On the other hand, the rapid reaction of alkyl and halogenated alkyl radicals with Fe(II) porphyrins to form iron–carbon bonded species appears to be a general phenomenon as exemplified by the ĊCl$_3$, ĊHCl$_2$, ĊH$_2$Cl, ĊH$_3$, and CF$_3$ĊHCl radicals.

C.   Demetallation

Silver(II) and (III) form stable complexes with porphyrins in aqueous solution, but Ag(I) porphyrins are unstable and demetallate rapidly. The mechanistic details of this process have been investigated[113] by using the pulse radiolysis technique to reduce Ag(II)TPPS and Ag(III)TPPS rapidly to Ag(I)TPPS.

The complex Ag(II)TPPS is reduced rapidly by the hydrated electron (pH 8.7; $k = 1.6 \times 10^{10}$ dm^3 mol^{-1} s^{-1}) and by (CH$_3$)$_2$ĊOH ($k = 6 \times 10^8$ dm^3 mol^{-1} s^{-1}) to form an intermediate assigned[113] as Ag(I)TPPS. The rapid reduction is followed by two first-order processes with rate constants of 5×10^4 and 1.3×10^3 s^{-1} to yield the final product, free porphyrin H$_2$TPPS. The following mechanism was proposed[113] to account for these observations.

$$\text{Ag(II)TPPS} + e_{aq}^- \ [\text{or } (\text{CH}_3)_2\dot{\text{C}}\text{OH}] \rightarrow \text{Ag(I)TPPS} \tag{52}$$

$$\text{Ag(I)TPPS} + \text{H}_2\text{O} \rightarrow \text{HAg(I)TPPS} + \text{OH}^- \tag{53}$$

$$\text{HAg(I)TPPS} + \text{H}_2\text{O} \rightarrow \text{H}_2\text{TPPS} + \text{Ag}^+ + \text{OH}^- \tag{54}$$

This stepwise mechanism is similar to that proposed for the acid-catalysed demetallation of Zn and Mg porphyrins,[114,115] but occurs much more rapidly and involves protonation directly by water molecules.

The free porphyrin H_2TPPS also reacts rapidly with e_{aq}^- ($k = 1.5 \times 10^{10}$ dm^3 mol^{-1} s^{-1}). However, γ-radiolysis studies showed that Ag(II)TPPS is reduced quantitatively to H_2TPPS and Ag^+, which indicates that $(H_2TPPS)^-$ also reduces Ag(II)TPPS.

V. CONCLUDING REMARKS

The metalloporphyrins behave much as expected. Thus, if the central metal atom is oxidisable or reducible, then redox changes occur there, either by direct reaction between the redox agent and the metal centre or indirectly through formation of a porphyrin radical anion or cation followed by intramolecular transfer, which seems generally to be quite efficient. When the metal atom is not redox active, for example in the case of zinc, there is clear evidence that porphyrin radical ions are the one-electron redox products.

The situation is much more complex in the case of the metalloproteins. Although there is abundant evidence that protein radicals can mediate the reduction of the metal centre by free-radical redox agents, both the efficiency and the rate with which this intramolecular electron transfer takes place are quite variable and the governing factors remain to be elucidated. This is particularly true when the reductant is the hydrated electron, because it reacts with the peptide chain, with the sulphur-containing residues cysteine and cystine (especially the disulphide bridge RSSR), and with histidyl residues.[116] Thus it is to be expected that e_{aq}^- will attack the metalloprotein rather indiscriminately. It is all the more remarkable, therefore, that almost 100% reduction of the metal centre has been observed in the reaction of e_{aq}^- with [2Fe–2S] and 2[4Fe–4S] ferredoxins.[78] The mechanisms of electron transfer from protein to the metal centre have yet to be delineated, but they could conceivably include any of the following processes: direct tunnelling, multistep transfer from one electron sink to another, and migration via a conduction channel such as may be provided by hydrogen bonds. If such pathways are available in the native protein, then it would not be surprising if they could be blocked or made less effective by small conformational changes induced in the protein structure by the experimental conditions.

A wider study of more selective reducing radicals should be informative. One example where this is evident is in the reduction of ascorbate oxidase by nitroaromatic radicals,[85] described in Section III,D, when the extent of reduction of the copper atoms is almost stoichiometric. Indeed, as these workers conclude,[85] "caution should be exercised in the use of e_{aq}^- and CO_2^{-} to probe electron-transfer processes inside multicentred redox proteins."

References

[1] Matheson, M. S.; Dorfman, L. M. "Pulse Radiolysis"; M.I.T. Press: Cambridge, Massachusetts, 1969.

[2] Spinks, J. W. T.; Woods, R. J. "An Introduction to Radiation Chemistry"; 2nd ed., Wiley-Interscience: New York, 1976.

[3] Buxton, G. V. "The Study of Fast Processes and Transient Species by Electron Pulse Radiolysis"; Baxendale, J. H., Busi, F., Eds.; Reidel Publ.: Dordrecht, Netherlands, 1982.

[4] Ross, A. B.; Neta, P. *Natl. Stand. Ref. Data Ser. (U.S., Natl. Bur. Stand.)* **1982,** *NSRDS-NBS 70.*

[5] Ross, A. B.; Neta, P. *Natl Stand. Ref. Data Ser. (U.S., Natl. Bur. Stand.)* **1979,** *NSRDS-NBS 65.*

[6] Buxton, G. V.; Sellers, R. M. *Natl. Stand. Ref. Data Ser. (U.S., Natl. Bur. Stand.)* **1972,** *NSRDS-NBS 68.*

[7] Anbar, M.; Bambenek, M.; Ross, A. B. *Natl. Stand. Ref. Data Ser. (U.S., Natl. Bur. Stand.)* **1973,** *NSRDS-NBS 43.*

[8] Anbar, M.; Farhataziz; Ross, A. B. *Natl. Stand. Ref. Data Ser. (U.S., Natl. Bur. Stand.)* **1975,** *NSRDS-NBS 51.*

[9] Farhataziz; Ross, A. B. *Natl. Stand. Ref. Data Ser. (U.S., Natl. Bur. Stand.)* **1977,** *NSRDS-NBS 59.*

[10] Klug-Roth, D.; Fridovich, I.; Rabani, J. *J. Am. Chem. Soc.* **1973,** *95,* 2786–2790.

[11] Fielden, E. M.; Roberts, P. B.; Bray, R. C.; Lowe, D. J.; Mantner, G. N.; Rotilio, G.; Calabrese, L. *Biochem. J.* **1974,** *139,* 49–60.

[12] O'Neill, P.; Fielden, E. M. "Chemical and Biochemical Aspects of Superoxide Dismutase"; Bannister, J. V.; Hill, H. A. O., Eds.; Elsevier/North Holland: Amsterdam, 1980, pp. 357–363.

[13] Fee, J. A. "Superoxide and Superoxide Dismutases"; Michelson, A. M.; McCord, J. M.; Fridovich, I., Eds.; Academic Press: New York, 1977, pp. 173–192.

[14] McAdam, M. E.; Fielden, E. M.; Lavelle, F.; Calabrese, L.; Cocco, D.; Rotilio, G. *Biochem. J.* **1977,** *167,* 271–274.

[15] O'Neill, P.; Fielden, E. M.; Cocco, D.; Rotilio, G.; Calabrese, L. *Biochem. J.* **1982,** *205,* 181–187.

[16] Malinowski, D. P.; Fridovich, I. "Chemical and Biochemical Aspects of Superoxide and Superoxide Dismutase"; Bannister, J. V.; Hill, H. A. O., Eds.; Elsevier/North Holland: Amsterdam, 1980, pp. 299–317.

[17] Simic, M. G.; Hoffman, M. Z. *J. Am. Chem. Soc.* **1977,** *99,* 2370–2371.

[18] Natarajan, P.; Raghavan, N. V. *J. Am. Chem. Soc.* **1980,** *102,* 4518–4519.

[19] Lavelle, F.; McAdam, M. E.; Fielden, E. M.; Puget, K.; Michelson, A. M. *Biochem. J.* **1977,** *161,* 3–11.

[20] Fee, J. A.; McClune, G. J.; O'Neill, P.; Fielden, E. M. *Biochem. Biophys. Res. Commun.* **1981,** *100,* 377–384.

[21] Pasternack, R. F.; Skowronek, W. R. *J. Inorg. Biochem.* **1979,** *11,* 261–267.

[22] Weinstein, J.; Bielski, B. H. J. *J. Am. Chem. Soc.* **1980,** *102,* 4916–4919.

[23] Ilan, Y.; Rabani, J.; Fridovich, I.; Pasternack, R. F. *Inorg. Nucl. Chem. Lett.* **1981,** *17,* 93–96.

[24] Solomon, D.; Peretz, P. P.; Faraggi, M. *J. Phys. Chem.* **1982,** *86,* 1842–1849.

[25] Peretz, P.; Solomon, D.; Weinraub, D.; Faraggi, M. *Int. J. Radiat. Biol.* **1982,** *42,* 449–456.

[26] Rohrbach, D. F.; Deutsch, E.; Heineman, W. R.; Pasternack, R. F. *Inorg. Chem.* **1977,** *16,* 2650–2652.
[27] McClune, G. J.; Fee, J. A.; McCluskey, G. A.; Groves, J. T. *J. Am. Chem. Soc.* **1977,** *99,* 5220–5222.
[28] Diguiseppi, J.; Fridovich, I. *Arch. Biochem. Biophys.* **1980,** *203,* 145–150.
[29] Bull, C.; Fee, J. A.; O'Neill, P.; Fielden, E. M. *Arch. Biochem. Biophys.* **1982,** *215,* 551–555.
[30] Butler, J.; Koppenol, W. H.; Margoliach, E. *J. Biol. Chem.* **1982,** *257,* 10747–10750.
[31] Koppenol, W. H.; van Buuren, K. J. H.; Butler, J.; Braams, R. *Biochim. Biophys. Acta* **1976,** *449,* 157–168.
[32] McCord, J. M.; Fridovich, I. *J. Biol. Chem.* **1969,** *244,* 6049–6055.
[33] Sutton, H. C.; Sangster, D. F. *J. Chem. Soc., Faraday Trans. 1* **1982,** *78,* 695–711.
[34] Sutton, H. C.; Roberts, P. B.; Winterbourne, C. C. *Biochem. J.* **1976,** *155,* 503–510.
[35] Winterbourne, C. C. "Chemical and Biochemical Aspects of Superoxide and Superoxide Dismutase"; Bannister, J. V.; Hill, H. A. O., Eds.; Elsevier/North Holland: Amsterdam, 1980, p. 372.
[36] Fee, J. A. *In* "Proceedings of the Third International Symposium on Oxidase and Related Oxidation-Reduction Systems"; King, T. E.; Mason, H. S.; Morrison, M., Eds.; Pergamon: Oxford, 1979.
[37] Samuni, A.; Chevion, M.; Czapski, G. *J. Biol. Chem.* **1981,** *256,* 12632–12635.
[38] Wilting, J.; Braams, R.; Nauta, H.; van Buuren, J. J. H. *Biochim. Biophys. Acta* **1972,** *283,* 543–547.
[39] Land, E. J.; Swallow, A. J. *Biophys. Biochim. Acta* **1974,** *368,* 86–96.
[40] Shafferman, A.; Stein, G. *Biochim. Biophys. Acta* **1975,** *416,* 287–317.
[41] Simic, M. G.; Taub, I. A. *Faraday Discuss. Chem. Soc.* **1977,** *63,* 270–278.
[42] Land, E. J.; Swallow, A. J. *Arch. Biochem. Biophys.* **1971,** *145,* 365–372.
[43] Butler, J.; Jayson, G. G.; Swallow, A. J. *Biochim. Biophys. Acta* **1975,** *408,* 215–222.
[44] Simic, M. G.; Taub, I. A.; Tocci, J.; Hurwitz, P. A. *Biochem. Biophys. Res. Commun.* **1975,** *62,* 161–167.
[45] Seki, H.; Ilam, Y. A.; Ilam, Y.; Stein, G. *Biochim. Biophys. Acta* **1976,** *440,* 573–586.
[46] Ilan, Y.; Shafferman, A.; Feinberg, B. A.; Lau, Y.-K. *Biochim. Biophys. Acta* **1979,** *548,* 565–578.
[47] Land, E. J.; Swallow, A. J. *Ber. Bunsenges. Phys. Chem.* **1975,** *79,* 436–437.
[48] Seki, H.; Imamura, M. *Biochim. Biophys. Acta* **1981,** *635,* 81–89.
[49] Shafferman, A.; Stein, G. *Science* **1974,** *183,* 428–430.
[50] Lichtin, N. N.; Shafferman, A.; Stien, G. *Biochim. Biophys. Acta* **1974,** *357,* 386–398.
[51] Van Leeuwen, J. W.; Raap, A.; Koppenol, W. H.; Nauta, H. *Biochim. Biophys. Acta* **1978,** *503,* 1–9.
[53] Ilan, Y.; Shafferman, A. *Biochim. Biophys. Acta* **1979,** *548,* 161–165.
[53] Ilan, Y.; Shafferman, A.; Stein, G. *J. Biol. Chem.* **1976,** *251,* 4336–4345.
[54] Heijman, M. G. H.; Nauta, H.; Levine, Y. K. *Biochim. Biophys. Acta* **1982,** *704,* 560–563.
[55] Isied, S. S.; Worosila, G.; Atherton, S. J. *J. Am. Chem. Soc.* **1982,** *104,* 7659–7661.
[56] Veerman, E. C. I.; Van Leeuwen, J. W.; van Buuren, K. J. H.; Van Gelder, B. F. *Biochim. Biophys. Acta* **1982,** *680,* 134–141.
[57] Fauvodon, V.; Ferradini, C.; Pucheault, J.; Gilles, L.; LeGall, J. *Biochem. Biophys. Res. Commun.* **1978,** *84,* 435–440.
[58] Van Leeuwen, J. W.; Van Dijk, H. J.; Grande, J.; Veeger, D. *Eur. J. Biochem.* **1982,** *127,* 631–637; **1983,** *130,* 619.
[59] Whitburn, K. D.; Shieh, J. J.; Sellers, R. M.; Hoffman, M. Z. *J. Biol. Chem.* **1982,** *257,* 1860–1869.
[60] Kobayashi, K.; Hayashi, K. *J. Biol. Chem.* **1981,** *256,* 12350–12354.

[61] Tamura, M.; Yamazaki, I. *J. Biochem.* (*Tokyo*) **1972**, *71*, 311–319.

[62] Raap, I. A.; Van Leeuwen, J. W.; Rollema, H. S.; de Bruin, S. H. *Eur. J. Biochem.* **1978**, *88*, 555–563.

[63] Van Leeuwen, J. W.; Butler, J.; Swallow, A. J. *Biochim. Biophys. Acta* **1981**, *667*, 185–196.

[64] Ilan, Y. A.; Samuni, A.; Chevion, M.; Czapski, G. *J. Biol. Chem.* **1978**, *253*, 82–86.

[65] Brill, A. S.; Castleman, B. W.; Knight, M. E. *Biochemistry* **1976**, *15*, 2309–2316.

[66] Hasinoff, B. B.; Pecht, I. *Biochim. Biophys. Acta* **1983**, *743*, 310–315.

[67] Klapper, M. H.; Faraggi, M. *Q. Rev. Biophys.* **1979**, *12*, 465–519.

[68] Simic, M. G. *J. Agric. Food Chem.* **1978**, *26*, 6–14.

[69] Neta, P. *J. Phys. Chem.* **1981**, *85*, 3678–3684.

[70] Brault, D.; Bizet, C.; Morliere, P.; Rougee, M.; Land, E. J.; Santus, R.; Swallow, A. J. *J. Am. Chem. Soc.* **1980**, *102*, 1015–1020.

[71] Evers, E. L.; Jayson, G. G.; Swallow, A. J. *J. Chem. Soc., Faraday Trans. 1*, **1978**, *74*, 418–426.

[72] Brault, D.; Morliere, P.; Rougee, M. *Proc. Int. Conf. Coord. Chem., 19th* **1978**, *1*, 112–113.

[73] Harrington, P. C.; Wilkins, R. G. *J. Biol. Chem.* **1979**, *254*, 7505–7508.

[74] Wiltig, J.; Raap, A'; Braams, R.; de Bruin, S. H.; Rollerma, H. S.; Janssen, L. H. M. *J. Biol. Chem.* **1974**, *249*, 6325–6330.

[75] Lovenberg, W. (Ed.) "Iron-Sulfur Proteins", Vols. 1–3; Academic Press: New York, 1973–1977.

[76] Averill, B. A.; Orme-Johnson, W. H. *Met. Ions Biol. Syst.* **1978**, *7*, 127–185.

[77] Maskiewicz, R.; Bielski, B. H. J. *Biochim. Biophys. Acta* **1981**, *638*, 153–160.

[78] Adzamli, I. K.; Ong, H.; Sykes, A. G.; Buxton, G. V. *J. Inorg. Biochem.* **1982**, *16*, 311–317.

[79] Butler, J.; Henderson, R. A.; Armstrong, F. A.; Sykes, A. G. *Biochem. J.* **1979**, *183*, 471–474.

[80] Butler, J.; Sykes, A. G.; Buxton, G. V.; Harrington, P. C.; Wilkins, R. G. *Biochem. J.* **1980**, *189*, 641–644.

[81] Stombaugh, N. A.; Sundquist, J. E.; Burris, R. H.; Orme-Johnson, W. H. *Biochemistry* **1976**, *15*, 2633–2641.

[82] Henderson, R. A.; Sykes, A. G. *Inorg. Chem.* **1980**, *19*, 3103–3105.

[83] Cammack, R. *Biochem. Biophys. Res. Commun.* **1973**, *54*, 548–554.

[84] Faraggi, M.; Pecht, I. *Biochem. Biophys. Res. Commun.* **1971**, *45*, 842–848.

[85] O'Neill, P.; Fielden, E. M.; Finazzi-Agro, A.; Avigliano, L. *Biochem. J.* **1983**, *209*, 167–174.

[86] Pecht, I.; Goldberg, M. *In* "Fast Processes in Radiation Chemistry and Biology"; Adams, G. E.; Fielden, E. M.; Michael, B. D., Eds.; Wiley: New York, 1975, pp. 277–284.

[87] Faraggi, M.; Pecht, I. *J. Biol. Chem.* **1973**, *248*, 3146–3149.

[88] Guissani, A.; Henry, Y.; Gilles, L. *Biophys. Chem.* **1982**, *15*, 177–190.

[89] Pecht, I.; Faraggi, M. *Nature* (*London*), *New Biol.* **1971**, *233*, 116–118.

[90] Adams, G. E.; McNaughton, G. S.; Michael, B. D. "The Chemistry of Ionization and Excitation"; Johnson, G. R. A.; Scholes, G., Eds.; Taylor & Francis: London, 1967, pp. 281–293.

[91] Mondovi, B.; Avigliano, L. *In* "Copper Proteins and Copper Enzymes"; Lontie, R., Ed.; CRC Press: Boca Raton, Florida, 1984.

[92] Chauvet, J. P.; Viovy, R.; Land, E. J.; Santus, R. C. *C. R. Hebd. Seances Acad. Sci., Ser. D* **1979**, *288*, 1423.

[93] Steenken, S. *J. Phys. Chem.* **1979**, *83*, 595–598.

[94] Bonnet, R.; Ridge, R. J.; Land, E. J.; Sinclair, R. S.; Tait, D.; Truscott, T. G. *J. Chem. Soc., Faraday Trans. 1* **1982,** *78,* 127–136.

[95] Neta, P.; Scherz, A.; Levanon, H. *J. Am. Chem. Soc.* **1979,** *101,* 3624–3629.

[96] Neta, P.; Grebel, V.; Levanon, H. *J. Phys. Chem.* **1981,** *85,* 2117–2119.

[97] Levanon, H.; Neta, P. *Chem. Phys. Lett.* **1980,** *70,* 100–103.

[98] Clarke, D. A.; Dolphin, D.; Grigg, R.; Johnson, A. W.; Pinnock, H. A. *J. Chem. Soc. C* **1968,** 881.

[99] Baral, S.; Neta, P. *J. Phys. Chem.* **1983,** *87,* 1502–1509.

[100] Endicott, J. F.; Ferraudi, G. J. *J. Am. Chem. Soc.* **1977,** *99,* 243–245.

[101] Tait, A. M.; Hoffman, M. Z.; Hayon, E. *Int. J. Radiat. Phys. Chem.* **1976,** *8,* 691–696.

[102] Grate, J. H.; Schrauzer, G. N. *J. Am. Chem. Soc.* **1979,** *101,* 4601–4611.

[103] Mulac, W. A.; Meyerstein, D. *J. Am. Chem. Soc.* **1982,** *104,* 4124–4128.

[104] Elroi, H.; Meyerstein, D. *J. Am. Chem. Soc.* **1978,** *100,* 5540–5548.

[105] Blackburn, R.; Kyaw, M.; Phillips, G. O.; Swallow, A. J. *J. Chem. Soc., Faraday Trans. 1* **1975,** *71,* 2277–2287.

[106] Blackburn, R.; Erkol, A. Y.; Phillips, G. O.; Swallow, A. J. *J. Chem. Soc., Faraday Trans. 1* **1974,** *70,* 1693–1701.

[107] Schrauzer, G. N.; Deutsch, E. *J. Am. Chem. Soc.* **1969,** *91,* 3341–3350.

[108] Tait, A. M.; Hoffman, M. Z.; Hayon, E. *J. Am. Chem. Soc.* **1976,** *98,* 86–93.

[109] Blackburn, R.; Kyaw, M.; Swallow, A. J. *J. Chem. Soc., Faraday Trans. 1* **1977,** *73,* 250–255.

[110] Brault, D.; Neta, P. *J. Am. Chem. Soc.* **1981,** *103,* 2705–2710.

[111] Lexa, D.; Mispelter, J.; Saveant, J.-M. *J. Am. Chem. Soc.* **1981,** *103,* 6806–6812.

[112] Brault, D.; Neta, P. *J. Phys. Chem.* **1982,** *86,* 3405–3410.

[113] Kumar, A.; Neta, P. *J. Phys. Chem.* **1981,** *85,* 2830–2832.

[114] Snellgrove, R.; Plane, R. A. *J. Am. Chem. Soc.* **1968,** *90,* 3185–3194.

[115] Shears, B.; Shah, B.; Hambright, P. *J. Am. Chem. Soc.* **1971,** *93,* 776–778.

[116] Braams, R. *Radiat. Res.* **1966,** *27,* 319.

ADVANCES IN INORGANIC AND BIOINORGANIC MECHANISMS, VOL. 3

Structure and Functional Properties of Peroxidases and Catalases

Jane E. Frew and Peter Jones

Radiation and Biophysical Chemistry Laboratory
University of Newcastle upon Tyne
Newcastle upon Tyne, England

I. INTRODUCTION

The occurrence in biological tissues of powerful catalysts for the decomposition of added hydrogen peroxide to oxygen and water (catalase enzymes) has been known for more than 150 years. An extensive range of peroxidase enzymes has also been characterized, which act as catalysts for the peroxide oxidation of a variety of donor (reducing) substrates. New enzymes possessing activities of these types continue to be discovered. The study of all such enzymes has formed an important aspect of the biological chemistry of iron, since, until comparatively recently, known members were all heme proteins. The characterization of the selenoenzyme glutathione peroxidase has added a new dimension to concepts of peroxide metabolism, and the existence of other, nonheme, enzymes is now becoming evident.

A number of other factors have influenced developments in this field. The discovery of superoxide dismutases has stimulated many areas of investigation concerned with the biochemistry of the molecular (H_2O_2 and H_2O) and free-radical (O_2^- and $OH \cdot$) products of oxygen reduction. In the present context attention has focused on the formation of hydrogen peroxide *in vivo* as a result of oxygen reduction via the superoxide level. From the characterization of the peroxisome as an organelle within liver cells and from other studies of intracellular enzyme localization, the concept of compartmentalization in the actions of peroxide metabolizing enzymes *in vivo* has emerged, particularly in mammalian systems.

The discovery of mixed-function oxidases, particularly the heme monooxygenase (cytochrome P-450) enzymes, has led to the recognition of the importance of oxygen activation, by reduction to the peroxide level, when binding occurs within the active sites of enzymes. Information derived from studies of the unusual species formed as intermediates in peroxidase and catalase actions has had a seminal influence on concepts of mechanisms in cytochrome P-450 reactions. Thus, in the past decade, the metabolism of hydrogen peroxide has become more clearly seen in the broader context of the biological chemistry of oxygen.

Although the development of mechanistic concepts of heme peroxidase functions has predominantly been concerned with H_2O_2 as the acceptor substrate, organic hydroperoxides have been used as mechanistic probes in a number of investigations. These studies were not originally considered to have any direct physiological relevance. Nonenzymatic lipid peroxidation has been intensively investigated with respect to both its physiological and technological aspects, and the phenomenon has been implicated in a variety of metabolic disorders ranging from aging to carcinogenesis. The formation of organic hydroperoxides from diverse biological molecules as a consequence of, for example, the action of ionizing radiation has also been examined. The finding that glutathione peroxidase could efficiently utilize a range of organic hydroperoxides as substrates has extended consideration of peroxide metabolism from hydrogen peroxide into a much wider field. This development has been powerfully reinforced by the further discovery that peroxidase action is an important early stage in the biosynthesis of prostaglandins from arachidonic acid. Thus, the biochemistry of the peroxidase enzymes is now very much concerned with the metabolism of hydroperoxides in general.

Early concepts of the *in vivo* action of the enzymes stressed the toxicity of hydrogen peroxide, and hence a primarily cellular detoxification role was proposed. It is becoming clear, particularly from the developing studies of mammalian enzymes, that despite the toxicity hazards, peroxide formation and metabolism are additionally important both in biosynthesis (thyroid hormones and prostaglandins) and in the deterrence and destruction of attacking organisms. The latter action is seen at its most spectacular in the bombardier beetle (*Brachynus*) and in a more subtle, and intrinsically important, form in the phagocytosis of invading microorganisms by mammalian leukocytes.

Understanding of the intimate catalytic mechanisms, particularly of the heme enzymes, has now reached a level of sophistication that was impossible before both the application of X-ray crystallography to the determination of the structures of some of the native enzymes and the application of a variety of spectroscopic probes in studies of some of the key catalytic intermediates.

Sections II, III, and IV are concerned with the occurrence, function, structure, and catalytic mechanisms of heme enzymes. In Section V a discussion of nonheme enzymes is presented.

II. HEME ENZYMES—OCCURRENCE, CHARACTERIZATION, AND FUNCTION

A. Introduction

In this section a comprehensive list of peroxidases and catalases is not presented, but rather the wide range of biological sources from which enzymes have been obtained and the very diverse chemical transformations that their specific catalytic reactions elicit are illustrated.

The *in vivo* functions of peroxidases are not well established, and classification indicates the tissue or organism of origin, with, in some cases, an indication of a predominant, *in vitro* catalytic property. The problem of classification is further complicated because most, if not all, hemoproteins may be converted to forms which show some "peroxidase" activity, e.g., metmyoglobin and and methemoglobin.[1] Mammalian cytochrome oxidase shows a "cytochrome *c* peroxidase" activity,[2] fragments of nonperoxidase hemoproteins, for example "microperoxidase" from cytochrome *c*, are catalytically effective, and even protein-free heme complexes show significant peroxidase- and catalase-like activities.[3]

Many of the enzyme preparations present problems of heterogeneity. In some cases the isoenzyme constituents arise from biosynthesis of quite distinct proteins, but it is evident that modification during extraction is a significant source of heterogeneity. The problems are particularly severe for those enzymes that are glycoproteins.

B. Plant peroxidases

1. Horseradish peroxidase

The root of the horseradish yields the most extensively studied enzyme of the peroxidase family.[4] Horseradish peroxidase (HRP) preparations of various levels of purity and activity are available from commercial suppliers and have applications in a variety of clinical biochemical assays. As many as 20 isoenzymes have been reported on the basis of isoelectric focusing studies.[5] Six basic isoenzymes (E1–E6) have recently been purified and crystallized, and preliminary crystallographic data for E4 have been reported.[6] The neutral type C isoenzyme is well characterized and is the predominant component of high-quality commercial preparations.[7,8] The enzyme is a glycoprotein (MW $\sim$42,000; protein MW 33,890 from sequence studies) containing one protoferriheme [Fe(III)–protoporphyrin IX] prosthetic group (Fig. 1), eight carbohydrate chains bound to asparagine residues, four disulfide bridges, and two calcium ions per molecule. The amino acid sequence has been determined.[9] The molar extinction coefficient of purified enzyme at the Soret band maximum (402.5 nm) is $1.02 \times 10^5\ M^{-1}\ cm^{-1}$, with a purity index *RZ* (the ratio of heme absorbance at 403 nm to protein absorbance at 280 nm) of 3.55. ORD and CD spectra suggest a high degree ($\sim$43%) of protein-helical structure. The Fe(III) center in the native enzyme is predominantly (80%) high spin. NMR studies have established that the fifth proximal ligand to iron is a neutral histidylimidazole, in which the N1

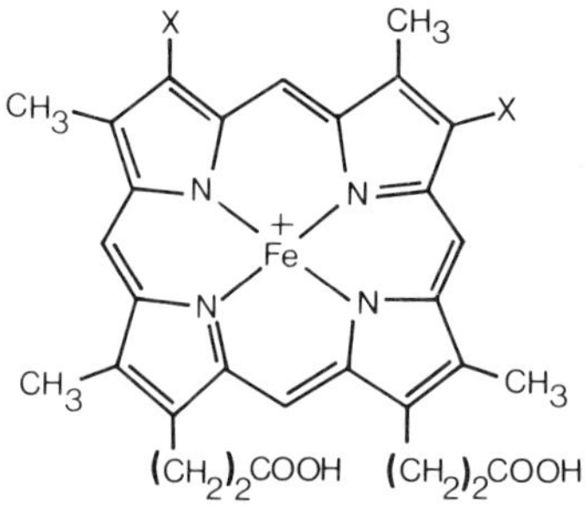

Fig. 1. Structure of some ferriheme [Fe(III)–porphyrin] complexes: X = H, deuteroferriheme; X = CH=CH$_2$, protoferriheme; X = CH$_2$CH$_3$, mesoferriheme.

histidyl proton remains bound and is only slowly exchangeable by deuterium from solvent.[10]

The nature of the ligand (if any) at the sixth (distal) coordination position of iron is not conclusively determined. A water molecule has often been assumed, but a number of studies using NMR methods did not reach a clear resolution of this problem.[11–13] More recent studies by NMR,[14] ^{17}O superhyperfine interaction in ESR,[15] spectra of HRP formed by photodissociation of the NO complex,[16] and analysis of resonance Raman spectra[17,18] strongly suggest a pentacoordinate iron center. The enzyme undergoes an "alkaline transition" above pH 11, which leads to an inactive, low-spin Fe species in which a nitrogenous ligand from the protein is bound to the sixth coordination position of iron.[19]

The function of HRP in the horseradish is not known, but *in vitro* the enzyme oxidizes a wide range of donor substrates, with a particularly high activity toward phenols and amines. In addition to H$_2$O$_2$ a range of organic hydroperoxides can be utilized; reaction with some peroxybenzoic acids occurs even more rapidly than with H$_2$O$_2$.[20,21]

2. Turnip peroxidases

Seven peroxidase isoenzymes have been isolated from the turnip (*Brassica napus*), of which five have been purified to homogeneity.[22] They appear to be genuinely different enzymes and not degradation products. The turnip peroxidase (TUP) enzymes are all glycohemoproteins with isoelectric points ranging from 3.3 (TUP$_1$) to 11.6 (TUP$_7$) and all contain protoferriheme as the prosthetic group. Molecular weights range from 51,000 (TUP$_1$) to 38,700 (TUP$_7$), and extinction coefficients at the Soret band maximum are $\varepsilon_{403} = 1.15 \times 10^5$ M^{-1} cm^{-1} for TUP$_1$ (*RZ* = 3.0) and $\varepsilon_{405} = 8.7 \times 10^4$

M^{-1} cm^{-1} for TUP$_7$ ($RZ = 2.8$). The enzymes undergo alkaline transitions (for TUP$_1$, pK ~10; for TUP$_7$, pK ~9) that convert the high-spin Fe(III) native enzymes to low-spin forms.[23]

The more acidic isoenzymes (TUP$_1$, TUP$_2$, TUP$_3$) appear to be closely related to HRP-C in structure, but the basic isoenzyme (TUP$_7$) possesses a distinct amino acid composition and sequence.[24,25] TUP$_7$, which is formed only during winter (when it constitutes 40% of the peroxidase activity), is a relatively inefficient peroxidase but acts as an oxygenase toward the plant hormone indole-3-acetate. It has been proposed that this oxygenase function may represent the *in vivo* activity of TUP$_7$.[26]

C. Yeast peroxidase

The intermembrane space of the mitochondria of aerobically grown baker's and brewer's yeasts contains a cytochrome *c* peroxidase (CcP) that is a monomeric hemoprotein (protein MW 33,419) containing one high-spin Fe(III) protoferriheme prosthetic group per molecule.[27,28] Both the amino acid sequence and the crystal structure of the enzyme have been determined (Section III,B).

In solution there is evidence that different anions can effect interconversion of the enzyme between forms that differ in spectra and reactivity.[29] Acetate-containing solutions yield species with absorption maxima at 620 nm (un-reactive) and 645 nm (reactive). In phosphate solutions only the reactive form is present (for purified enzyme $RZ = 1.23$ and $\varepsilon_{408} = 1.0 \times 10^5$ M^{-1} cm^{-1}).

The classification of the enzyme is based on the efficiency of its catalysis of the oxidation of yeast and mammalian ferrocytochromes *c* by H$_2$O$_2$ and a range of hydroperoxides, although other donors (ascorbate, pyrogallol, ferrocyanide) are also oxidized. Ferricytochrome *c* inhibits the oxidation of ferrocytochrome *c* by the enzyme,[30] and complexes between both ferri- and ferrocytochrome *c* and CcP have been studied.[31] The *in vivo* role of the enzyme is not known although it has been suggested that it may be involved in restricting the accumulation of peroxides.

D. Bacterial peroxidase

The denitrifying bacterium and opportunistic pathogen *Pseudomonas aeruginosa* contains a membrane-bound peroxidase that is unusual in a number of respects.[32] The extracted protein (*Pseudomonas* cytochrome c_{551} peroxidase) is a monomer (MW 43,200) that contains *two* molecules of mesoheme (Fig. 1) as prosthetic groups.[33] In the enzyme as prepared (now termed the resting enzyme[34]), both hemes are in the Fe(III) form. The two heme groups show quite different properties; one is a high-spin heme that is accessible to ligands and behaves as a typical peroxidase heme prosthetic group.[35] The other is low spin and inaccessible to ligands, and behaves as a hexacoordinate cytochrome unit, which is capable of undergoing electron transfer.[36]

The resting enzyme reacts with H_2O_2 to form a complex but is not per-oxidatically active. Enzyme activation involves reduction of the low-spin (cytochrome-like) heme to the Fe(II) state.[37]

The basis for the enzyme classification is the high peroxidatic activity displayed toward the ferrocytochrome c_{551} from the bacterium. However, the enzyme is also markedly reactive in catalyzing the peroxide oxidation of azurin, a Type 1 (blue) copper protein from the same organism.[38]

E.   Fungal peroxidase

The fungal chloroperoxidase is a glycohemoprotein isolated from the mold *Caldariomyces fumago*.[39] The enzyme contains one protoferriheme group per molecule and 25–30% carbohydrate (MW 42,000; $\varepsilon_{398} = 8.6 \times 10^4 \, M^{-1} \, cm^{-1}$, $RZ > 1.4$). Preliminary X-ray data on the crystalline enzyme have been reported.[40] The reduced Fe(II) form of the enzyme forms a CO complex with a Soret absorption band at 446 nm, and on this basis a possible structural similarity with cytochrome P-450 enzymes was suggested,[41] although a search for a proximal thiolate ligand to the heme of chloroperoxidase has not been successful.

The enzyme catalyzes the halogenation reactions involved in the bio-synthesis of caldariomycin (2,2-dichloro-1,3-cyclopentadione). Classification is based on the facility for catalyzing oxidation of Cl^- by H_2O_2, an attribute shared only by myeloperoxidase (all peroxidases oxidize I^-; some also oxidize Br^-). However, chloroperoxidase has been found to display a quite remarkable range of catalytic functions:

1. In the absence of halide and donor substrates, H_2O_2 undergoes catalytic decomposition (catalase-like action). In the presence of ethanol and formate, catalase-like peroxidatic oxidation of these donors occurs.[42]

2. Hydroperoxides and peroxyacids (except peroxyacetic acid) undergo catalytic decomposition,[43] a reaction unique to chloroperoxidase and of a very interesting mechanism (Section IV,E).

3. In the absence of halide, typical horseradish peroxidase substrates are oxidized using H_2O_2 or various hydroperoxides.[44]

4. In the presence of halide (Cl^-, Br^-, I^-), carbon–halogen bond formation is catalyzed with suitable halogen acceptors (most nucleophiles with an activated position available for attack by electrophilic halogen—β-ketoacids and β-diketones are extremely good acceptors).[42]

5. N-Demethylation reactions are catalyzed using hydroperoxide oxidants (reactions that parallel those of cytochrome P-450 using O_2 and NADPH).[45]

6. Chloroperoxidase catalyzes the dismutation of both chlorite and chlorine dioxide.[46]

F. Algal peroxidase

Many marine organisms, particularly marine algae, possess considerable bromoperoxidase activity.[47,48] The bromoperoxidase from *Penicillus capitatus* is the best characterized[49]: it contains no carbohydrate and exists in solution as a dimer of identical subunits, each containing one tightly bound protoferriheme prosthetic group (dimer MW 97,600; Soret band $\varepsilon_{413} = 2.12 \times 10^5\ M^{-1}\ cm^{-1}$, $RZ = 1.2$).

Bromoperoxidase can use Br^- (but not Cl^-) to form carbon–halogen bonds with a halogen acceptor. Bromine is the major halogen constituent of most halometabolites found in marine organisms, many of which show antimicrobial and antialgal activity.[50]

G. Protozoan peroxidase

A hemoprotein peroxidase obtained from the cytosol of *Euglena gracilis* Z has been partially characterized (Soret band 407 nm, MW $\sim 76,000$). The enzyme specifically requires L-ascorbic acid as natural electron donor.[51] Of artificial donor substrates, only pyrogallol and D-araboascorbic acid show significant reactivity. In addition to H_2O_2, cumene hydroperoxide and *t*-butyl hydroperoxide are good oxidizing substrates. *Euglena* cells lack catalase, and the peroxidase appears to be the sole agent decomposing H_2O_2 in this organism.[52]

H. Mammalian peroxidases

1. Lactoperoxidase

The first mammalian secretory peroxidase was found in cow's milk,[53,54] and since then similar enzymes have been found in goat and sheep milk and in other exocrine secretions (saliva and tears). The peroxidase activity of human milk is not a true secretory peroxidase but instead derives from myeloperoxidase (present in leukocytes).[55]

Bovine lactoperoxidase is a glycohemoprotein consisting of a single polypeptide chain[56] (MW 78,000; earlier work suggested two nearly identical subunits), 8–10% carbohydrate, and a tightly bound heme prosthetic group that has proved difficult to identify but that has recently been established to be protoferriheme ($\varepsilon_{412} = 1.1 \times 10^5 \ M^{-1} \ cm^{-1}$, $RZ = 0.92$).[57] Crystalline enzyme has been obtained.

Lactoperoxidase displays a range of peroxidatic functions. The facile oxidation of iodide has suggested a possible role in biological iodination. The enzyme has also been implicated in melanin formation. More recently, an antimicrobial role for lactoperoxidase in saliva has been suggested on the basis of the oxidation of thiocyanate (SCN^-) to hypothiocyanate ($OSCN^-$).[58,59] $OSCN^-$ (or its conjugate acid[60]) exerts antimicrobial activity[61] by inhibition of glyceraldehyde 3-phosphate dehydrogenase in the glycolytic pathway.[62]

2. Thyroid peroxidase

Thyroid peroxidase is a tightly bound membrane glycohemoprotein in the endoplasmic reticulum of the thyroid gland. The enzyme has been obtained from several mammals by proteolytic digestion,[63] solubilization by nonionic detergents,[64] or a combination of these techniques. The difficulties of extraction have hampered detailed characterization.

There is considerable evidence that thyroid peroxidase plays an important role in the biosynthesis of thyroxine.[65,66] In this process the enzyme successively catalyzes two reactions: (1) the iodination of several tyrosine residues of thyroglobulin and (2) the oxidative coupling of the hormonogenic diiodotyrosine residues of thyroglobulin to form thyroid hormones (Fig. 2). Iodide stimulates the coupling reaction, which may derive from a regulatory ligand action.[67] Different enzyme intermediates may be involved in the two reactions.[65]

3. Myeloperoxidase

Myeloperoxidase is a complex glycohemoprotein that is located in azurophilic granules within mammalian leukocytes (polymorphonuclear neutrophils and monocytes). Both the canine enzyme (isolated from the pus of infected dog uteri)[68] and the human enzyme (isolated from blood leukocytes)[69] are tetramers containing two heavy (MW ~57,000) and two light

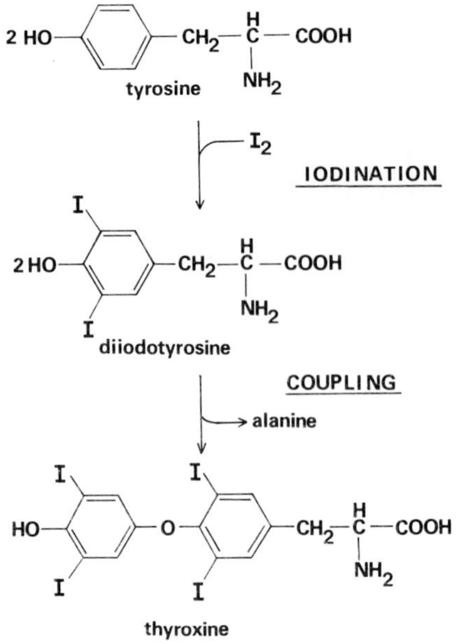

Fig. 2. Iodination and coupling reactions in the formation of thyroxine.

(MW 10,000–14,000) polypeptide chain subunits. The heavy subunits each contain a tightly bound high-spin ferriheme prosthetic group (the porphyrin structure is not yet known) and are covalently linked by a single disulfide bond.[70] The purified human enzyme has $\varepsilon_{403} = 1.78 \times 10^5 \ M^{-1} \ cm^{-1}$ and $RZ = 0.81$.[71] Reduction and alkylation of the human enzyme (treatment with dithiothreitol followed by iodoacetamide) cleaved the molecule in half; the product hemimyeloperoxidase retained the full activity of the tetramer. On the basis of these studies, Andrews and Krinsky[70] have proposed a model of the quaternary structure (Fig. 3). Electrophoretic studies have indicated that human myeloperoxidase consists of a number of isoenzymes; a study using enzyme prepared from the blood of a single donor suggests the existence of three isoenzymes.[72]

The primary function of the leukocyte is the destruction of invading microorganisms. This process proceeds by phagocytosis (engulfment of microorganism into the cell within a phagosomal space), followed by the attack upon the invader by a range of cidal agents generated by the leukocyte. The oxidative aspect of this process[73] originates from an intense burst of respiration (50-fold increase) within the phagocyte in which oxygen is reduced to O_2^- and H_2O_2. These species function as precursors of more reactive oxidants, which are the true microbicidal agents and which include

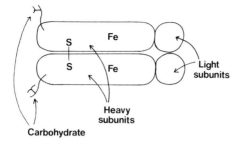

Fig. 3. Quaternary structure of myeloperoxidase. (From Ref. [70].)

hypochlorite, formed by the myeloperoxidase-catalyzed oxidation of chloride,[74] and hydroxyl radical, formed in a catalyzed Haber–Weiss reaction. Myeloperoxidase and chloride may effectively catalyze the latter process via the following sequence[75,76]:

$$H_2O_2 + Cl^- \xrightarrow{\text{myeloperoxidase}} HOCl + OH^-$$

$$HOCl + O_2^- \longrightarrow OH\cdot + O_2 + Cl^-$$

$$\overline{H_2O_2 + O_2^- \longrightarrow OH\cdot + O_2 + OH^-}$$

$$\text{(Haber-Weiss reaction)}$$

The microbicidal activity of the myeloperoxidase/H_2O_2/chloride system on *Escherichia coli* is associated with loss of iron from the organism.[77]

4. Prostaglandin hydroperoxidase

The early stages of prostaglandin biosynthesis (Fig. 4) involve two processes: (1) oxygenation of arachidonic acid to form the endoperoxide–hydroperoxide species PGG_1 and (2) reduction of the hydroperoxide function of PGG_1 to an alcohol, forming PGH_1. The enzymatic activity associated with catalysis of process (1) is termed cyclooxygenase; that associated with process (2) is termed prostaglandin hydroperoxidase. A single molecular species appears to possess both catalytic functions.[78] Both activities of the protein require heme as a cofactor but they show distinctly different substrate and inhibitor specificities. Cyclooxygenase activity is specifically inhibited by nonsteroidal antiinflammatory drugs such as aspirin. The enzyme obtained from bovine or ram seminal vesicular gland microsomes is a glycoprotein with two identical heme-containing subunits; an absorption at

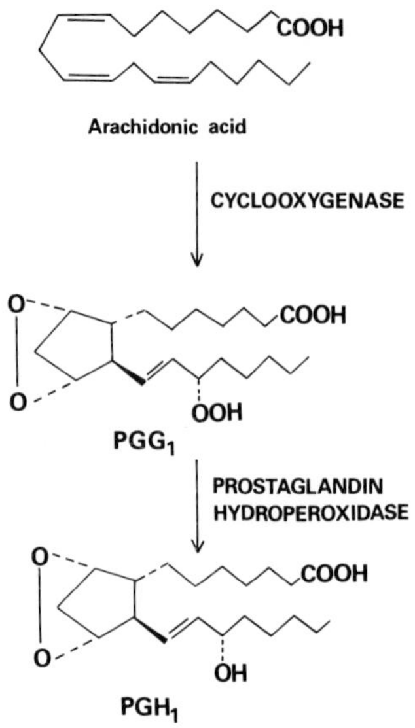

Fig. 4. Cyclooxygenase and hydroperoxidase actions in the formation of prostaglandins from arachidonic acid.

412 nm has been attributed to the heme–enzyme complex.[79] The enzyme can effectively utilize alkyl hydroperoxides (preferably lipid hydroperoxides) as oxidizing substrate, and a wide range of species can act as donor substrate *in vitro*.[80,81]

I. Catalases

Catalases are almost ubiquitous in aerobic organisms.[82] The enzymes from various sources are tetramers of identical subunits (tetramer MW $2.1–3.8 \times 10^5$), each subunit containing a protoferriheme prosthetic group (although the enzymes from *Neurospora crassa* may contain a ferric chlorin[83]). Catalase from beef liver is commercially available (a proportion of the prosthetic groups of enzyme from this source may be damaged by porphyrin oxidation either before or during extraction); other major sources include mammalian erythrocytes and aerobic bacteria. The crystal structures of the enzymes from beef liver and from the fungus *Penicillium vitale* have been determined (Section III,A).

The ability of animal and vegetable tissues to decompose H_2O_2 to H_2O and O_2 was first observed by Thenard (1818) and this function ("catalatic" activity) has formed the basis for the classification of the catalase enzymes.

This reaction is undoubtedly a major function of the enzyme *in vivo*, notably in the liver cell peroxisome, an intracellular organelle that contains both catalase and hydrogen peroxide-producing oxidases.[84] It is probable that the peroxidatic action of catalases in catalyzing the H_2O_2 oxidation of methanol and ethanol may also be significant *in vivo*.

Catalases are highly selective for H_2O_2 as oxidizing substrate but will also utilize alkyl and acyl hydroperoxides with small end groups. The quaternary structure of the catalase tetramer is essential for catalatic activity. In Swiss-type acatalasemia, enzyme is synthesized that is active but that possesses a very unstable quaternary structure.[85]

III. HEME ENZYMES—X-RAY STRUCTURE DETERMINATIONS

A. Catalases

X-Ray crystallographic determinations of the structures of bovine liver catalase (to 2.5 Å resolution)[86] and the fungal catalase from *P. vitale* (to 3.5 Å resolution)[87] have been reported. Both enzymes are tetramers of identical (or nearly identical) subunits, although the fungal enzyme is a molecule (MW 280,000) appreciably larger than the bovine enzyme.

The complete sequence of the 506 amino acid residues in each subunit of beef liver catalase has been determined.[88] The molecule has 222 symmetry, and is dumbbell shaped with a length of 105 Å and a waist of 50 Å.[86] The four heme groups are buried 20 Å below the molecular surface and 23 Å from the center of the molecule. The smallest iron–iron distance is 30.6 Å. The structure determination is in agreement with the amino acid sequence data, although no appreciable electron density was observed from 19 residues in each subunit. About 26% of the residues are in helical structures and 12% in sheet structures. The four SH groups in each subunit are too far apart to be involved in intrasubunit or intersubunit S–S bond formation.

The structure of each subunit (Fig. 5) is composed of four domains: (1) residues 1–75 at the amino terminus form an arm extending from the globular region of the subunit, which contains two helices, both of which are involved in intersubunit contacts; (2) residues 76–320 form an eight-stranded, antiparallel sheet β barrel, and most residues on the distal side of the heme are from this region; (3) residues 321–436 form an outer layer to each subunit

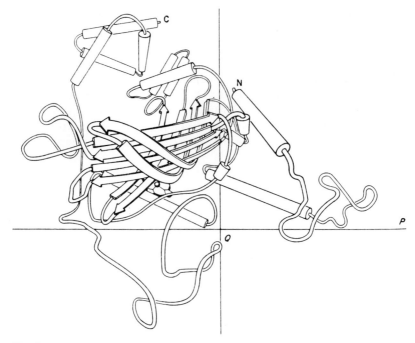

Fig. 5. Diagrammatic representation of the structure of a subunit of beef liver catalase. (From Ref. [86].)

(the "wrapping domain"), which contains the essential helix with the proximal heme ligand, Tyr 357; and (4) residues 437–506 are folded into a four-helical domain, which is not involved in intersubunit contacts but which contributes to limitation of the accessibility of the active site.

Most of the intersubunit contacts in the quaternary structure of the tetramer are confined to domains (1) and (3). The most remarkable feature is that the first 25 residues of the N-terminal arm are inserted into a hole in an adjacent (Q-axis-related) subunit. This interaction has important implications for the biosynthesis of catalase and the stability of the tetrameric molecule.

Each heme site is accessible from the surface via a channel that is 30 Å long and with a maximum width of 15 Å at its mouth in the restricted waist of the molecule. The channel is lined with polar residues at the entrance and with hydrophobic residues as it descends toward the heme group. Unlike most hemoproteins, both propionyl groups of the heme are buried and directed toward the molecular center where they are neutralized by the guanidinium groups of arginine residues. The proximal side of the heme is crowded and the guanidinium group of Arg 353 is only 3.5 Å away from the phenolic oxygen of the (presumably deprotonated) proximal ligand, Tyr 357. The distal side of the heme is less confined and, most interestingly, there is no distal ligand to iron (contrast yeast cytochrome *c* peroxidase and metmyoglobin, in which this position is occupied by a water molecule). Residues of likely catalytic importance on the distal side are His 74 and Asn 147. A water molecule may be hydrogen bonded to these two residues. On separation of the tetramer into monomers, pyrrole rings I and IV of the heme become partially exposed to solvent.

The catalase from *P. vitale* contains ~650 amino acid residues per subunit (the amino acid sequence is unknown).[87] The structure shows a close similarity in the polypeptide chain folding to the bovine liver enzyme except that the chain is extended at the carboxy terminus by 150 residues, to form an additional domain containing a β sheet of five parallel strands flanked on each side by two helices, with a topology closely resembling flavodoxin. The quaternary structure again shows the remarkable feature of the amino-terminal section of one subunit passing through a loop formed by an irregular segment of a neighboring subunit.

Interestingly, bovine erythrocyte catalase contains a polypeptide chain of at least 517 residues, in which the carboxy terminus is extended beyond that of the liver enzyme.[88] Sequence studies have suggested that the liver enzyme, at some point in its history, also has a C-terminal extension that is lost either through a different processing of the molecule in liver or by partial degradation in the first stages of catabolism.

B.   Yeast cytochrome *c* peroxidase

The 2.5 Å resolution electron density map could account for 266 of the expected 293 amino acid residues of CcP,[89] but the model has since been extended to include 288 residues after incorporation of data from the complete primary sequence of the molecule.[90] The locations of residues 1–5 are still uncertain. The molecule is a compact ellipsoid with ~50% helical structure in 10 helices (A–J) and little β structure. The most striking feature is the division of the molecule into two domains, connected by helix E. The heme is bound in a crevice between the two domains, sandwiched between helix B and the carboxy terminus of helix F, with the propionate residues extended toward the surface. Only an edge of pyrrole ring IV of the heme is accessible to the external milieu. The imidazole NE2 nitrogen of His 174 from the C-terminus of helix F provides the proximal heme ligand. The sixth coordination position of iron is occupied by a water molecule and the iron atom is displaced 0.3–0.5 Å below the heme plane toward the proximal ligand.

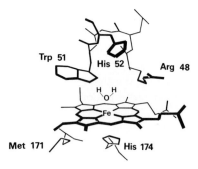

Fig. 6. Amino acid residues in the heme environment of yeast cytochrome *c* peroxidase. (From Ref. [89].)

Residues from helix B lie close to the distal water molecule (Fig. 6). Trp 51 lies 3.5 Å above and parallel to the most buried surface of the heme, and together Trp 51 and the distal His 52 form a tight pocket around the sixth coordination position of the heme. The guanidinium group of Arg 48 lies 3.6 Å above the heme propionate on pyrrole ring IV; the most direct route for ligands entering the heme pocket is through an opening near this guanidinium group. The surface of the molecule surrounding the propionate edge of the heme contains a preponderance of negatively charged residues, a feature that is considered to be important in the interaction with cytochrome *c* as donor substrate (Section IV,E).

IV. HEME ENZYMES—CATALYTIC MECHANISMS

A. General features of the mechanisms

The enzymes of the catalase–peroxidase family have many common features in their reaction chemistry, and it is likely that their actions all involve variations on a general catalytic mechanism. The central feature of this mechanism is the formation of a spectroscopically distinct intermediate, Compound I, by the action of oxidizing substrate on the native enzyme. Compound I of HRP was first detected by Theorell.[91] It was at first supposed that Compound I intermediates are enzyme–substrate complexes, but many lines of evidence have demonstrated that they are oxidized derivatives of the enzymes, which retain both oxidizing equivalents of a hydroperoxide progenitor molecule.[92] Thus, formally, they are Fe(V) forms of the enzymes.

Donor substrates reduce Compound I either directly to native enzyme or in two stages via a second intermediate, Compound II, which contains one oxidizing equivalent above native enzyme and is therefore an Fe(IV) form of enzyme. Overall, five oxidation states of peroxidases and catalases have been characterized, and are related to enzyme species as follows: Fe(VI), Compound III; Fe(V), Compound I; Fe(IV), Compound II; Fe(III), native enzyme; and Fe(II), ferroenzyme.

The major species in peroxidatic and catalatic actions are native enzyme and Compounds I and II, so that the catalytic cycles have the general form:

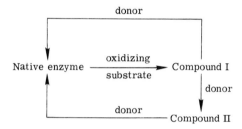

B.   Formation of Compound I intermediates

In the absence of donor substrates, peroxidases (other than chloroperoxidase) can be completely converted to Compound I with a variety of oxidizing substrates (although it is not a trivial matter to obtain preparations adequately free from adventitious donor impurities). With catalases and chloroperoxidase it is necessary to use a substrate that is a good oxidizing substrate but an ineffective donor in order to obtain complete conversion to Compound I. Peroxyacetic acid has proved suitable in both systems.[93,94]

Stopped-flow spectrophotometry has provided a suitable technique for study of the kinetics of formation of Compound I, both in terms of time resolution and because the absorption changes in the Soret band region (Fig. 7) are sufficiently large at convenient ($\leq 1\ \mu M$) enzyme concentrations. Reactions are accurately second order (first order with respect to both enzyme and oxidizing substrate), and the rate constant for most enzymes with H_2O_2 is near $10^7\ M^{-1}\ s^{-1}$. At the extremes of pH the rate constant decreases; the effects are associated with the alkaline transitions of peroxidases at high pH and with inhibiting protonation of protein residues at low pH.[95–98] The latter effect with HRP has been shown to be complicated in

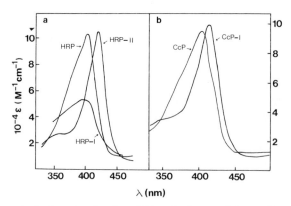

Fig. 7. Spectra in the Soret band region of (a) native horseradish peroxidase, Compound I, and Compound II; (b) native yeast cytochrome c peroxidase and Compound I.

an unusual way by the binding of some anions (nitrate, perchlorate, acetate), not at the heme iron but to the protonated inhibitory protein residue, thus influencing the observed rate-controlling pK_a.[99,100] The reaction is inhibited in an apparently more straightforward way by the binding of typical ligands (F^-, CN^-) to the heme iron, although these reactions are again complicated in a rather subtle fashion since it is the un-ionized conjugate acid forms of the ligands that react with the enzymes.[101,102]

The high rate constants and low activation energies (≤ 4 kcal mol^{-1}) for the pH-independent reactions between active enzyme and H_2O_2[103] have led to suggestions that the reactions may be diffusion controlled, but a number of lines of evidence, including the absence of viscosity effects, indicate that they are chemically controlled and involve the reversible formation of a precursor complex[92]:

$$E + H_2O_2 \rightleftharpoons (E \cdot H_2O_2) \rightarrow \text{Compound I} + H_2O$$

Limiting kinetics have not been observed in direct studies of Compound I formation, although the steady-state turnover of H_2O_2 by catalases shows limiting kinetics at high $[H_2O_2]$ with $K_m \sim 1\ M$.[104] It is unlikely that such high $[H_2O_2]$ has physiological relevance in general, although the bombardier beetle secretes $\sim 25\%\ H_2O_2$ (i.e., $> 7\ M$).[105]

HRP reacts more rapidly with unhindered peroxybenzoic acids (rate constants $\sim 10^8\ M^{-1}\ s^{-1}$) than with H_2O_2, and in the case of m-chloroperoxybenzoic acid, the rate constant shows the classical dependence on reciprocal solvent viscosity of a diffusion-controlled reaction.[106] The pH dependences of reactions of peroxybenzoic acids with peroxidases and of peroxyacetic acid with catalases indicate that the neutral, un-ionized form of the peroxyacid is predominantly the reactive form.[94,107,108]

C.   Structures of Compound I intermediates

A variety of chemical studies have established that formation of Compound I intermediates has the following reaction stoichiometry:

$$\text{Fe(III)} \quad + \text{ROOH} \rightleftharpoons \text{ "Fe(V)"} \quad + \text{ROH}$$

$$\text{native enzyme} \qquad\qquad \text{Compound I}$$

Measurements have involved titration studies of the formation of Compound I, release of ROH,[21] and reduction of Compound I, and studies of proton balance.[109] The spectra and reactivities of Compound I species are independent of the nature of the oxidizing substrate. These findings have motivated many studies aimed at elucidating the structures of the intermediates. Magnetic susceptibility and Mössbauer spectroscopic data have shown that the heme iron center in Compound I is invariably in the ferryl $[S = 1, \text{Fe(IV)}]$ state with two unpaired electrons.[110,111] Compound I species must therefore store an additional oxidizing equivalent somewhere other than at the iron center. Recent work on different enzymes has shown that evolution has provided more than one solution to this problem.

Compound I of CcP was early recognized as different from that of other peroxidases because its absorption spectrum in the Soret band region (Fig. 7) closely resembled that of HRP Compound II (historically the CcP intermediate has been called Compound ES in order to emphasize this difference), although both intermediates contain three unpaired electrons.[112] However, Compound I from CcP shows a narrow free-radical-like EPR signal that is observable at 77 K.[113] Compound I from HRP was thought to be EPR silent and the broad, stoichiometric EPR signal has only recently been observed.[114,115] EPR studies of radical signals associated with chloroperoxidase Compound I have also been reported.[116]

The character of the CcP Compound I EPR signal implies that the second oxidizing equivalent is well removed from the heme center, that is, it exists as a free radical formed from an amino acid. A tryptophanyl radical at Trp 51 has been suggested,[117] but a more recent analysis implicates a sulfur-based radical, and a thioether cation radical based on the sulfur atom of Met 171 has been proposed.[118]

In contrast, the second oxidizing equivalent in Compound I of HRP, several other peroxidases, and the catalases is now generally believed to occur as a π-cation radical of the porphyrin ligand of iron.[116,119,120] This assignment was originally made on the basis of marked similarities between the electronic spectra of these species and those of synthetic porphyrin π-cation radicals.[121] The spectra were further classified into two groups, which arise from different ground state symmetries of the radicals. Direct evidence for the occurrence of the porphyrin π-cation radical in HRP Compound I has been obtained by proton and ^{14}N ENDOR studies and assignments have been confirmed using enzyme reconstituted from apoprotein and selectively deuterated heme complexes.[119] NMR,[122] ESR, and resonance Raman[123] studies were consistent with this interpretation but did not provide direct proof of the structure.

A number of suggestions that the high-oxidation-state iron center in Compound I might be stabilized by an oxide or hydroxide ligand first received experimental support from isotopic tracer studies of the decomposition of m-chloroperoxybenzoic acid by chloroperoxidase.[124] In the decomposition of hydrogen peroxide by catalase[125] and chloroperoxidase, the oxygen molecule produced via

$$2H_2O_2 \rightleftharpoons 2H_2O + O_2$$

derives from a single molecule of H_2O_2. In contrast, with m-chloroperoxybenzoic acid decomposing by chloroperoxidase catalysis according to

$$2ROOH \rightleftharpoons 2ROH + O_2$$

where

double isotopic labeling experiments, using mixtures of $R^{16}O^{16}OH$ and $R^{18}O^{18}OH$, showed that scrambling occurred such that in each O_2 product molecule the two atoms derived from different peroxyacid molecules. This observation is only readily explained if chloroperoxidase Compound I retains one oxygen atom of its parent hydroperoxide molecule and if oxidation of the second molecule involves a "three-oxygen" intermediate:

$$\underset{\text{Compound I}}{\overset{O}{\underset{\|}{Fe\,''(V)''}}} + ROO^- \longrightarrow \underset{\substack{\text{''three-oxygen}\\ \text{intermediate''}}}{\overset{(O-O-O-R)^-}{\underset{|}{Fe\,''(V)''}}} \longrightarrow \underset{\substack{\text{native}\\ \text{enzyme}}}{Fe(III)} + O_2 + {}^-OR$$

The mechanism bears a marked resemblance to that which can occur in the thermal decomposition of peroxyacids when nucleophilic peroxyanion attacks the electrophilic outer peroxidic oxygen of a peroxyacid molecule.[126]

Direct proof of the retention of one oxygen atom from hydroperoxide in the coordination sphere of Fe(IV) has come from ^{17}O ENDOR measurements on HRP Compound I formed using $H_2^{17}O_2$.[127] The core structure of HRP Compound I is a triplet $(S = 1)$ oxyferryl center $[Fe(IV){=}O]$ whose axis lies normal to the porphyrin cation radical plane and with substantial covalency in the iron–oxygen bond through $d_\pi - p_\pi$ bonding.

An additional type of Compound I structure arises in the *P. aeruginosa* cytochrome c_{551} peroxidase.[37,128] In this case the cytochrome-like heme center provides a store for the second oxidizing equivalent, so that formation of Compound I in this case involves the following sequence (h.s., high spin; l.s., low spin):

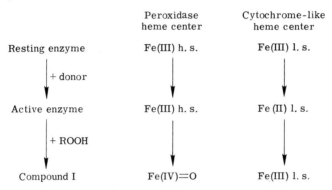

	Peroxidase heme center	Cytochrome-like heme center
Resting enzyme	Fe(III) h. s.	Fe(III) l. s.
+ donor ↓	↓	↓
Active enzyme	Fe(III) h. s.	Fe (II) l. s.
+ ROOH ↓	↓	↓
Compound I	Fe(IV)=O	Fe(III) l. s.

In summary, the oxidative activation of peroxidases and catalases involves not only the iron center of the heme enzymes but also either the porphyrin ligand, or an amino acid residue, or another redox-active heme cofactor as an accessible site for storage of one oxidizing equivalent from the oxidizing substrate.

D. Compound II and Compound III intermediates

HRP Compound II was the first intermediate to be observed because early peroxidase preparations contained sufficient adventitious donor to reduce Compound I rapidly to Compound II. Compound II species are low-spin Fe(IV) derivatives of the enzymes, containing two unpaired electrons but exhibiting no EPR signal. In addition to their formation by one-electron reduction of Compound I, a number of additional routes to Compound II have been described and are summarized in Fig. 8.

Compound III species were first discovered as the products of reaction of hydrogen peroxide with Compound II species, but additional routes have been described (Fig. 8), notably that involving reaction of O_2 with ferroperoxidases (leading to the alternative description of peroxidase Compound III as oxyperoxidase to emphasize the formal similarity with oxyhemoglobin). Compound III is diamagnetic and shows no EPR signal. A slow

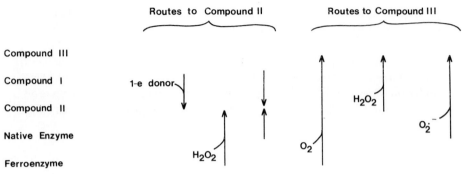

Fig. 8. Reaction pathways for the formation of Compound II and Compound III.

interprotein conproportionation reaction between HRP Compound I and native enzyme yields Compound II.[129] Interprotein redox reactions can also occur between Compound III and ferroperoxidase (1 molecule Compound III + 3 molecules ferroperoxidase → 4 molecules of native enzyme).[130]

E. Donor substrate oxidation

The structures of the Compound I intermediates suggest, and extensive kinetic and isotopic tracer studies confirm, that a variety of donor oxidation mechanisms may occur in these systems.

1. One-electron equivalent oxidation of donor substrate either by electron abstraction from a donor (D) or by hydrogen atom abstraction from a donor (DH):

$$
\begin{array}{ccc}
\text{DH} & \text{Compound I} & \text{D} \\
& & \\
\text{D}\cdot & \text{Compound II} & \text{D}^+
\end{array}
$$

2. Consecutive one-electron oxidation of donor in which the transient appearance of Compound II is observed:

$$
\begin{array}{cc}
\text{Compound I} & \text{D} \\
& \text{D}^+ \\
\text{Compound II} & \text{D} \\
& \text{D}^+ \\
\text{Native enzyme} & \text{D}^+
\end{array}
$$

3. Two-electron oxidation of donor. (a) One criterion for this classification has been failure to detect the transient appearance of Compound II. This may merely imply that the rate constant for Compound II oxidation of donor is much larger than that for Compound I, which may be tested by independent studies of Compound II + donor reactions (commonly donor oxidation by Compound II occurs much more slowly than with Compound I). (b) The product of one-electron oxidation of donor may remain trapped within the active site so as to undergo a second one-electron oxidation:

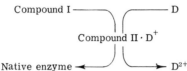

(c) Substrate may be oxidized by hydride transfer from donor. (d) Substrate may be oxidized by oxygen atom transfer to donor.

All these possibilities have been either demonstrated or proposed in various systems. The reactions of HRP have been most intensively studied. HRP Compound I is reduced to Compound II by O_2^-, but Compound II is not reduced to native enzyme.[131] Mechanism (2) is established as the normal donor oxidation process for HRP with its most favored phenol and aromatic amine substrates.[132] However, reduction of HRP Compound I to native enzyme by both sulfite and iodide occurs too rapidly to involve Compound II as an intermediate.[133,134] In the decomposition of H_2O_2 by catalase, catalysis is inhibited by Compound II formation. Thus the oxidation of H_2O_2 by catalase Compound I involves either mechanism (3b) or (3c). The occurrence of donor oxidation by mechanism (3d) in the decomposition of *m*-chloroperoxybenzoic acid by chloroperoxidase (Section IV,C) has been important in focusing attention on the relationship between heme peroxidases and heme monooxygenases (cytochromes P-450).

Kinetics of the oxidation of donors by both Compound I and Compound II are complicated by the influence of ionization on both substrate and protein and suggest the occurrence of acid catalysis effects.[101] A protein residue with $pK_a = 5.1$ (probably distal His 42) influences donor oxidation by HRP Compound I. Reactions of Compound II are more complex; the reactivity of many donors is influenced by a protein residue with $pK_a = 8.6$,[101] although the reaction with I^- is accurately proportional to $[H^+]$ over a very wide range of pH (down to pH 3).[135] This latter behavior cannot reflect reaction of HI but must be associated with protonation of a protein residue with a very low pK_a. The one-electron oxidations of phenols and anilines by HRP Compound I show marked substituent effects in which increased electron-donating power of the substituent enhanced the rate. The rate constant variations obey Hammett relationships with large negative ρ values.[132]

The oxidation of ferrocytochrome c by CcP Compound I is of particular interest as an interprotein electron transfer reaction and because the crystal structures of both enzyme and cytochrome are known. Steady-state kinetic studies of the overall reaction have yielded conflicting results, mainly because the reaction is very sensitive to the nature of the buffer, pH, and ionic strength. A recent study[136] at pH 7 in a cacodylate/KNO_3 buffer system of low ionic strength concluded that only a single binding site for cytochrome c is involved (two sites of very different affinity were proposed in earlier work), and that the reaction involves a random-order mechanism, which is unusual in peroxidase chemistry:

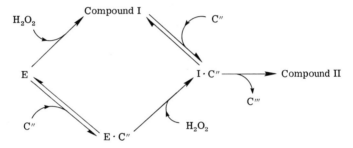

where C'' and C''' are reduced and oxidized cytochrome c.

A hypothetical model of the CcP–cytochrome c electron transfer complex, based on optimization of the fit between the two known protein structures assumes that a ring of aspartate residues on the molecular surface of CcP is complementary to a similar array of highly conserved lysine residues surrounding the exposed edge of the cytochrome c heme crevice.[90] Cytochrome c lysine modification,[137] CcP aspartate modification,[138] and covalent interprotein cross-linking studies[139] have given support to this model. The two heme groups in the model are parallel (although not co-planar), with an edge separation of 16.5 Å. A system of inter- and intra-molecular π–π and hydrogen bonding interactions forms a bridge between the hemes.[90] The route for ligands to and from the heme crevice in CcP remains open in the model complex, enabling hydroperoxides to react with CcP even while it is bound to cytochrome c, as is required by kinetic data.[140]

Because the two oxidizing equivalents of CcP Compound I are spatially separated, the question of intramolecular electron transfer between the radical and heme sites is important. Control of electron flow may be influenced by different conformations of the radical site that have very different reduction potentials.[141]

F. Halogenation

It has been (and remains) a matter of contention whether the peroxidase catalysis of halogenation reactions involves (1) generation of free halogen (or hypohalite) followed by nonenzymatic halogenation of substrate or (2) formation of an enzyme-bound halogenating intermediate. There is appreciable evidence that both types of process occur. Recent work has shown that, with chloroperoxidase, the rate of oxidation of chloride to chlorine is considerably slower than the rate of enzymatic chlorination of acceptor substrates, and the involvement of an Fe(III) hypochlorite halogenating intermediate has been proposed.[142] Chlorination reactions of myeloperoxidase and bromination reactions of lactoperoxidase probably involve free hypohalite halogenating agents.

Although HRP cannot oxidize chloride ion using H_2O_2, it can catalyze chlorination reactions in cases in which chlorite serves both as halogen donor and as oxidizing substrate. In the absence of halogen acceptor, the reaction of HRP with sodium [^{36}Cl]chlorite at pH 10.7 forms a ^{36}Cl-labeled intermediate that is very similar to Compound II.[143] This intermediate (Compound X) halogenates donor substrates with ^{36}Cl transfer. It has been suggested that Compound X is an Fe(IV) form of the enzyme with a "chlorine oxide" ligand at the heme iron.

G. Roles of the proteins in catalysis

Determination of the structure of CcP made possible the first attempt to offer a detailed analysis of the roles of the protein in peroxidase action. Poulos and Kraut used computer modeling to study the interaction of a peroxide group (O–O bond length 1.48 Å) at the heme site of CcP.[117] With one oxygen atom (O2) fixed at the site of the water ligand in native enzyme, a preferred orientation of peroxide within the restricted active site was clear, in which two hydrogen bonds from Arg 48 were possible to the second oxygen atom (O1). Furthermore, His 52, which is poorly positioned for hydrogen bonding with water in the native enzyme, is well positioned to form an additional hydrogen bond to O1. With an alkyl hydroperoxide (R–O1–O2–H) the alkyl group extended out over pyrrole IV of the heme group into the solvent.

The proposed mechanism of Compound I formation (Fig. 9) involves the following stages[117]: (1) diffusion of un-ionized ROOH (as required by kinetic data) into the active site results in proton transfer to His 52, entry of peroxyanion via ligation of O2 into the iron coordination sphere, and hydrogen bonding of His 52 to O1 (Fig. 9b); (2) heterolysis of the peroxide bond

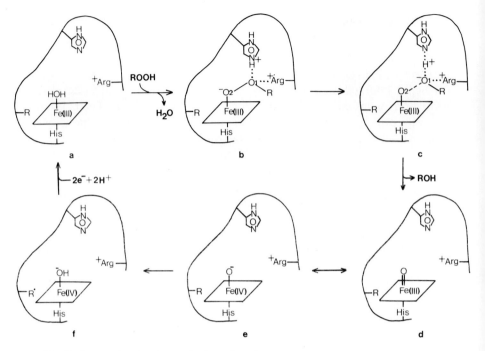

Fig. 9. Proposed molecular mechanism of the formation of yeast cytochrome *c* peroxidase Compound I. (From Refs. [117 and 118].)

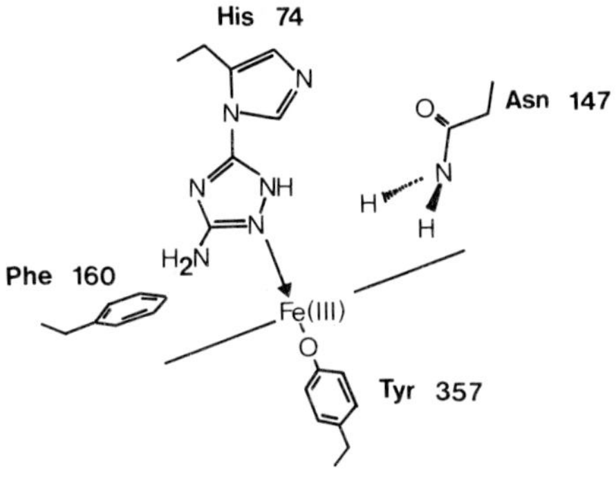

Fig. 10. Proposed binding of 3-amino-1,2,4-triazole inhibitor in the active site of beef liver catalase. (From Ref. [146].)

occurs via the transition state (Fig. 9c), in which the developing negative charge on O1 is stabilized by interaction with the positive charge on Arg 48, leads to release of ROH, and yields initially an "oxene" adduct of the Fe(III) enzyme (Fig. 9d); and (3) electron transfer from the protein residue R yields Compound I as a $[Fe(IV)=O + R^+]$ species. In Fig. 9e the group R is not specified; Poulos and Kraut propose that R is Trp 51, whereas Hoffman *et al.* suggest Met 171, which lies 5.6 Å below the proximal surface of the heme.[118]

On the basis of sequence homologies between CcP and horseradish and turnip enzymes, it was suggested that this group of enzymes catalyzes the reduction of hydroperoxides by virtually identical mechanisms. However, the formation of HRP Compound I requires deprotonation of a group with $pK_a \leq 3$ (compared with $pK_a = 5.5$ in CcP[95]), which appears too low for a distal histidine and which has been assigned to Asp 43.[144] The role of the protein in HRP Compound I formation may involve Asp 43 and Arg 38

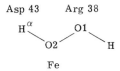

with loss of proton α to Asp 43 accompanying ligation of HO_2^- via O2 to iron. Heterolysis of the O–O bond is accompanied by rotation of the carboxylic acid group of Asp 43 to transfer proton α to the O1 of departing OH^-. It is possible that a push–pull (concerted) mechanism occurs: α H^+ to Asp 43, Arg 38 H^+ to O1.

A detailed proposal for the formation of catalase Compound I, based on the X-ray structure determinations, has not as yet been presented, although the likely catalytic importance of the distal residues His 74 and Asn 147 has been emphasized.[86] Acid–base catalysis by distal residues has been implicated on the basis of kinetic studies.[3] Model-building studies have shown that the irreversible catalase inhibitor 3-amino-1,2,4-triazole[145] combining with His 74 can form an adduct (Fig. 10) in which the N2 atom of aminotriazole occupies the sixth coordination position of heme iron.[146] The comparable rates of reaction of peroxidases and catalases with H_2O_2 may indicate that a structure as complex as that of active catalases is not essential for the efficient formation of a Compound I intermediate, but it is certainly important in determining the high selectivity of catalases for H_2O_2 as both oxidizing and reducing substrate.

Whereas in HRP Compound I the protoporphyrin IX π-cation radical has a $^2A_{2u}$ ground state, the ground state in catalase Compound I is $^2A_{1u}$; the difference can be attributed to differences in the proximal ligation of the hemes. A modified HRP, formed from deuteroferriheme (Fig. 1) and apoprotein, yields a Compound I with a $^2A_{1u}$ ground state but possesses no catalase activity.[147] Other aspects of the protein environment of the heme must play the major role in determining the reactivities of Compound I species toward donors, but these problems are as yet largely unresolved.

H. Model systems

Phenomenological, iron-containing models of catalases and peroxidases abound, although, in general, their catalytic reactions bear little mechanistic similarity to the enzymatic processes. In the catalytic action of iron(III) salts in decomposing H_2O_2 at acid pH, it is widely held that the initial action of H_2O_2 upon aquo iron(III) is reductive[148]

$$Fe^{3+} + HO_2^- \rightarrow Fe^{2+} + HO_2^{\cdot}$$

and that this process initiates a complex chain reaction in which $HO_2^{\cdot}/O_2^{\cdot-}$, $OH\cdot$, and Fe^{2+} are chain carriers. Similarly, catalysis by the Fe(III)–edta complex also involves a free-radical chain mechanism.[149]

It is only with Fe(III)–porphyrins that redox chemistry comparable to that of the enzymes appears. The Fe(III) complex of tetramesitylporphyrin is oxidized by *m*-chloroperoxybenzoic acid at $-78°C$ in methylenechloride–methanol solvent to an Fe(IV)–porphyrin π-cation radical species that will epoxidize norbornene at $-80°C$.[150] The oxygen of the epoxidizing species is exchangeable with added $H_2{}^{18}O$.

In aqueous solution, water-soluble Fe(III)–porphyrins display both catalase- and peroxidase-like activities. An extensive study of the reactions of deuteroferriheme (Fig. 1) has been made since it shows less tendency to be autoxidized and to dimerize than does protoferriheme. On reaction with hydroperoxides[151] or chlorite[152] an oxidized intermediate is formed that is formally an Fe(IV)–porphyrin. It is likely that a Compound I analog is first formed but that it undergoes rapid conproportionation with Fe(III)–porphyrin. The intermediate shows marked peroxidatic properties, and is reduced by anilines at rates that are closely comparable to those of HRP Compound I.[153] A similar relationship holds between the rates of reduction of the model intermediate by phenolate anions and those of HRP Compound I by phenols.[154]

Although the rate of formation of the intermediate is markedly slower than that of Compound I at physiological pH, it is comparable at high pH,[155] and an important role of the protein in peroxidases and catalases may be to permit the efficient use of molecular H_2O_2 as substrate at physiological pH, rather than HO_2^- since pK_a $(H_2O_2) \sim 11.6$. Mechanisms of Compound I formation based on protein structure studies (Section IV,G) are in accord with this view. The overall catalytic rate constants for mono-meric ferrihemes vary inversely with $[H^+]$, and approach the (pH-inde-pendent) rate constant for catalase at high pH. Since the model intermediates are restricted to one-electron oxidation, it was suggested that the oxidation of H_2O_2 by catalase Compound I may occur by a type (3b) mechanism (Section IV,E), in which the intensely hydrophobic tunnel into the protein active site acts to trap the intermediate superoxide species.[156]

I. Oxidase functions of peroxidases

Oxidase properties of peroxidase have been studied, in particular with the horseradish enzyme. Two mechanisms (pathways) have been proposed, and the possibility that both may occur simultaneously in the oxidation of indole acetate has been discussed[157]:

1. Traces of hydroperoxide initiate reaction by oxidation of enzyme to Compound I. Radicals formed by donor (AH) oxidation react with O_2 to form peroxyl radicals AO_2^- and then hydroperoxide $AO_2H \cdot$, which may initiate further reaction.[158]

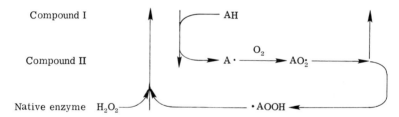

2. Native enzyme is initially reduced by donor to ferroenzyme, which reacts with O_2 to form Compound III. Donor oxidation by Compound III yields donor radicals and the superoxide ion.

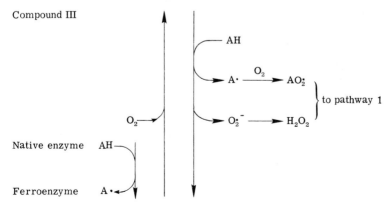

In suitable cases the final oxidation products are those expected from the cleavage of an intermediate dioxetane.[159]

$$\underset{R_2}{\overset{R_1}{\text{HC}}}-\underset{X}{\overset{O}{\text{C}}} + O_2 \xrightarrow{\text{HRP}} \left[\underset{R_2}{\overset{R_1}{\text{C}}}\underset{O-O}{\overset{O^-}{\underset{|}{\text{C}}}}-X \right] \longrightarrow \underset{*O}{\overset{R_1}{\underset{\underset{}{}}{\text{C}}}}\overset{R_2}{\underset{}{}} + \underset{O}{\overset{O^-}{\underset{||}{\text{C}}}}-X$$

In this process one of the products may be generated in an electronically excited state, and a number of chemiluminogenic reactions of this type have been reported, e.g., isobutyraldehyde yields formic acid and triplet acetone.[160] The peroxidase-catalyzed chemiluminogenic oxidation of luminol by H_2O_2 also requires oxygen and probably involves intermediate dioxetane formation.[161]

V. NONHEME ENZYMES

A. Selenium glutathione peroxidase

Selenium-containing peroxidases are widely distributed in mammalian tissues, although the level of activity shows large species differences in all organs. The molecules are tetramers of identical subunits (tetramer MW 76,000–100,000), each subunit containing one essential atom of selenium in the form of selenocysteine (or a derivative of selenocysteine).[162] No other unusual component or cofactor has been identified. In rat liver 70% of the activity is found in the cytosol and 30% in the mitochondrial matrix (none in the catalase-containing peroxisomes).[163] The structure of the crystalline enzyme from bovine erythrocytes (MW 84,000) has been determined.[164] The selenocysteine groups are located in shallow depressions on the surface of the molecule (minimum Se–Se separation 20.7 Å) in an environment largely composed of hydrophobic aromatic side chains. A histidine residue in this region may have a role in catalysis. It is probable that each active site is built up from segments contributed from two symmetry-related subunits.

The enzymes differ markedly in reactivity from the heme enzymes in two respects: (1) Virtually any hydroperoxide is an effective oxidizing substrate (ranging from H_2O_2 to "DNA hydroperoxide").[165,166] This presumably reflects the open access to the selenocysteine in the active site. (2) The enzymes specifically require reduced glutathione (GSH) as the physiological donor substrate (the intracellular concentration of GSH can reach 10 mM). Many thiols show some donor substrate reactivity *in vitro*, although all thiols with a positively charged amino group near the SH group are very poor substrates.[167] They may be repelled by a positively charged group in the active site that normally interacts with the carboxylate group of GSH. Typical donor substrates of heme peroxidases are not accepted.[168] The overall reaction catalyzed by selenium glutathione peroxidase is shown in Fig. 11.

Despite extensive studies, the detailed catalytic mechanism is not completely defined. Removal of selenium by cyanide treatment inactivates the enzyme. ESCA studies have shown that, in the enzyme, selenium undergoes a reversible, substrate-induced redox change.[169] Difference Fourier synthesis between crystalline oxidized and substrate-reduced enzyme showed no conformational changes or large side chain movements. The difference map is consistent with the loss of one or more selenium-bound oxygen atoms in the transition from oxidized to reduced enzyme.[164]

On the basis that the reduced enzyme is a selenol species (R–SeH) it has been proposed that enzyme oxidation involves the formation of either a seleninyl (R–SeO–OH) or a selenenyl (R–Se–OH) derivative. However, in the crystallographic studies it was observed that neither oxidized nor reduced enzyme binds mercury compounds at the selenium site, which appears incompatible with selenol present in the reduced form but favors transition from seleninic to selenenic acid upon substrate-induced reduction.[164] These results suggest a general mechanism of the form

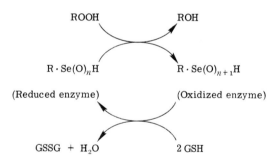

where $n = 0$ for a selenol, $n = 1$ selenenic acid, and $n = 2$ seleninic acid.

Kinetic data imply that the catalytic mechanism is a ping-pong process[170] of the form

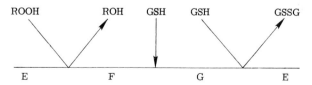

where the intermediate G may be a selenosulfide species. Rate constants for

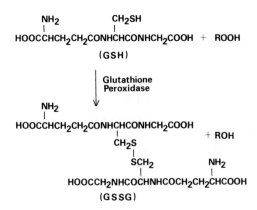

Fig. 11. Oxidation of glutathione catalyzed by glutathione peroxidase.

the oxidation of enzyme by hydroperoxide are high (6×10^7 to 1.8×10^8 M^{-1} s^{-1} for H_2O_2 depending on the medium, decreasing to $\sim 10^7$ for cumene hydroperoxide or t-butyl hydroperoxide).

Selenium glutathione peroxidase was first reported by Mills as an enzyme in erythrocytes capable of protecting hemoglobin from oxidative breakdown.[168] Nutritional studies have established that selenium is an essential trace element. Biochemical and biological studies have indicated that glutathione peroxidase is a major defense system against oxidative damage to intracellular components, proteins, and polyunsaturated fatty acids.[171] The compartmentalization of H_2O_2 catabolism in liver cells has been extensively discussed.[172]

B.   Nonselenium glutathione peroxidase

A striking and unexpected recent discovery has been the demonstration of important non-selenium-dependent glutathione peroxidase activity in a variety of tissues.[173,174] The activity has been shown to be associated with some isoenzymes of the glutathione S-transferase group of enzymes.[175] Dual function occurs in the bovine liver enzyme but not in the enzymes of the bovine lens. The bovine retina, a tissue exposed to high oxidative stress, contains no selenium glutathione peroxidase, and activity is related solely to the more anionic ($pI = 6.34$) glutathione S-transferase isoenzyme.[176] Nonselenium glutathione activity of the glutathione S-transferases differs notably in specificity for oxidizing substrate from the selenoenzymes in that H_2O_2 is not an acceptable substrate, although a wide range of organic hydroperoxides are efficient oxidants.[177] It has recently been suggested that the intermembrane space of cardiac mitochondria contains yet another type of nonselenium glutathione peroxidase that is membrane bound and can react with both H_2O_2 and cumene hydroperoxide.[178]

C. Pseudocatalase

There have been a number of reports that some microorganisms, which are unable to synthesize heme, can produce proteins with catalase-like properties. A manganese-containing pseudocatalase protein from *Lactobacillus plantarum* has been purified and characterized.[179] The molecule is a hexamer of noncovalently associated subunits (hexamer MW 172,000) containing one (1.12 ± 0.37) atom of manganese per subunit. The optical spectrum is similar to that of manganese superoxide dismutase and indicates the presence of Mn(III) in the resting enzyme. At pH 7 and 25°C, K_m for the reaction with H_2O_2 is 250 mM and the turnover number is 3.9×10^5 s^{-1} (compared with 3.5×10^6 s^{-1} for bovine liver catalase). In contrast to the heme enzyme, manganese pseudocatalase is not inhibited by CN$^-$ or N$_3^-$. A mechanism involving cycling of the enzyme between the Mn(III) and Mn(IV) states during catalysis has been proposed.

References

[1] George, P.; Irvine, D. H. *Nature (London)* **1951,** *168,* 184.

[2] Orii, Y. *J. Biol. Chem.* **1982,** *257,* 9246.

[3] Jones, P.; Wilson, I. *Met. Ions Biol. Syst.* **1978,** *7,* 185.

[4] For an extensive review see Dunford, H. B. *Adv. Inorg. Biochem.* **1982,** *4,* 41.

[5] Delincée, H.; Radola, B. J. *Biochim. Biophys. Acta* **1970,** *200,* 404.

[6] Aibara, S.; Kobayashi, T.; Morita, Y. *J. Biochem. (Tokyo)* **1981,** *90,* 489.

[7] Shannon, L. M.; Kay, E.; Lew, J. Y. *J. Biol. Chem.* **1966,** *241,* 2166.

[8] Paul, K. G.; Stigbrand, T. *Acta Chem. Scand.* **1970,** *24,* 3607.

[9] Welinder, K. G. *FEBS Lett.* **1976,** *72,* 19.

[10] La Mar, G. N.; Chacko, V. P.; de Ropp, J. S. *In* "The Biological Chemistry of Iron"; Dunford, H. B.; Dolphin, D.; Raymond, K. N.; Sieker, L. C., Eds.; Reidel Publ.: Dordrecht, Netherlands, 1982, p. 357.

[11] Lanir, A.; Schejter, B. *Biochem. Biophys. Res. Commun.* **1975,** *62,* 199.

[12] Vuk-Pavlović, S.; Benko, B. *Biochem. Biophys. Res. Commun.* **1975,** *66,* 1154.

[13] Williams, R. J. P.; Wright, P. E.; Mazza, G.; Ricard, J. *Biochim. Biophys. Acta* **1975,** *412,* 127.

[14] Gupta, R. K.; Mildvan, A. S.; Schonbaum, G. R. *Arch. Biochem. Biophys.* **1980,** *202,* 1.

[15] Gupta, R. K.; Mildvan, A. S.; Schonbaum, G. R. *Biochem. Biophys. Res. Commun.* **1979,** *89,* 1334.

[16] Kobayashi, K.; Tamura, M.; Hayashi, K.; Hori, H.; Morimoto, H. *J. Biol. Chem.* **1980,** *255,* 2239.

[17] Teraoka, J.; Kitagawa, T. *J. Biol. Chem.* **1981**, *256*, 3969.
[18] Spiro, T. G.; Stong, J. D.; Stein, P. *J. Am. Chem. Soc.* **1979**, *101*, 2648.
[19] Morishima, I.; Ogawa, S.; Inubushi, T.; Yonezawa, T.; Iizuka, T. *Biochemistry* **1977**, *16*, 5109.
[20] Davies, D. M.; Jones, P.; Mantle, D. *Biochem. J.* **1976**, *157*, 247.
[21] Schonbaum, G. R.; Lo, S. *J. Biol. Chem.* **1972**, *247*, 3353.
[22] Mazza, G.; Charles, C.; Bouchet, M.; Ricard, J.; Reynaud, J. *Biochim. Biophys. Acta* **1968**, *167*, 89.
[23] Job, D.; Ricard, J.; Dunford, H. B. *Arch. Biochem. Biophys.* **1977**, *179*, 95.
[24] Mazza, G.; Welinder, K. G. *Eur. J. Biochem.* **1980**, *108*, 473.
[25] Mazza, G.; Welinder, K. G. *Eur. J. Biochem.* **1980**, *108*, 481.
[26] Ricard, J.; Mazza, G.; Williams, R. J. P. *Eur. J. Biochem.* **1972**, *28*, 566.
[27] Yonetani, T.; Ray, G. S. *J. Biol. Chem.* **1965**, *240*, 4503.
[28] For a review of early literature see Yonetani, T. *Adv. Enzymol.* **1970**, *33*, 309.
[29] Mathews, R. A.; Wittenberg, J. B. *J. Biol. Chem.* **1979**, *254*, 5991.
[30] Nicholls, P.; Mochan, E. *Biochem. J.* **1971**, *121*, 55.
[31] Erman, J. E.; Vitello, L. B. *J. Biol. Chem.* **1980**, *255*, 6224.
[32] Ellfolk, N.; Soininen, R. *Acta Chem. Scand.* **1970**, *24*, 2126.
[33] Ellfolk, N.; Soininen, R. *Acta Chem. Scand.* **1971**, *25*, 1535.
[34] Ellfolk, N.; Rönnberg, M.; Aasa, R.; Andréasson, L.-E.; Vänngård, T. *Biochim. Biophys. Acta* **1983**, *743*, 23.
[35] Rönnberg, M.; Ellfolk, N. *Acta Chem. Scand.* **1973**, *27*, 35.
[36] Rönnberg, M.; Ellfolk, N. *Biochim. Biophys. Acta* **1978**, *504*, 60.
[37] Rönnberg, M.; Araiso, T.; Ellfolk, N.; Dunford, H. B. *Arch. Biochem. Biophys.* **1981**, *207*, 197.
[38] Rönnberg, M.; Araiso, T.; Ellfolk, N.; Dunford, H. B. *J. Biol. Chem.* **1981**, *256*, 2471.
[39] Morris, D. R.; Hager, L. P. *J. Biol. Chem.* **1966**, *241*, 1763.
[40] Rubin, B.; Van Middlesworth, J.; Thomas, K.; Hager, L. P. *J. Biol. Chem.* **1982**, *257*, 7768.
[41] Hollenberg, P. F.; Hager, L. P. *J. Biol. Chem.* **1973**, *248*, 2630.
[42] Thomas, J. A.; Morris, D. R.; Hager, L. P. *J. Biol. Chem.* **1970**, *245*, 3129.
[43] Hager, L. P.; Doubek, D. L.; Silverstein, R. M.; Lee, T. T.; Thomas, J. A.; Hargis, J. H.; Martin, J. C. *In* "Oxidases and Related Redox Systems", Vol. 1; King, T. E.; Mason, H. S.; Morrison, M., Eds.; University Park Press: Baltimore, 1973.
[44] Thomas, J. A.; Morris, D. R.; Hager, L. P. *J. Biol. Chem.* **1970**, *245*, 3129.
[45] Kedderis, G. L.; Koop, D. R.; Hollenberg, P. F. *J. Biol. Chem.* **1980**, *255*, 10174.
[46] Shahangian, S.; Hager, L. P. *J. Biol. Chem.* **1981**, *256*, 6034.
[47] Hewson, W. D.; Hager, L. P. *J. Phycol.* **1980**, *16*, 340.
[48] Ahern, T. J.; Allan, G. G.; Medcalf, D. G. *Biochim. Biophys. Acta* **1980**, *616*, 329.
[49] Manthey, J. A.; Hager, L. P. *J. Biol. Chem.* **1981**, *256*, 11232.
[50] Kurata, K.; Amiya, T. *Bull. Jpn. Soc. Sci. Fish.* **1975**, *41*, 657.
[51] Shigeoka, S.; Nakano, Y.; Kitaoka, S. *Arch. Biochem. Biophys.* **1980**, *201*, 121.
[52] Shigeoka, S.; Nakano, Y.; Kitaoka, S. *Biochem. J.* **1980**, *186*, 377.
[53] Theorell, H.; Åkeson, Å. *Ark. Kemi, Mineral. Geol.* **1943**, *17B*, No. 7.
[54] Paul, K. G.; Ohlsson, P. I.; Henriksson, A. *FEBS Lett.* **1980**, *110*, 200.
[55] Moldoveanu, Z.; Tenovuo, J.; Mestecky, J.; Pruitt, K. M. *Biochim. Biophys. Acta* **1982**, *718*, 103.
[56] Sievers, G. *FEBS Lett.* **1981**, *127*, 253.
[57] Sievers, G. *Biochim. Biophys. Acta* **1979**, *579*, 181.
[58] Oram, J. D.; Reiter, B. *Biochem. J.* **1966**, *100*, 382.

[59] Aune, T. M.; Thomas, E. L. *Eur. J. Biochem.* **1977**, *80*, 209.
[60] Thomas, E. L. *Biochemistry* **1981**, *20*, 3273.
[61] Klebanoff, S. J.; Clem, W. H.; Luebke, R. G. *Biochim. Biophys. Acta* **1966**, *117*, 63.
[62] Carlsson, J.; Iwami, Y.; Yamada, T. *Infect. Immun.* **1983**, *40*, 70.
[63] Hosoya, T.; Morrison, M. *J. Biol. Chem.* **1967**, *242*, 2828.
[64] Neary, J. T.; Koepsell, D.; Davidson, B.; Armstrong, A.; Strout, H. V.; Soodak, M.; Maloof, F. *J. Biol. Chem.* **1977**, *252*, 1264.
[65] Virion, A.; Pommier, J.; Deme, D.; Nunez, J. *Eur. J. Biochem.* **1981**, *117*, 103.
[66] Sugawara, M.; Hagen, G. A. *J. Lab. Clin. Med.* **1982**, *99*, 580.
[67] Michot, J.-L.; Osty, J.; Nunez, J. *Eur. J. Biochem.* **1980**, *107*, 297.
[68] Harrison, J. E.; Pabalan, S.; Schultz, J. *Biochim. Biophys. Acta* **1977**, *493*, 247.
[69] Olsson, I.; Olofsson, T.; Odeberg, H. *Scand. J. Haematol.* **1972**, *9*, 483.
[70] Andrews, P. C.; Krinsky, N. I. *J. Biol. Chem.* **1981**, *256*, 4211.
[71] Olsen, R. L.; Little, C. *Biochem. J.* **1983**, *209*, 781.
[72] Pember, S. O.; Fuhrer-Krüsi, S. M.; Barnes, K. C.; Kinkade, J. M., Jr. *FEBS Lett.* **1982**, *140*, 103.
[73] Babior, B. M. "Oxygen and Life", 2nd B.O.C. Priestley Conf., R. Soc. Chem. Spec. Publ. No. 39. 1981, p. 107.
[74] Klebanoff, S. J. "The Reticuloendothelial System: Biochemistry and Metabolism", Vol. 2; Sbarra, A. J.; Strauss, R. R., Eds.; Plenum: New York, 1980.
[75] Hill, H. A. O; Okolow-Zubkowska, M. J. "Oxygen and Life", 2nd B.O.C. Priestley Conf., R. Soc. Chem. Spec. Publ. No. 39. 1981, p. 98.
[76] Green, M. R.; Hill, H. A. O.; Okolow-Zubkowska, M. J.; Segal, A. W. *FEBS Lett.* **1979**, *100*, 23.
[77] Rosen, H.; Klebanoff, S. J. *J. Biol. Chem.* **1982**, *257*, 13731.
[78] Ohki, S.; Ogino, N.; Yamamoto, S.; Hayaishi, O. *J. Biol. Chem.* **1979**, *254*, 829.
[79] Van der Ouderaa, F. J.; Buytenhek, M.; Slikkerveer, F. J.; Van Dorp, D. A. *Biochim. Biophys. Acta* **1979**, *572*, 29.
[80] Egan, R. W.; Gale, P. H.; Baptista, E. M.; Kennicott, K. L.; Vanden Heuvel, W. J. A.; Walker, R. W.; Fagerness, P. E.; Kuehl, F. A., Jr. *J. Biol. Chem.* **1981**, *256*, 7352.
[81] Egan, R. W.; Gale, P. H.; Baptista, E. M.; Kuehl, F. A., Jr. *Prog. Lipid Res.* **1981**, *20*, 173.
[82] For an extensive review of earlier work see Deisseroth, A.; Dounce, A. L. *Physiol. Rev.* **1970**, *50*, 319.
[83] Jacob, G. S.; Orme-Johnson, W. H. *Biochemistry* **1979**, *18*, 2975.
[84] Sies, H. *Angew. Chem., Int. Ed. Engl.* **1974**, *13*, 706.
[85] Aebi, H.; Wyss, S. R.; Scherz, B.; Skvaril, F. *Eur. J. Biochem.* **1974**, *48*, 137.
[86] Murthy, M. R. N.; Reid, T. J., III; Sicignano, A.; Tanaka, N.; Rossmann, M. G. *J. Mol. Biol.* **1981**, *152*, 465.
[87] Vainshtein, B. K.; Melik-Adamyan, W. R.; Barynin, V. V.; Vagin, A. A.; Grebenko, A. I. *Nature (London)* **1981**, *293*, 411.
[88] Schroeder, W. A.; Shelton, J. R.; Shelton, J. B.; Robberson, B.; Apell, G.; Fang, R. S.; Bonaventura, J. *Arch. Biochem. Biophys.* **1982**, *214*, 397.
[89] Poulos, T. L.; Freer, S. T.; Alden, R. A.; Edwards, S. L.; Skogland, U.; Takio, K.; Eriksson, B.; Xuong, Ng. H.; Yonetani, T.; Kraut, J. *J. Biol. Chem.* **1980**, *255*, 575.
[90] Poulos, T. L.; Kraut, J. *J. Biol. Chem.* **1980**, *255*, 10322.
[91] Theorell, H. *Enzymologia* **1941**, *10*, 250.
[92] Jones, P.; Dunford, H. B. *J. Theor. Biol.* **1977**, *69*, 457.
[93] Araiso, T.; Rutter, R.; Palcic, M. M.; Hager, L. P.; Dunford, H. B. *Can. J. Biochem.* **1981**, *59*, 233.

[94] Jones, P.; Middlemiss, D. N. *Biochem. J.* **1972**, *130*, 411.
[95] Loo, S.; Erman, J. E. *Biochemistry* **1975**, *14*, 3467.
[96] Job, D.; Ricard, J.; Dunford, H. B. *Can. J. Biochem.* **1978**, *56*, 702.
[97] Maguire, R. J.; Dunford, H. B.; Morrison, M. *Can. J. Biochem.* **1971**, *49*, 1165.
[98] Dolman, D.; Newell, G. A.; Thurlow, M. D.; Dunford, H. B. *Can. J. Biochem.* **1975**, *53*, 495.
[99] Araiso, T.; Dunford, H. B. *Biochem. Biophys. Res. Commun.* **1980**, *94*, 1177.
[100] Araiso, T.; Dunford, H. B. *J. Biol. Chem.* **1981**, *256*, 10099.
[101] Dunford, H. B.; Stillman, J. S. *Coord. Chem. Rev.* **1976**, *19*, 187.
[102] Millar, F.; Wrigglesworth, J. M.; Nicholls, P. *Eur. J. Biochem.* **1981**, *117*, 13.
[103] Hewson, W. D.; Dunford, H. B. *Can. J. Chem.* **1975**, *53*, 1928.
[104] Jones, P.; Suggett, A. *Biochem. J.* **1968**, *110*, 617.
[105] Aneshansley, D. J.; Eisner, T.; Widom, J. M.; Widom, B. *Science* **1969**, *165*, 61.
[106] Dunford, H. B.; Hewson, W. D. *Biochemistry* **1977**, *16*, 2949.
[107] Job, D.; Jones, P. *Eur. J. Biochem.* **1978**, *86*, 565.
[108] Frew, J. E.; Jones, P. *Biochim. Biophys. Acta* **1983**, *742*, 1.
[109] Yamada, H.; Yamazaki, I. *Arch. Biochem. Biophys.* **1974**, *165*, 728.
[110] Maeda, Y.; Morita, Y. *Biochem. Biophys. Res. Commun.* **1967**, *29*, 680.
[111] Moss, T. H.; Ehrenberg, A.; Bearden, A. J. *Biochemistry* **1969**, *8*, 4159.
[112] Yonetani, T. *J. Biol. Chem.* **1966**, *241*, 2562.
[113] Yonetani, T. *In* "The Enzymes", Vol. 13; Boyer, P. D., Ed.; Academic Press: New York, 1976, p. 345.
[114] Aasa, R.; Vänngård, T.; Dunford, H. B. *Biochim. Biophys. Acta* **1975**, *391*, 259.
[115] Schulz, C. E.; Devaney, P. W.; Winkler, H.; Debrunner, P. G.; Doan, N.; Chiang, R.; Rutter, R.; Hager, L. P. *FEBS Lett.* **1979**, *103*, 102.
[116] Rutter, R.; Hager, L. P. *J. Biol. Chem.* **1982**, *257*, 7958.
[117] Poulos, T. L.; Kraut, J. *J. Biol. Chem.* **1980**, *255*, 8199.
[118] Hoffman, B. M.; Roberts, J. E.; Kang, C. H.; Margoliash, E. *J. Biol. Chem.* **1981**, *256*, 6556.
[119] Roberts, J. E.; Hoffman, B. M.; Rutter, R.; Hager, L. P. *J. Biol. Chem.* **1981**, *256*, 2118.
[120] Browlett, W. R.; Stillman, M. J. *Biochim. Biophys. Acta* **1981**, *660*, 1.
[121] Dolphin, D.; Forman, A.; Borg, D. C.; Fajer, J.; Felton, R. H. *Proc. Natl. Acad. Sci. U.S.A.* **1971**, *68*, 614.
[122] La Mar, G. N.; de Ropp, J. S.; Smith, K. M.; Langry, K. C. *J. Biol. Chem.* **1981**, *256*, 237.
[123] Felton, R. H.; Romans, A. Y.; Yu, N.-T.; Schonbaum, G. R. *Biochim. Biophys. Acta* **1976**, *434*, 82.
[124] Hager, L. P.; Doubek, D. L.; Silverstein, R. M.; Hargis, J. H.; Martin, J. C. *J. Am. Chem. Soc.* **1972**, *94*, 4364.
[125] Jarnagin, R. C.; Wang, J. H. *J. Am. Chem. Soc.* **1958**, *80*, 786.
[126] For a review see Brown, S. B.; Jones, P.; Suggett, A. *Prog. Inorg. Chem.* **1970**, *13*, 159.
[127] Hoffman, B. M. *In* "The Biological Chemistry of Iron"; Dunford, H. B.; Dolphin, D.; Raymond, K. N.; Sieker, L. C., Eds.; Reidel Publ.: Dordrecht, Netherlands, 1982, p. 391.
[128] Araiso, T.; Rönnberg, M.; Dunford, H. B.; Ellfolk, N. *FEBS Lett.* **1980**, *118*, 99.
[129] Santimone, M. *Biochimie* **1975**, *57*, 265.
[130] Wittenberg, J. B.; Noble, R. W.; Wittenberg, B. A.; Antonini, E.; Brunori, M.; Wyman, J. *J. Biol. Chem.* **1967**, *242*, 626.
[131] Bielski, B. H. J.; Comstock, D. A.; Haber, A.; Chan, P. C. *Biochim. Biophys. Acta* **1974**, *350*, 113.

[132] Job, D.; Dunford, H. B. *Eur. J. Biochem.* **1976,** *66,* 607.
[133] Roman, R.; Dunford, H. B. *Biochemistry* **1972,** *11,* 2076.
[134] Roman, R.; Dunford, H. B. *Can. J. Chem.* **1973,** *51,* 588.
[135] Roman, R.; Dunford, H. B.; Evett, M. *Can. J. Chem.* **1975,** *53,* 1563.
[136] Kang, D. S.; Erman, J. E. *J. Biol. Chem.* **1982,** *257,* 12775.
[137] Smith, M. B.; Millett, F. *Biochim. Biophys. Acta* **1980,** *626,* 64.
[138] Waldmeyer, B.; Bechtold, R.; Bosshard, H. R.; Poulos, T. L. *J. Biol. Chem.* **1982,** *257,* 6073.
[139] Bisson, R.; Capaldi, R. A. *J. Biol. Chem.* **1981,** *256,* 4362.
[140] Mochan, E.; Nicholls, P. *Biochem. J.* **1971,** *121,* 69.
[141] Ho, P. S.; Hoffman, B. M.; Kang, C. H.; Margoliash, E. *J. Biol. Chem.* **1983,** *258,* 4356.
[142] Libby, R. D.; Thomas, J. A.; Kaiser, L. W.; Hager, L. P. *J. Biol. Chem.* **1982,** *257,* 5030.
[143] Shahangian, S.; Hager, L. P. *J. Biol. Chem.* **1982,** *257,* 11529.
[144] Dunford, H. B.; Araiso, T. *Biochem. Biophys. Res. Commun.* **1979,** *89,* 764.
[145] Margoliash, E.; Novogrodsky, A.; Schejter, A. *Biochem. J.* **1960,** *74,* 339.
[146] Reid, T. J., III; Murthy, M. R. N.; Sicignano, A.; Tanaka, N.; Musik, W. D. L.; Rossmann, M. G. *Proc. Natl. Acad. Sci. U.S.A.* **1981,** *78,* 4767.
[147] Di Nello, R. K.; Dolphin, D. *J. Biol. Chem.* **1981,** *256,* 6903.
[148] Baxendale, J. H. *Adv. Catal.* **1952,** *4,* 343.
[149] Walling, C.; Partch, R. E.; Weil, T. *Proc. Natl. Acad. Sci. U.S.A.* **1975,** *72,* 140.
[150] Groves, T. J.; Haushalter, R. C.; Nakamura, M.; Nemo, T. E.; Evans, B. J. *J. Am. Chem. Soc.* **1981,** *103,* 2884.
[151] Jones, P.; Mantle, D.; Davies, D. M.; Kelly, H. C. *Biochemistry* **1977,** *16,* 3974.
[152] Kelly, H. C.; Parigi, K. J.; Wilson, I.; Davies, D. M.; Jones, P.; Roettger, L. *J. Inorg. Chem.* **1981,** *20,* 1086.
[153] Frew, J. E.; Jones, P. *J. Inorg. Biochem.* **1983,** *18,* 33.
[154] Jones, P.; Mantle, D.; Wilson, I. *J. Inorg. Biochem.* **1982,** *17,* 293.
[155] Kelly, H. C.; Davies, D. M.; King, M. J.; Jones, P. *Biochemistry* **1977,** *16,* 3543.
[156] Jones, P. *In* "The Biological Chemistry of Iron"; Dunford, H. B.; Dolphin, D.; Raymond, K. N.; Sieker, L. C., Eds.; Reidel Publ.: Dordrecht, Netherlands, 1982, p. 427.
[157] Smith, A. M.; Morrison, W. L.; Milham, P. J. *Biochemistry* **1982,** *21,* 4414.
[158] Yamazaki, I. *In* "Molecular Mechanisms of Oxygen Activation"; Hayaishi, O., Ed.; Academic Press: New York, 1974, p. 538.
[159] Zinner, K.; Vidigal-Martinelli, C.; Durán, N.; Marsaioli, A. J.; Cilento, G. *Biochem. Biophys. Res. Commun.* **1980,** *92,* 32.
[160] Faria Oliveira, O. M. M.; Haun, M.; Durán, N.; O'Brien, P. J.; O'Brien, C. R.; Bechara, E. J. H.; Cilento, G. *J. Biol. Chem.* **1978,** *253,* 4707.
[161] Misra, H. P.; Squatrito, P. M. *Arch. Biochem. Biophys.* **1982,** *215,* 59.
[162] Forstrom, J. W.; Zakowski, J. J.; Tappel, A. L. *Biochemistry* **1978,** *17,* 2639.
[163] Flohé, L.; Schlegel, W. *Hoppe-Seyler's Z. Physiol. Chem.* **1971,** *352,* 1401.
[164] Ladenstein, R.; Epp, O.; Bartels, K.; Jones, A.; Huber, R.; Wendel, A. *J. Mol. Biol.* **1979,** *134,* 199.
[165] Christophersen, B. O. *Biochim. Biophys. Acta* **1969,** *176,* 463.
[166] For a review see Sies, H.; Wendel, A.; Bors, W. *In* "Metabolic Basis of Detoxication"; Jakoby, W. B.; Bend, J. R.; Caldwell, J., Eds.; Academic Press: New York, 1982, p. 307.
[167] Flohé, L.; Günzler, W.; Jung, G.; Schaich, E.; Schneider, F. *Hoppe-Seyler's Z. Physiol. Chem.* **1971,** *352,* 159.
[168] Mills, G. C. *J. Biol. Chem.* **1957,** *229,* 189.
[169] Wendel, A.; Pilz, W.; Ladenstein, R.; Sawatzki, G.; Weser, U. *Biochim. Biophys. Acta* **1975,** *377,* 211.

[170] Flohé, L.; Loschen, G.; Günzler, W. A.; Eichele, E. *Hoppe-Seyler's Z. Physiol. Chem.* **1972**, *353*, 987.

[171] For a review see Flohé, L. *In* "Free Radicals in Biology", Vol. 5; Pryor, W. A., Ed.; Academic Press: New York, 1982, p. 223.

[172] Jones, D. P.; Eklöw, L.; Thor, H.; Orrenius, S. *Arch. Biochem. Biophys.* **1981**, *210*, 505.

[173] Lawrence, R. A.; Burk, R. F. *Biochem. Biophys. Res. Commun.* **1976**, *71*, 952.

[174] Prohaska, J. R.; Ganther, H. E. *Biochem. Biophys. Res. Commun.* **1977**, *76*, 437.

[175] Irwin, C.; O'Brien, J. K.; Chu, P.; Townsend-Parchman, J. K.; O'Hara, P.; Hunter, F. E., Jr. *Arch. Biochem. Biophys.* **1980**, *205*, 122.

[176] Saneto, R. P.; Awasthi, Y. C.; Srivastava, S. K. *Biochem. J.* **1982**, *205*, 213.

[177] Shrene, M. R.; Morrissey, P. G.; O'Brien, P. J. *Biochem. J.* **1979**, *177*, 761.

[178] Katki, A. G.; Myers, C. E. *Biochem. Biophys. Res. Commun.* **1980**, *96*, 85.

[179] Kono, Y.; Fridovich, I. *J. Biol. Chem.* **1983**, *258*, 6015.

Mechanistic Studies Involving Hydroxylamine

Karl Wieghardt

Anorganische Chemie I
Ruhr-Universität Bochum
Bochum, Federal Republic of Germany

I. INTRODUCTION

The chemistry of nitrosyl coordination complexes of the transition metals continues to attract considerable interest. Numerous structure determinations have been carried out[1] that have established on a very firm basis the diverse structural chemistry of these complexes. In recent years, interest has focused on the reactivity of coordinated nitrogen monoxide, and many interesting reactions and their mechanisms have been described.[2−4] The examination of reactions of nitrosyl complexes has been prompted by the widespread occurrence of nitrogen oxides as pollutants in the atmosphere. A number of different reduced intermediates of the kind H_xNOH_y ($x = 0$–3 and $y = 0$ or 1) have been identified, and the exact nature of these is described in this article. In some cases hydroxylamine and/or coordination compounds containing hydroxylamine ligands have been characterized as products of such reactions. The reverse reaction, i.e., preparation of nitrosyl complexes via oxidation of hydroxylamine, has also been used since the early work of Nast *et al.*[5]

Detailed understanding of the structure and reactivity of nitrosyl complexes is in marked contrast to the limited information available on the chemistry of coordination compounds containing hydroxylamine ligands. This is rather surprising because well-characterized hydroxylamine complexes such as $[Ni(H_2NOH)_6]SO_4$ have been known since 1899,[6] although the X-ray structure was not known until 1974. By this time a number of structures of the much more elusive dinitrogen complexes had already been published.

As has been the case in many areas of coordination chemistry, in which relevance in biochemistry has caused renewed interest in the study of reactions of many known compounds, the unequivocal detection of hydroxylamine as an intermediate in the nitrogen cycle (Fig. 1) has stimulated inorganic chemists to seek a better understanding of the diverse and interesting chemistry of hydroxylamine with transition metals.

In the enzymatic oxidation of ammonia to nitrite and/or nitrate (or, reversibly, in the enzymatic reduction of NO_2^- or NO_3^- to NH_3), it is established that hydroxylamine is an intermediate[7] (Fig. 1).

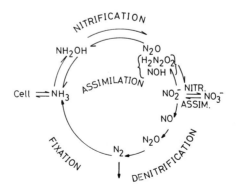

Fig. 1. The nitrogen cycle. (From Ref. [7].)

The bacterium *Nitrosomonas europae* oxidizes ammonia to nitrite in aerobic soils.[8,9] The enzyme hydroxylamine oxidoreductase isolated from *N. europae* catalyzes the rapid aerobic oxidation of hydroxylamine to nitrite in the presence of a sacrificial electron acceptor. This enzyme has a molecular weight of 220,000 and contains approximately 21 C-type hemes and three residues of an unusual prosthetic group termed P-460. The enzyme is a complex protein with an $(\alpha\beta)_3$ subunit structure. The P-460 moiety is believed to be a heme also.[10,11] It is generally accepted that the oxidation of NH_4^+ to nitrite proceeds via two-electron changes. The first oxidation product of H_2NOH, a species with nitrogen in the formal oxidation state of $+1$, has not been identified unambiguously, although the nitrosyl anion NO^- appears to be a very likely candidate. The elucidation of mechanisms for the interconversions of coordinated ammonia to hydroxylamine and the development of models for the sequence

$$NH_3 \rightleftharpoons H_2NOH \rightleftharpoons NO^-(?) \rightleftharpoons NO_2^- \rightleftharpoons NO_3^-$$

have been a challenge to inorganic chemists.

This article deals with the coordination chemistry of hydroxylamine ligands. The available structural data of such complexes are comprehensively summarized, covering the literature through 1983. Synthetic routes to hydroxylamine complexes are reviewed more selectively. The complicated and diverse redox chemistry of hydroxylamine with transition metals is then described, with emphasis on nitrosylation reactions using hydroxylamine and on the reductive conversion of coordinated nitrosyls yielding hydroxylamine or amine species. The chemistry of uncoordinated hydroxylamine has been reviewed by Stedman.[12]

The term "hydroxylamine-containing compound" will be used in a rather broad sense. Complexes with ligands having nitrogen–oxygen single bonds and N and/or O donors are the subject of this article. The coordination chemistry of oximes ($R_2C{=}N{-}OH$) will be taken into consideration only if O,N coordination of the ligand prevails. A specific example may highlight the problem.[13] Nitrosobenzene (with nitrogen in the $+1$ state, sp^2 hybridized, and doubly bonded to oxygen) reacts with an oxomolybdenum(IV) complex to yield a product with an O,N-coordinated Ph–NO moiety. The same product is obtained by reacting a molybdenum(VI) complex with *N*-phenylhydroxylamine (in which nitrogen is in the -1 state, is sp^3 hybridized, and is singly bonded to oxygen) [reaction (1)].

$$(1)$$

In principle, two mesomeric formulas (**A** and **B**) may describe the bonding situation in the complex.

A	**B**

The question arises whether this is a hydroxylamine-containing complex or "side-on" coordinated nitrosobenzene. Structure determination by X-ray diffraction clearly shows that in this instance the description as a molybdenum(VI) complex (**B**) with an O,N-coordinated N-phenylhydroxylamido (−2) ligand is preferred. There are other complexes, well characterized by X-ray structure determinations, in which the nitroso description (**C**) fits the bonding situation more appropriately.

C

The ambiguity of a nomenclature based upon formal oxidation states is circumvented by referring to three-membered rings that contain a nitrogen, an oxygen, and a metal ion as metallaoxaziridines.[13]

II. SOME PROPERTIES AND REACTIONS OF UNCOORDINATED HYDROXYLAMINE

Hydroxylamine in the absence of water forms colorless needles (mp 33°C) that explode when heated rapidly. In aqueous solution, hydroxylamine behaves as a weak base, and many salts of the hydroxylammonium cation are known.

$$H_3NOH^+ \rightleftharpoons NH_2OH + H^+ \qquad pK_a = 5.9 \qquad (2)$$

Structure determinations (X-ray and neutron diffraction) have been carried out on $[H_3NOH]Cl$[14] and $[H_3NOH]ClO_4$.[15] An N–O distance of ∼1.40 Å in the H_3NOH^+ cation is typical for a covalent nitrogen–oxygen single bond. This distance compares well with that found for the neutral hydroxylamine molecule of 1.47(3) Å.[16] The latter value has been obtained from film data with visually estimated X-ray intensities. A redetermination of the structure of this fundamental molecule by neutron diffraction methods is needed.

In alkaline aqueous solution, hydroxylamine behaves as a weak acid. It probably deprotonates at the O atom. The dissociation constant has been determined at $25°C$ ($pK = 13.7$).[17]

$$H_2NOH + OH^- \rightleftharpoons H_2NO^- + H_2O \qquad (3)$$

It is therefore rather surprising that very few, if any, genuine salts containing the H_2NO^- anion have been prepared as crystalline solids. No structural data on such compounds have been published to date. The reactions of calcium or zinc hydroxide with water-free H_2NOH are believed to produce the explosive salts[18] $Ca(ONH_2)_2$ and $Zn(ONH_2)_2 \cdot 3H_2NOH$, respectively.

Further deprotonation of hydroxylamine leads to the anions hydroxylamide (-2) (HNO^{2-}) and finally NO^{3-}, neither of which are present in aqueous solution. No simple salts of these anions are known, but they may well be stabilized on coordination to suitable transition metal centers.

The behavior of hydroxylamine in alkaline aqueous solution in the presence and absence of oxygen is quite complicated. Two reactions, the autoxidation and the alleged disproportionation of H_2NOH, are briefly discussed here since they have important implications on a series of reactions involving H_2NOH and transition metals. The autoxidation of H_2NOH has been studied by Hughes and co-workers,[17,19] who found that it produces peroxonitrite via the nitroxyl ion. The rate of autoxidation is markedly affected by adventitious or added heavy metal ions.

$$NH_2O^- + O_2 \rightarrow H_2O_2 + NO^- \qquad (4)$$

$$NO^- + O_2 \rightarrow ONOO^- \qquad (5)$$

In the presence of catalytic metal ions, nitrite is the main product, and the hydroxylamine is largely consumed via reaction with peroxonitrite.[20]

$$ONOO^- + NH_2O^- \xrightarrow{Cu^{2+}} NO_2^- + NO^- + H_2O \qquad (6)$$

Hydroxylamine decomposes slowly in alkaline solutions, the rate increasing with increasing pH. In the absence of oxygen, the reaction products are N_2, N_2O, and NH_3. The reaction has been formulated in a completely formal sense by Hieber and Nast,[5,21] involving nitroxyl (HNO) as a hypothetical intermediate, in reactions (7) and (8):

$$2H_2NOH \rightarrow H_2O + NH_3 + HNO \qquad (7)$$

$$H_2NOH + HNO \rightarrow N_2 + 2H_2O \qquad (8)$$

The distribution of the final gaseous products N_2 and N_2O is dependent upon pH. Reduction of HNO to N_2 and dimerization to N_2O were believed to predominate in alkaline and acidic solutions, respectively,[22] and the above scheme has been accepted in textbooks. On the other hand, careful investigation has shown that the rate is highly dependent upon the purity of the reagents employed.[23] Moews and Audrieth[24] and Cooper *et al.*[25] reported that hydroxylamine solutions at pH 9 are stable for 1 week. The very slow decomposition was found to be first order in H_2NOH,[23] which obviously rules out reaction (7) as an elementary process. Bonner *et al.*[22] studied the reaction of H_2NOH with NO, which produces equimolar amounts of N_2 and N_2O at pH > 13, whereas at pH 8 the product is almost entirely N_2O. Nitroxyl (HNO) and the radical NHOH are intermediates. In the same study, the disproportionation of H_2NOH at pH 6–13.5 was reinvestigated, and tracer experiments seem to lend support to the hypothesis that nitroxyl is in fact a primary product of H_2NOH disproportionation.

III. STRUCTURES OF TRANSITION METAL COMPLEXES CONTAINING HYDROXYLAMINE LIGANDS

A. Structure types

The structural chemistry of hydroxylamine compounds is rather complex. The reason appears to be the inherent ambiguity concerning the mode of coordination, which is due to the relative mobility of the protons of H_2NOH. Thus, the neutral molecule H_2NOH and its tautomer H_3NO (ammine oxide) are known to coordinate as monodentate ligands through nitrogen or oxygen as in **I** and **II**, respectively. A third conceivable mode, **III**, with bidentate O,N coordination has been discussed in a number of papers, but has not been unambiguously established by an X-ray or neutron diffraction study.

$$M-N\begin{smallmatrix}OH\\ \\H\\ \\H\end{smallmatrix} \qquad M-O-N\begin{smallmatrix}H\\ \\ \\ \\H\end{smallmatrix} \qquad M\begin{smallmatrix}O-H\\ |\ \ H\\N\\ \ \ H\end{smallmatrix}$$

I   **II**   **III**

Structures **IV** and **V** are related deprotonated forms. The hydroxylamide (−1) anion may coordinate in the sideways-on fashion (O,N coordination, η^2 mode) to a transition metal (type **V**). Indeed, the majority of structure determinations have revealed structure type **V** for a variety of transition metals in high and low oxidation states.

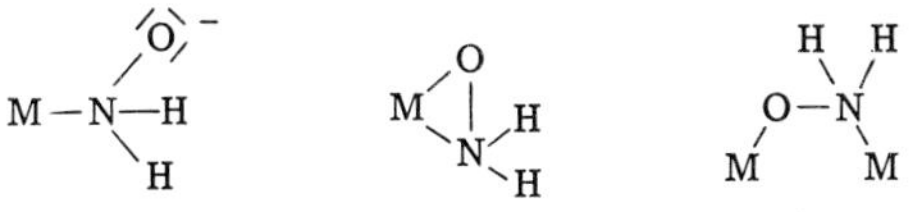

IV   V   VI

Although well-characterized complexes with coordinated hydroxylamide (-1) functioning as bridging ligand between two metals (type **VI**) have not been established, this mode of coordination has been proposed for some polymeric materials, and is certainly a distinct possibility.

Coordinatively saturated hydroxylamine complexes of type **I** deprotonate in alkaline aqueous solutions, e.g., *cis*- and *trans*-$[Pt(NH_3)_3(H_2NOH)_2]^{2+}$ according to reactions (9) and (10).[26]

$$[Pt(NH_3)_2(H_2NOH)_2]^{2+} + OH^- \rightleftharpoons [Pt(NH_3)_2(H_2NO)(H_2NOH)]^+ + H_2O \quad (9)$$

$$[Pt(NH_3)_2(H_2NO)(H_2NOH)]^+ + OH^- \rightleftharpoons [Pt(NH_3)_2(H_2NO)_2]^0 \quad (10)$$

Complexes with N-bound, monodentate hydroxylamide (-1) ligands are thought to be formed (type **IV**). Alexander's base is frequently formulated as $Pt(NH_2OH)_4(OH)_2$; it is insoluble in aqueous solution, and is prepared from alkaline solutions of $[Pt(H_2NOH)_4]Cl_2$.[26] Therefore, it is most probable that in fact a neutral species $[Pt(H_2NOH)_2(H_2NO)_2] \cdot 2H_2O$ (type **IV**) precipitates. In 2 *M* NaOH the precipitate dissolves and the anions $[Pt(H_2NO)_3(NH_2OH)]^-$ and $[Pt(H_2NO)_4]^{2-}$ may exist in solution.[26]

Structures types **VII–IX** may be anticipated for the hydroxylamide (-2) anion, HNO^{2-}.

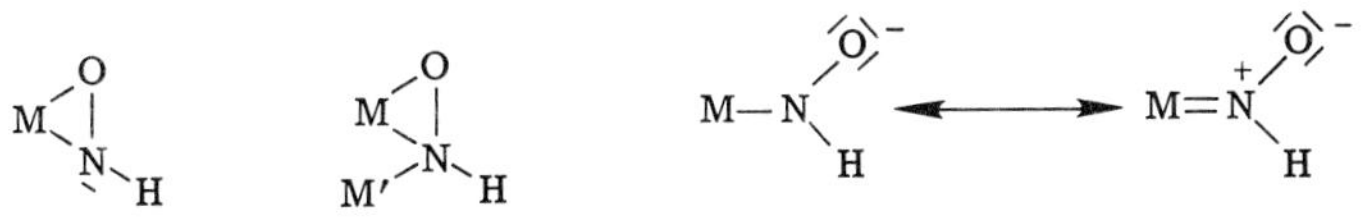

VII   VIII   IX

Transition metal complexes of types **VII** and **VIII** have been characterized, but in the literature they are referred to as nitroso alkane or nitroso arene complexes when the proton is substituted by alkyl or aryl groups as in **X** or **XI**. In general, these compounds are prepared from the reaction of organic nitroso derivatives with low-valent transition metal complexes (via an oxidative addition). The situation is somewhat confusing since genuine nitroso complexes (with retention of the nitrogen–oxygen double bond when coordinated to a metal) have also been prepared and their structures have been determined.[27–29]

$$\underset{\textbf{X}}{\underset{M}{\overset{H\diagdown N\diagup\!\!\diagup O}{|}}} \qquad \underset{\textbf{XI}}{\underset{M}{\overset{R\diagdown N\diagup\!\!\diagup O}{|}}}$$

R = alkyl or aryl

No example of type **IX** is known and only one genuine complex of type **VII** has been found to date.[30] Complexes of type **VIII** have been characterized only when H is substituted by a methyl or a phenyl group.[31,32]

The structural chemistry of sideways-on-bound RNO species is not quite comparable with the bonding situation of η^2-coordinated olefine complexes, in which there is a dependence on the nature of the olefin and the metal center. The carbon-to-carbon distance could range from that of a C–C single bond to that of a nearly undistorted C–C double bond. The data compiled in Table 1 do suggest that the concept of distinct formal oxidation states holds, and N–O distances may be assigned a single- or double-bond character. Therefore, differentiation between hydroxylamido (-2) and nitroso complexes by nomenclature does appear to reflect the true bonding situation, and may not be a purely semantic question.

Interestingly, the formally tautomeric form of the hydroxylamide (-2) anion, the so-called hydroxyimide species NOH^{2-}, does prevail when bonded to a transition metal ion.[33] It is bound as a μ_3-NOH ligand and has an N–O single bond, as in **XII**.

$$\underset{\textbf{XII}}{\overset{M\diagdown}{\underset{M\diagup}{M-N-OH}}}$$

Complexes containing the NO^{3-} anion have not been characterized. The deprotonated form of **XII** yields a μ_3-nitrosyl complex having an N–O double bond.[33]

Monoximes are derivatives of hydroxylamine. Their known structural chemistry with transition metals is represented by **XIII** and **XIV**.

$$\underset{\textbf{XIII}}{\overset{M\diagdown O}{\underset{N\diagup\!\diagdown C\diagup\overset{R}{\diagdown R}}{|}}} \qquad \underset{\textbf{XIV}}{\overset{M\diagdown\,\,\,M}{\underset{C\diagup\overset{R}{\diagdown R}}{O-N\diagdown}}}$$

TABLE 1

Compilation of structural data for complexes with coordinated hydroxylamine ligands

Complex[a]	M–N (Å)	M–O (Å)	N–O (Å)	Structure	Ref.
$[Ni(NH_2OH)_6]SO_4$	2.12(1)	–	1.44(1)	I	[34]
$[Pd(PhNO)(Pt\text{-}Bu_3)]_3$	2.14(2), 2.09(1)	2.043(4)	1.35(1)	VIII	[31]
$[MoO(PhNO)(dipic)(hmpa)]$	2.036(5)	1.944(4)	1.416(7)	VII	[37]
$[Fe_2(CO)_6(3\text{-}Cl,2\text{-}CH_3C_6H_3NO)_2]$	2.02(1)	1.91(1)	1.40(1)	VIII	[32]
$[Mo(NO)(terpy)(CN)(HNO)]\cdot 2H_2O$	2.044(7)	2.022(6)	1.422(9)	VII	[30]
$[UO_2(NH_2O)_2(H_2O)_2]2H_2O$	2.431(7)	2.297(7)	1.434(9)	V	[38]
$[UO_2(NH_2O)_2(NH_3O)_2]\cdot 2H_2O$	2.410(7)	2.330(8)	1.421(12)	V, II	[35]
	–	2.477(9)	1.422(12)		
$[UO_2(NH_2O)_2(H_2O)_2]\cdot H_2O$	2.421(7)	2.317(8)	1.433(10)	V	[39]
$[UO_2(NH_2O)_2(C_2H_6O_2)]$	2.41(2)	2.30(1)	1.41(2)	V	[40]
$[VO(dipic)(NH_2O)(H_2O)]$	2.007(3)	1.903(3)	1.371(4)	V	[41]
$Cs[V(dipic)(NH_2O)(NO)(H_2O)]\cdot 2H_2O$	2.054(17)	2.015(12)	1.42(2)	V	[42]
$[Mo(bpy)(NH_2O)_2(NO)]Cl$	2.110(5)	2.030(6)	1.41(1)	V	[43]
$[MoO_2(CH_3NHO)_2]$	2.12(1)	1.962(10)	1.43(1)	V	
$(phen)[Mo(phen)_2(NO)(NH_2O)]I_2\cdot H_2O$	2.06(2)	2.00(2)	1.33(3)	V	[44]
$[MoO_2(Et_2NO)_2]$	2.143(2)	1.970(2)	1.427(3)	V	[45]
$[V_2O_3(Et_2NO)_4]$	2.078(2)	1.863(2)	1.402(2)	V	
$[MoO[H(CH_3)NO]_2L]$	2.100(10)	1.961(8)	1.400(10)	V	[46]
$[MoO[H(CH_3)NO]LL']\cdot H_2O$	2.114(4)	1.976(3)	1.376(7)	V	
$[MoO[(CH_3)_2NO](BenzH)Benz]$	2.172(4)	1.969(3)	1.399(4)	V	[47]
$[MoO[H(CH_3)NO]_2L'']$	2.11(1)	1.961(7)	1.41(1)	V	[48]
$(Et_2HNOH)_2[Mo_2O_4(Et_2NO)_2(C_2O_4)_2]$	2.149(4)	2.010(5)	1.406(5), 1.427(9)	V	[49]
$[Ti(Et_2NO)_4]$	2.108(5)	1.980(3)	1.402(7)	V	[50]
$[MoO_2(C_5H_{10}NO)_2]$	2.145(7)	1.96(1)	1.44(1)	V	[51]
$[MoS_2(C_5H_{10}NO)_2]$	2.151(4)	1.956(4)	1.427(5)	V	
$[MoOS(C_5H_{10}NO)_2]$	2.144(5)	1.970(5)	1.433(6)	V	[52]
$[[(C_5H_{10}NO)_2Mo(\mu_2\text{-}S)_2Cu(\mu_2\text{-}Cl)]_2]$	2.162(4)	1.957(4)	1.433(5)	V	[53]
$[MoS_2(Et_2NO)_2]$	2.173(3)	1.962(2)	1.426(3)	V	[54]
$[PdCl(PPh_3)(tmpo)]$	2.093(4)	2.023(4)	1.372(6)	V	[55]
$[(C_6H_5)_4P]_2[Mo(NO)(H_2NO)(N_3)_4]\cdot H_2O$	2.125(5)	2.048(5)	1.404(7)	V	[57]
$[(C_6H_5)_4P]_2[Mo(NO)(NH_2O)(NCS)_4]$	2.102(9)	2.053(8)	1.398(12)	V	[58]
$[Mn(CO)_3(C_9H_{18}NO)]$	1.981(3)	1.839(3)	1.413(3)	V	[59]
$[(\eta^5\text{-}C_5H_4Me)_3Mn_3(NO)_3(NOH)]BF_4$	1.873(3)	–	1.393(4)	XII	[33]
$[Ru_3(CO)_{10}(NOCH_3)]$	2.027(3)	–	1.433(6)	XII	[60]
NH_2OH	–	–	1.47(3)	–	[36]
$[NH_3OH]Cl$	–	–	1.369(7)	–	[14]
$[NH_3OH]ClO_4\ (-150°C)$	–	–	1.41(1)	–	[15]

[a] Abbreviations: t-Bu, t-butyl; Ph, phenyl; dipic, pyridine-2,6-dicarboxylate; hmpa, hexamethyl-phosphoramide; terpy, terpyridine; bpy, bipyridine; Et, ethyl; phen, phenanthroline; tmpo, 2,2,6,6-tetramethylpiperidine *N*-oxide; Me, methyl; BenzH (Benz), benzohydroximate (benzohydroxamate).

Finally, it should be pointed out that structures **I–XII** are also adopted by N-mono- and N-disubstituted organic hydroxylamines or by O-substituted derivatives, as in **I** and **XII**.

B. X-Ray structure determinations

In this section a few benchmark structures of complexes containing ligands bound to transition metals as in **I–XIV** will be discussed in more detail. A complete list of structures of complexes with hydroxylamine ligands is given in Table 1. As already mentioned, the first X-ray determination of a genuine hydroxylamine complex was published in 1974 by A. White for $[Ni(H_2NOH)_6](SO_4)$. This is an octahedral complex of type **I** with a mean N–O distance of 1.44 Å.[34] Some earlier structures with nitrosobenzene ligands may also be looked at as N-substituted hydroxylamido (-2)-containing complexes.[32] In 1979 Adrian and van Tets published a careful neutron diffraction study on $[UO_2(NH_2O)_2(NH_3O)_2] \cdot 2H_2O$,[35] which contains sideways-on-coordinated hydroxylamido (-1) ligands (type **V**) and two O-coordinated neutral molecules of the tautomeric ammine oxide form of hydroxylamine (type **II**). The structure is shown in Fig. 2. The only characterized complex with a hydroxylamido (-2) ligand of type **VII** is $[Mo(NO)(terpy)(CN)(HNO)] \cdot H_2O$,[30] the structure of which is shown in Fig. 3.

It is remarkable that the N–O distances do not vary extensively in going from coordinated H_2NOH (1.44 Å) to η^2-H_2NO^- (~ 1.42 Å) or to

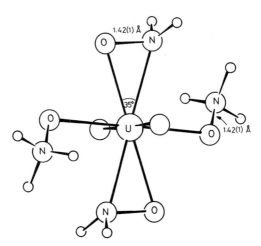

Fig. 2. The structure of the molecule $[UO_2(NH_3O)_2(H_2NO)_2]$. (From Ref. [35].)

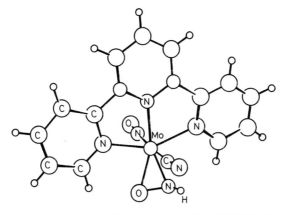

Fig. 3. The structure of the molecule [Mo(NO)(terpy)(CN)(HNO)]. (From Ref. [30].)

TABLE 2

Summary of structural data of complexes with alkylnitroso, arylnitroso, and monoxime ligands

Complex	M–N (Å)	M–O (Å)	N–O (Å)	Structure	Ref.
[(C$_6$H$_5$NO)$_2$PdCl$_2$]	1.994(2)		1.209(3)	XI	[27]
[CoCl$_2$[p-NC$_6$H$_4$N(CH$_3$)$_2$]$_2$]	2.025 (av)		1.265(8) (av)	XI	[28]
[OsCl$_2$(CO)(HNO)(PPh$_3$)$_2$]	1.915(6)		1.193(7)	X	[29]
[CuL(ClO$_4$)(H$_2$O)]$_2$	1.987(5)		1.325(6)	XIV	[61]
[Fe$_2$(CO)$_6$(C$_6$H$_{14}$N$_2$O)]	1.99(1)	1.96(1)	1.38(1)	XIV	[62]
[Fe$_2$(CO)$_6$(C$_6$H$_{12}$N$_2$O)]	1.982(5)	1.952(4)	1.351(6)	XIV	[63]
[Mo(C$_5$H$_5$)(CO)$_2$[(CH$_3$)$_2$CNO]]	2.089(12)	2.139(9)	1.336(27)	XIII	[64]
[(C$_6$H$_5$)$_4$P]$_2$[Mo(NO)[ONC(CH$_3$)$_2$](NCS)$_4$]	2.086(8)	2.090(6)	1.358(10)	XIII	[56]
[(C$_6$H$_5$)$_4$P]$_2$[Mo(NO)[ONC(C$_2$H$_5$)$_2$](NCS)$_4$]	2.064(6)	2.090(4)	1.333(7)	XIII	[58]

η^2-HNO^{2-} (1.42 Å). The distances are characteristic of N–O single bonds and agree well with N–O bond lengths found in uncoordinated hydroxyl-amine (1.47 Å)[36] or in hydroxylammonium salts (1.37–1.44 Å).[14,15] Furthermore, the nitrogens have a formal coordination number 4 in complexes of types **I, II,** and **V** (sp^3 hybridization), which is to be compared with the structure of [Mo(NO)(terpy)(CN)(HNO)], in which the hydrogen of the HNO ligand is clearly not in the plane defined by the three-membered Mo–N–O ring. This again implies sp^3 hybridization of the nitrogen, under-lining the hydroxylamido (-2) designation rather than coordinated nitroxyl:

$$\text{Mo} \leftarrow \begin{array}{c} \text{O} \\ \| \\ \text{N} \\ \diagdown \\ \text{H} \end{array}$$

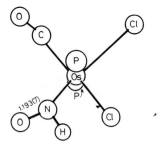

Fig. 4. Inner coordination sphere of [OsCl₂(CO)(HNO)(PPh₃)₂]. (From Ref. [29].)

An *N*-phenylhydroxylamido (−2) ligand is O,N coordinated in a molybdenum(VI) complex[37] contrasting two structures with N-coordinated nitrosobenzene,[27,28] [(C₆H₅NO)₂PdCl₂] and [CoCl₂[*p*-ONC₆H₄N(CH₃)₂]₂] (Table 2). A genuine nitroxyl complex has been characterized by Ibers and co-workers,[29] [OsCl₂(CO)(HNO)[P(C₆H₅)₃]₂] (Fig. 4).

The majority of hydroxylamine complexes contain sideways-on hydroxylamido (−1) ligands (type **V**).[38–59] Such a bond appears to be thermodynamically the most stable and is prevalent among the complexes with metal ions in both high oxidation states [e.g., Ti(IV), V(V), Mo(VI), and U(VI)] and low-valent forms [e.g., V(I), Mo(II), or Mn(I)]. Complexes with one, two, and four O,N-bound R₂NO⁻ ligands have been isolated. An example of the latter, an eight-coordinate complex of titanium(IV), is shown in Fig. 5. A substantial number of these complexes have structures with a coordination number 7.[38–49] The metal is in a pentagonal bipyramid ligand environment. Due to the small chelate bite of the hydroxylamide (−1) ligands (∼40°), these moieties are ideally suited to occupy two coordination sites in equatorial positions. HNO²⁻ is isoelectronic with the peroxo group O₂²⁻, and from a structural point of view it is interesting to note that the structural chemistry of R₂NO⁻ with transition metals is generally very similar to that of the peroxo group.[65] This may be exemplified by two complexes of vanadium(V) (Fig. 6). The reaction of NH₄VO₃ in acidic aqueous solution with H₂O₂ in the presence of pyridine-2,6-dicarboxylate (dipic) yields the red monoperoxo complex NH₄[VO(O₂)(dipic)H₂O][66]; when hydroxylamine is used instead of H₂O₂ a colorless neutral complex is obtained [VO(H₂NO)(dipic)(H₂O)].[42] Both structures have been determined by X-ray diffraction studies.[41,67] Even some hydroxylamide (−1) analogs of the yellow diperoxo complexes of vanadium(V) have been characterized.[68] Thus the analog of the [O[VO(O₂)₂]₂]⁴⁻ anion[69] is [O[VO(Et₂NO)₂]₂]⁰[45] [Et₂NO = *N*,*N*-diethylhydroxylamide (−1)].

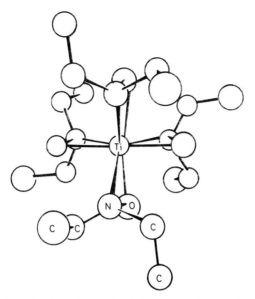

Fig. 5. The structure of the molecule [Ti(ONEt$_2$)$_4$]. (From Ref. [50].)

Recently, a few structures have been reported containing O-coordinated nitroxyl radicals.[70-72]

$$O\overset{|}{\underset{M}{}}=N\overset{R}{\underset{R}{}}$$

R$_2$NO = 2,2,6,6-tetramethyl-
piperidinyl-1-oxide

The N—O distances of ~1.27(1) Å are indicative of double bonds. This is

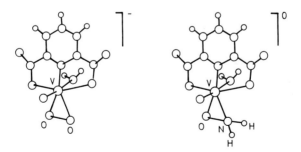

Fig. 6. The structures of the monoanion of [VO(O$_2$)(dipic)(H$_2$O)]$^-$ and of the neutral molecule [VO(H$_2$NO)(dipic)(H$_2$O)]. (From Refs. [41] and [67].)

in contrast to structures in which the reduced form, the monoanion hydroxyl-amide (-1), is O,N coordinated (type **V**) to transition metals,[55,59] the N–O distance then being ~ 1.41 Å.

IV. SURVEY OF COMPLEXES AND SYNTHETIC ROUTES TO HYDROXYLAMINE COMPLEXES

A. Survey of complexes

The information on the majority of known complexes of transition metals containing H_2NOH or its N- and/or O-substituted derivatives (Table 3),

TABLE 3

Derivatives of hydroxylamine used as ligands

Derivative	Ref.	Derivative	Ref.
$\begin{matrix} H_3C \\ \quad\diagdown \\ \quad\quad N{-}OH \\ \quad\diagup \\ H \end{matrix}$	[43, 46, 48, 78, 80, 89, 98, 109, 131, 139, 141, 156]	$(t\text{-Bu})_2NOH$ (phenyl)$\diagup^{N-OH}_{\diagdown H}$	[121–124] [31, 37, 88–90 119, 120]
$\begin{matrix} t\text{-Bu} \\ \quad\diagdown \\ \quad\quad N{-}OH \\ \quad\diagup \\ H \end{matrix}$	[37, 83, 120]	$\begin{matrix} \quad\quad CH_3 \\ \quad\quad\diagup \\ O{-}N{-}CH_3 \\ \quad\quad\diagdown \\ \quad\quad CH_3 \end{matrix}$	[81, 97, 110, 111, 147]
$(H_3C)_2NOH$ $(H_5C_2)_2NOH$	[47] [45, 49, 50, 51, 77, 85]	$X{-}\text{(phenyl)}{-}N{-}OH$ below: H $X = Cl, CN, CH_3,$ $N(CH_3)_2, OCH_3$	[37, 90, 119]
$\left(\text{(phenyl)}{-}CH_2\right)_2 NOH$	[49, 50]		
$\text{(piperidine)}{-}N{-}OH$	[51–53]	$H_2N{-}OCH_3$	[32, 60, 98–101, 112, 137, 138, 141, 146, 154, 155]
$\begin{matrix} H_3C \\ H_3C \end{matrix}\!\!\diagdown\!\!\text{(ring)}\!\!\diagup\!\!\begin{matrix} CH_3 \\ CH_3 \end{matrix}$ with N and OH	[55, 59, 125]	$(H_3C)_2C{=}N{-}OH$ $(H_5C_2)_2C{=}N{-}OH$	[77] [58]

such as trimethylamineoxide, *N*-alkyl- and *N,N*-dialkylhydroxylamines, and O-methylhydroxylamine, is summarized in Table 4. This table cannot be comprehensive with respect to the rich chemistry of platinum(II) and palladium(II) complexes with hydroxylamine. Extensive studies in this area have been carried out by Kharitonov and Sarukhanov.[73,74]

The exact structures of many of the simple complexes, some of which were prepared as early as 1870, are not yet established, e.g., $Mn(H_2NO)_2Cl_2$.[91-93] Infrared spectroscopy has been used as a means to distinguish the modes of coordination of H_2NOH with metals. These investigations have been carried out primarily by Russian workers; Kharitonov gave an authoritative account of these studies in 1971.[75] It must be kept in mind, however, that the O,N coordination of hydroxylamide (-1) anions was recognized only in 1977,[39,42] and since then only one paper has appeared that deals with IR characteristics of sideways-on-bound H_2NO^- ligands.[76] The $v(NO)$ stretching frequency is found for all types of coordination between 920 and 1050 cm^{-1}, indicating N–O single bonds.

B.   Syntheses

1.   Displacement reactions

Many of the simple homoleptic hydroxylamine complexes of type **I** are prepared from ethanolic (or aqueous) solutions of divalent metal halides and ethanolic solutions of H_2NOH. The desired complexes precipitate out. In general, only divalent metal centers that are stable with respect to redox reactions and that have labile aquo ligands undergo this type of reaction [Ni(II), Co(II), Mn(II), Zn(II)]. Complexes of $[Ni(H_2NOH)_6]^{2+}$ or of the explosive $[Cu(NH_2OH)_4]^{2+}$ can be obtained by this method.

The reaction of metal oxides or carbonates with hydroxylammonium salts in aqueous solution

$$M(II)O + 2[H_3NOH]^+X \rightarrow M(II)(H_2NOH)_2X_2 + H_2O \tag{11}$$

$$MCO_3 + 2[H_3NOH]X \rightarrow M(II)(H_2NOH)_2X_2 + H_2O + CO_2 \tag{12}$$

gives metal complexes of 1:2 stoichiometry, e.g., $Zn(H_2NOH)Cl_2$, $Mn(H_2NOH)_2Cl_2$, and $Co(H_2NOH)_2Cl_2$ are obtained. Furthermore, $[Co(NH_2OH)_6]^{3-}$ has been prepared from $Na_3[Co(CO_3)_3] \cdot 3H_2O$ and $[H_3NOH]Cl$ in aqueous solution.[105]

Platinum(II) hydroxylamine complexes are prepared by ingeneously devised substitution reactions starting with $PtCl_4^{2-}$ and H_2NOH in aqueous solution. Chernaev and Chugaev have used the kinetic trans effect to synthesize a large number of essentially square-planar complexes of platinum(II), with varying numbers (one to four) of N-coordinated H_2NOH ligands.

TABLE 4

Summary of transition metal complexes containing hydroxylamine or its derivatives

Complex[a]	Ref.	Complex[a]	Ref.
Titanium, zirconium, hafnium			
$[Ti(OEt)_x(ONEt_2)_{4-x}]$ $x = 1$–3	[77]	$[Zr(ONEt_2)_4]$	[50]
$[Ti(OPr^i)_x[ONC(CH_3)_2]_{4-x}]$ $x = 1$–3	[77]	$[Zr(ONBz_2)_4]$	[50]
$[Ti(ONBz_2)_4]$	[50]		
Vanadium, niobium, tantalum			
$[VO(dipic)(H_2NO)(H_2O)]$	[41, 42]	$[V(NO)(CN)_4(H_2NO)]^{3-}$	[79]
$[VO[H(CH_3)NO](2\text{-pic})_2]$	[78]	$[O[VO(ONEt_2)_2]_2]$	[45]
$[VO[H(CH_3)NO]_2(2\text{-pic})]$	[78]	$[V(dipic)(H_2NO)(H_2O_2]$	[80]
$[VO(H_2NO)_2(2\text{-pic})]$	[78]	$[V(dipic)[H(CH_3)NO](H_2O_2)]$	[80]
$[VO(bpy)(H_2NO)_2]^+$	[78]	$[(VO(Et_2NO)_2)_2(C_2O_4)]$	[49]
$[VO(phen)(H_2NO)_2]^+$	[78]	$[(VO(Bz_2NO)_2)_2(C_2O_4)]$	[49]
Chromium, molybdenum, tungsten			
$[Cr(tmno)_6](ClO_4)_3$	[81]	$[Mo(bpy)(H_2NO)_2(NO)]Cl$	[43]
$[Cr(dipic)(NO)(NH_2OH)_2]$	[82]	$[Mo(NO)(H_2NO)(terpy)(H_2O)](ClO_4)_2$	[86]
$[Cp_2(NO)_3Cr_2(t\text{-BuNHO})]$	[83]	$[Mo(NO)(HNO)(terpy)(CN)] \cdot 2H_2O$	[30, 86]
$[CrCl(H_2O)(NH_2OH)_3](OH)_2 \cdot C_2H_5OH$	[84]	$[Mo(NO)(H_2NO)(dipic)(H_2O)] \cdot 1.5H_2O$	[87]
$[Cr(NH_2OH)_5(OH)]Cl_2$	[84]	$[Mo(NO)(H_2NO)(dipic)X]^-$ $X = N_3^-,$	[87]
$[Cr(OH)_2(H_2O)(NH_2OH)_3]OH \cdot 1.5H_2O$	[84]	NCS^-, CN^-	
$[CrCl(SO_4)(NH_2OH)_4] \cdot 2H_2O$	[84]	$[Mo(NO)(H_2NO)(terpy)(H_2NOH)](ClO_4)_2$	[87]
$[MoXY(ONR_1R_2)]$ $X = O, S, Se; Y = O, S$	[43, 45, 51, 52, 54, 85]	$[Mo(NO)(H_2NO)_2(phen)]^+$	[87]
		$[Mo(NO)(H_2NO)(phen)_2]^{2+}$	[87]
		$[Mo(NO)(H_2NO)(2\text{-pic})_2]$	[87]

[MoO[(CH₃)NO]₂[HNC(S)N(CH₃)O]]	[46]	[Mo(NO)[(C₂H₅)₂CNO]CNO[(NCS)₄]²⁻	[58]
[MoO[(CH₃)NO][HNC(S)N(CH₃)O][H₂NC(S)N-(CH₃)O]]·H₂O	[46]	[MoO(RNO)[(CH₃)₂NCS₂]] R = t-Bu, Ph, C₆H₄–Cl, C₆H₄CN	[37]
[MoO(dipic)(PhNO)(hmpa)]	[88]	[MoO(RNO)(dipic)(hmpa)] R = Ph, C₆H₄Cl, C₆H₄CN	[37]
[MoO(dipic)(PhNO)(py)]	[88]	[MoO(ONPh)(S₂CNEt₂)₂]	[160]
[MoO(2-pic)(PhNO)₂]H	[88]	[Mo(ONPh)(S₂CNEt₂)₂]	[160]
[MoO[(CH₃)₂NO](Benz)(BenzH)]	[47]	[MoO₂[ON(CH₃)]]	[89]
[MoO[(CH₃)HNO]₂[ON(CH₃)CS₂]]	[48]	[MoO₂(ONPh)]	[89]
(Et₂NHOH)₂[Mo₂O₄(Et₂NO)₂(C₂O₄)₂]	[49]	[Mo₂O(NO)₂[ON(CH₃)]₂[OP(CH₃)₃]₂]	[89]
[Mo(NO)(H₂NO)(N₃)₄]²⁻	[57]	[MoO(S₂CNEt₂)₂(ONC₆H₄p-X)] X = CH₃, H, Cl	[90]
[Mo(NO)(H₂NO)Cl₄]²⁻	[57]	[WO(dipic)(PhNO)(hmpa)]	[88]
[Mo(NO)(H₂NO)(CN)₄]²⁻	[82]		
[Mo(NO)(NCS)₄(H₂NO)]²⁻	[58]		

Manganese, technetium, rhenium

Mn(H₃NO)₂Cl₂	[91–94]	[(η⁵-C₅H₄Me)₃Mn₃(NO)₃(NOH)]BF₄	[33]
Mn(H₃NO)₂SO₄H₂O	[95, 96]	Mn(H₂NOH)(NCS)₂C₂H₅OH	[103]
[Mn[(CH₃)₃NO]₄](ClO₄)₂	[81]	Mn(CO)₃(tmpo)	[59]
Mn[(CH₃)₃NO]₂Br₂	[97]	Re₂O₃Me₄(ONMe)₂	[89]
Mn[(CH₃)H₂NO]₂Cl₂	[98]		

Iron, ruthenium, osmium

[Fe₂(CO)₆(R–NO)₂] R = 3-Cl-2-CH₃-phenyl	[32, 99–101]	[RuCl(H₂NO)(H₂NOH)(PPh₃)₂]	[102]
[Fe(CN)₅(H₂NOH)]³⁻	[166, 167]	[RuCl₂(H₂NOH)₂(PPh₃)₂]	[102]
[Fe(CN)₅(R₂NOH)]³⁻	[168]	[Ru₃(CO)₁₀(NOCH₃)]	[60]

(continued)

229

TABLE 4 (continued)

Complex[a]	Ref.	Complex[a]	Ref.
Cobalt, rhodium, iridium			
[Co(H₂NOH)₆]³⁺	[91, 104–106]	[Co(tmno)₂X₂] X = Cl⁻, Br⁻, I⁻, SCN⁻, NO₃⁻	[97, 159]
[CoCl(en)₂(NH₂OH)]Cl₂	[104, 107]	[Co(NCS)₂(H₂NOH)2C₂H₅OH]	[103]
[CoCl(NH₂OH)₅Cl]Cl₂	[93]	[Co(H₂NOCH₃)₂Cl₂]	[98, 112]
[Co(H₂NOH)₄]Cl₂C₂H₅OH	[95]	[Co(H₂NOCH₃)₄](NCS)₂	[98, 112]
Co(NH₂OH)₂Cl₂	[94, 108]	RhCl₃(H₂NOH)(PPh₃)₂	[158]
Co(NH₂OH)₂SO₄	[108]	[IrCl₃(NH₂OH)(PPh₃)₂]	[113]
[Co(CH₃)](CH₃)NO][P(CH₃)₃]₂]₂	[109]	[Ir(NH₃)₅(NH₂OH)](ClO₄)₃	[114]
[Co(tmno)₄]X₂ X = Cl⁻, Br⁻, I⁻, SCN⁻, NO₃⁻	[81, 97, 110, 111, 159]		
Nickel, palladium, platinum			
[Ni(H₂NOH)₆]SO₄	[17, 34, 95, 108, 115, 116, 117]	[Ni(PhNO)(PPh₃)₂]	[119, 120]
		[Ni(t-Bu–NO)(PEt₃)₂]	[120]
[Ni(NH₂OH)₄]X₂ X = Cl⁻, Br⁻, I⁻, NO₃⁻, ClO₄⁻, NSC⁻	[93, 103, 106, 108, 118]	[Pd₃(PhNO)L₃] L = P(t-Bu)₃, PPh(t-Bu)₂	[31, 119]
		[Pd(p-ClC₆H₄NO)(t-BuNC)₂]	[119]
		[Pd(PhNO)(PPh₃)₂]	[119]
		[PdCl[ON(t-Bu)₂]]₂	[121]
[Ni(NH₂OH)₂X₂] X = Cl⁻, Br⁻	[95, 108]	[PdX[ON(t-Bu)₂]L] L = PPh₃, AsPh₃, P(OPh)₃; X = Cl, Br⁻, I⁻	[122]
Ni(NH₂OH)X₂ X = Cl⁻, NO₃⁻	[108]		
[Ni(H₂NOH)₃(H₂NO₂)]·2H₂O	[108]	[Pd(NH₂O)₂(NH₂OH)₂]·2H₂O	[157]
[Ni(H₂NOH)₃X]X X = Cl⁻, Br⁻	[108]	[Pd(NH₂OH)₄]F₂·4H₂O	[157]
Ni(H₂NOCH₃)₂X₂ X = Cl⁻, Br⁻	[98, 112]	[Pd(NH₂OH]₄]PtCl₄	[157]
[Ni(H₂NOCH₃)₄](NCS)₂	[98, 112]	[Pd(NH₂OH)₂Cl₂]	[157]
[Ni(tmno)₄](ClO₄)₂	[81]	[Pd(NH₂OH)₂(NH₃)₂]PtCl₄	[157]
[Ni(ArNO(t-BuNC)₂] Ar = p-XC₆H₄; X = NMe₂, OMe, Me, H, Cl, Br	[119]	[PdCl[ON(t-Bu)₂][ZCHC(O)Ph]] Z = PPh₂Me, PPh₃, AsPh₃, SMe₂	[123]

Compound	Ref.	Compound	Ref.
[Pd[ON(t-Bu)]₂](L–L)]X	[124]	[Pt(NH₃)(NH₂OH)py(NO₂)]₂PtCl₄	[135, 136]
L–L = bpy, phen, 2PPh₃; X = ClO₄⁻, BPh₄⁻		[Pt(H₂NOH)(py)Cl₂]	[135]
[PdCl(tmpo)]₂	[125]	[Pt(NH₂OH)₂(NO₂)₂]	[136, 140]
[PdCl(tmpoL)] L = phenacylide, PPh₃	[55, 125]	[Pt(H₂NOCH₃)₄]PtCl₄	[98, 137, 138]
[Pt(NH₂OH)₄]Cl₂	[106, 117, 126, 127, 128]	[Pt((CH₃)HNOH]₄PtCl₄	[131, 139]
[Pt(NH₂OH)₂(NH₂O)₂]·2H₂O	[106, 127, 128]	[Pt(H₂NOCH₃)₄]Cl₂	[141]
		[Pt(CH₃CHNOH)₄]Cl₂	[141]
[Pt(NH₂OH)₂X₂] X = Cl, I	[106, 127, 129, 130]	[Pt(dms)(H₂NOH)Cl₂]	[142]
		[PtCl₂(H₂NOH)(dmso)]	[143]
[Pt(NH₃)₂(NH₂OH)₂]Cl₂	[106, 131, 132, 133]	[Pt(NH₂OH)₂X₄] X = Cl⁻, Br⁻	[144, 145]
		[Pt(NH₂OH)₄X₂]X₂ X = Cl⁻, Br⁻	[144, 145]
[Pt(NH₃)(NH₂O)(NO₂)(py)]	[134]	[Pt A₄X₂]X₂ X = Cl⁻, Br⁻; A = NH₂OH,	[146]
[Pt(NH₂OH)(NH₂O)(NH₃)(NO₂)]	[134]	H₂NOCH₃	
[Pt(NH₂OH)₂(NH₂O)(NO₂)]	[134]	[PtA₂X₄]	[146]
[PtCl₂(NH₃)(NH₂OH)]	[134]	[Pt(PhNO)(PPh₃)₂]	[119]

Copper, silver, gold, uranium, zinc, cadmium

Compound	Ref.	Compound	Ref.
[Cu(NH₂OH)₄]Cl	[93]	[UO₂(H₂NO)₂(HOCH₂CH₂OH)]	[40]
[UO₂X₂L₂] L = (CH₃)₃NO;	[147]	Zn(H₃NO)₂Cl₂	[93, 152, 153]
X = diethyldithiocarbamate,			
diethyldiselenocarbamate		Zn(H₂NOH)₂X₂ X = Cl, Br	[108]
[UO₂(OAc)₂LH₂O]₂	[147]	Zn₂(H₂NOH)₃(H₂NO)X₃ X = Cl, Br	[108]
[UO₂(NO₃)₂L₂]	[147]	Cd(H₃NO)₂Cl₂	[92, 98, 108, 153]
[UO₂L₄(NO₃)₂]	[147]		
[UO₂(H₂NO)₂(NH₃)₂]·2H₂O	[148, 149]	M(H₂NOCH₃)₂X₂ X = Cl⁻, Br⁻, I⁻;	[154, 155]
[UO₂(H₂NO)₂(H₂O)₂]·H₂O	[38, 39, 76, 149, 150, 151]	M = Zn, Cd	
		M(CH₃)(HNOH)₂X₂ Z = Cl⁻, Br⁻, I⁻;	[156]
[UO₂(H₂NO)₂(NH₃O)₂]	[35]	M = Zn, Cd	

231

ᵃ Abbreviations: dms, dimethylsulfide; dmso, dimethyl sulfoxide; OPrⁱ, isopropyl; 2-pic, pyridine 2-carboxylate; tmdo, trimethylamine oxide; other abbreviations as in footnote to Table 1.

Mehrotra *et al.*[77] have reported an elegant synthesis, the potential of which has not yet been exploited. Titanium(IV) tetraalkoxides react with N,N-disubstituted alkylhydroxylamines, yielding O,N-coordinated[50] hydroxylamido (-1) complexes of titanium[77] and zirconium[50] and the corresponding alcohol, which can easily be distilled off, thus driving the reaction (13) to completion.

$$M(OR)_4 + n \; \overset{R_1}{\underset{R_2}{\diagdown}} NOH \longrightarrow [M(OR)_{4-n}(R_1R_2NO)_n] + nROH \tag{13}$$

$$M = Ti, Zr$$

The same reaction with monoximes has been used to generate oxime complexes of titanium(IV).

Complexes with metal centers in high oxidation states and with relatively labile terminal oxo ligands react with H_2NOH and with N-substituted or N,N-disubstituted hydroxylamines, forming compounds with O,N-coordinated hydroxylamido (-1) or hydroxylamido (-2) ligands[37,41–43,45,49, 51,52,88] according to reactions (14) and (15).

$$M{=}O + \overset{R_1}{\underset{R_2}{\diagdown}}\overset{+}{\underset{\underset{H}{|}}{N}}{-}OH \longrightarrow M\overset{O}{\underset{N\diagdown R_2}{\diagup|\diagdown R_1}} + H_2O \tag{14}$$

$$M = Mo(VI), V(V)$$

$$M{=}O + \overset{R}{\underset{H}{\diagdown}}N{-}OH \longrightarrow M\overset{O}{\underset{N\diagdown R}{\diagup|}} + H_2O \tag{15}$$

2. Oxidative addition reactions

Stetsenko *et al.*[144–146] have reported the preparation of H_2NOH complexes of platinum(IV). This is somewhat surprising because of the reducing ability of hydroxylamine: it can precipitate elemental platinum from solutions of Pt(II) complexes. The heterogeneous action of bromine in CCl_4 upon hydroxylamine complexes of platinum(II) leads to the bromination of these complexes:

$$[Pt(II)(NH_2OH)_4Cl_2] + Br_2 \rightarrow [Pt(IV)(H_2NOH)_4Br_2Cl_2] \tag{16}$$

Reaction (16) is in essence an oxidative addition reaction in which the square-planar platinum(II) species (d^8) is converted to an octahedral complex of platinum(IV) (d^6). The general applicability of this method has been demonstrated and many hydroxylamine platinum(IV) complexes have been isolated and characterized by their IR[145] and ESCA[144] spectra. In these reactions the N-coordinated hydroxylamine ligands are not involved in the redox process.

Because nitroso groups in compounds such as nitrosobenzene or nitroso alkyl derivatives are isoelectronic with oxygen, their reactions with metals in low formal oxidation states (i.e., zero) have been investigated.[119] Otsuka et al.[119] prepared a series of complexes by reacting $Ni^0(t$-$BuNC)_4$, Pd^0-$(t$-$BuNC)_2$, and $Pt^0(C_2H_4)(PPh_3)_2$ with a variety of arylnitroso compounds. Monomeric compounds with sideways-on coordinated hydroxylamido (-2) ligands as well as the trimeric species $[Pd_3(Pt$-$Bu_3)_3(HNO)_3]$ have been characterized. The $v(NO)$ stretching frequency at ~ 1035 cm^{-1} is thought to be indicative of an NO bond order of 1.

$$L_4Ni^0 \;+\; R-N{=}O \;\longrightarrow\; L_2Ni(II) \underset{N}{\overset{O}{\diagup}} \diagdown R \;+\; 2\,L \qquad (17)$$

3. Redox reactions with nitroxide radicals and nitro compounds

In 1967, in an attempt to study the coordination behavior of nitroxide radicals, Beck et al.[121] reacted the neutral radical di-*t*-butylnitroxide (in large excess) with $PdCl_2$ or $Na_2[PdCl_4]$ and isolated a diamagnetic compound, to which they assigned a dimeric structure, containing the O,N-bound di-*t*-butylhydroxylamide (-1) ligand.

The basis of this assignment was the $v(NO)$ stretching frequency at 1020 cm^{-1}, which indicates an N–O single bond. This was in fact the first time that this mode of coordination for a derivative of hydroxylamine had been proposed. The exact nature of the reaction which led to the above complex in good yields (70–80%) was not at all clear, since the oxidation state of palladium had not changed whereas the nitroxide radical obviously had been reduced to a hydroxylamide (-1) anion.

Subsequent work by Okunaka, Matsubayashi, and Tanaka[122–125] proved the O,N coordination of the hydroxylamido (-1) ligands correct, and a crystal structure of the monomeric $[PdCl(PPh_3)(tmpo)]$ complex was published in 1981.[31] Furthermore, the nature of the reaction was shown to be a disproportionation of the nitroxide radicals into cationic species, R_2NO^+, and hydroxylamide (-1) anions, R_2NO^-.[125]

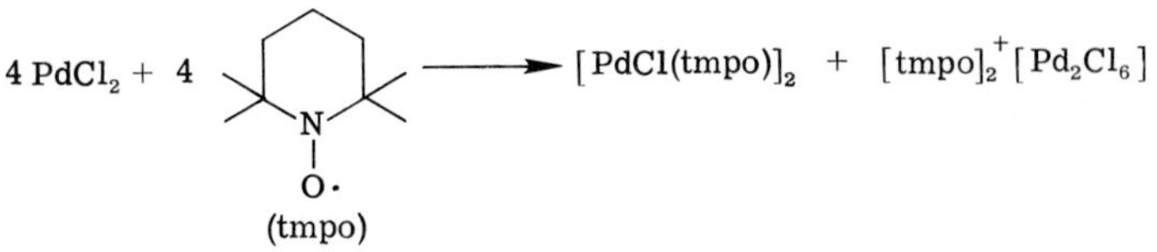

$$4\,PdCl_2 \;+\; 4 \quad \text{(tmpo)} \quad \longrightarrow \quad [PdCl(tmpo)]_2 \;+\; [tmpo]_2^+[Pd_2Cl_6] \tag{18}$$

(tmpo)

Organic nitro compounds give hydroxylamido (-2) complexes when reacted with an appropriate reductant and a low-valent metal complex. Thus, $Fe(CO)_5$ reacts with nitrobenzene upon irradiation with ^{60}Co γ rays according to reaction (19).[99–101]

$$2C_6H_5NO_2 + 2Fe(CO)_5 \rightarrow [(C_6H_5NO)Fe(CO)_3]_2 + 2CO_2 + 2CO \tag{19}$$

Monomeric and dimeric species have been isolated and in one instance a dimeric complex was characterized by X-ray crystallography, establishing the hydroxylamido (-2) nature of the coordinated nitrosobenzene.[32]

Berman and Kochi[120] have investigated the mechanism of reaction (20), in which the material balance requires an oxygen atom transfer from a nitrogen at nitroalkanes or of nitroarenes to a phosphorus center.

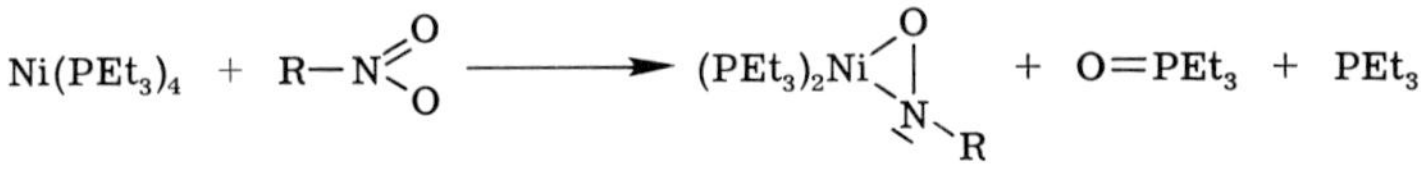

$$Ni(PEt_3)_4 \;+\; R-N\!\!\begin{smallmatrix}O\\\\O\end{smallmatrix} \quad \longrightarrow \quad (PEt_3)_2Ni\!\!\begin{smallmatrix}O\\|\\N\!-\!R\end{smallmatrix} \;+\; O{=}PEt_3 \;+\; PEt_3 \tag{20}$$

A careful kinetic study[120] of reaction (20) has led to a formulation of a mechanism for oxygen transfer that includes a rate-limiting electron transfer process according to reactions (21)–(23).

$$NiL_4 \overset{k_1}{\rightleftharpoons} NiL_3 + L \tag{21}$$

$$NiL_3 + RNO_2 \overset{k_2}{\longrightarrow} [Ni(I)L_3, RNO_2^{-}] \tag{22}$$

$$[NiL_3, RNO_2^{-}] \overset{fast}{\longrightarrow} L_2Ni(ONR) + LO \tag{23}$$

The collapse of the ion pair, which is invoked here, via a cyclic intermediate (A or B) is an attractive route to the elimination of phosphine oxide and the concomitant formation of an η^2-hydroxylamido (-2) complex:

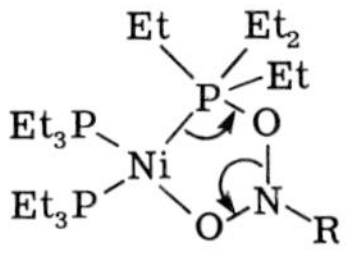

A 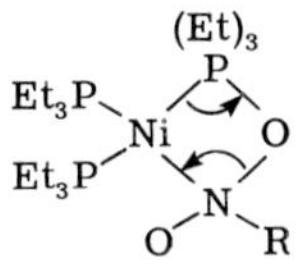   **B**

$Mo(CO)_2(S_2CNEt_2)_2$ has been shown to react with nitrobenzene in CH_2Cl_2, yielding $MoO(ONPh)(S_2CNEt_2)_2$, which can be further reduced to $Mo(ONPh)(S_2CNEt_2)_2$ by triphenylphosphine (generating $Ph_3P=O$).[160] η^2 coordination of the ONPh ligand has been suggested for both complexes from ${}^1H$ NMR evidence. $Mo(ONPh)(S_2CNEt_2)_2$ can also be prepared by the displacement of two carbon monoxide ligands in $Mo(CO)_2(S_2CNEt_2)_2$ and oxidative addition of nitrosobenzene.[160]

4. Insertion of nitric oxide into metal–carbon bonds

Insertion of nitric oxide into the metal–carbon bond of transition metal alkyls can lead to η^2- or bridged hydroxylamido (-2) complexes (**1a, 1b**) provided that the starting transition metal alkyl is paramagnetic[89,102] by one electron.

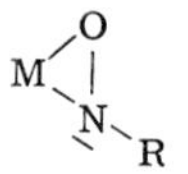

1a 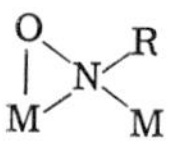　　　　**1b**

Complexes of type **1a** or **1b** have been referred to as nitroalkane[102] or metallaoxaziridines,[13] but, as stated earlier, the N–O distances in these complexes are as expected for single bonds, and the designation as η^2- hydroxylamido (-2) ligands is also possible.

The paramagnetic complex $[Co(CH_3)_2[P(CH_3)_3]_3]$ (one unpaired electron; d^7) reacts with nitric oxide to form an unstable diamagnetic nitrosyl complex, which then forms the (probably) dimeric complex **2a**.[109]

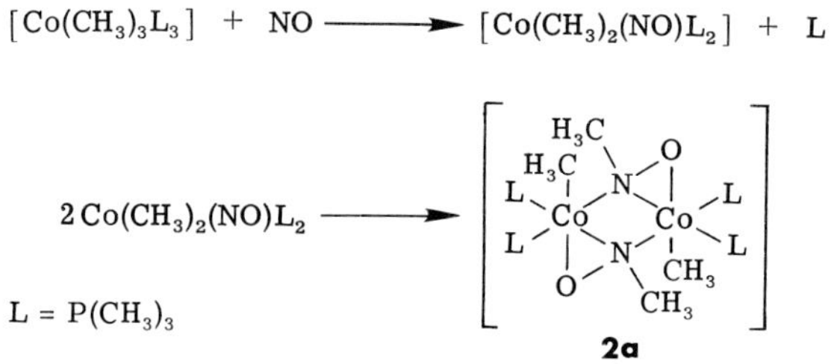

$$[Co(CH_3)_3L_3] \; + \; NO \longrightarrow [Co(CH_3)_2(NO)L_2] \; + \; L$$

$$2\,Co(CH_3)_2(NO)L_2 \longrightarrow$$

$L = P(CH_3)_3$

2a (24)

The reaction of $ReO(CH_3)_4$ with NO was found to give *cis*-trimethyl-dioxorhenium(VII) and azomethane. An unstable, diamagnetic intermediate with structure **1a** has been proposed.[161]

In the reaction of paramagnetic $[(CH_3)_2Nb(Cp)_2]$ (Cp, cyclopentadienyl) with NO at low temperatures two intermediates have been identified spectroscopically: a nitrosyl complex, and its isomer, a complex containing

$\eta^2 N$-methylhydroxylamido (-2), which in turn decomposes via nitrene transfer (dimerization) to yield azomethane and an oxomethyl complex.[102]

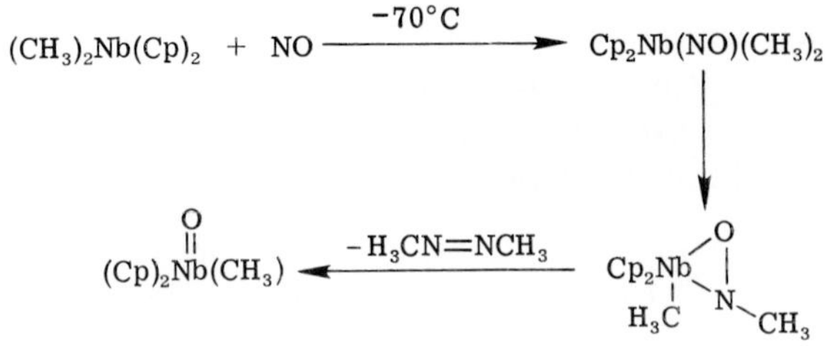

$$(CH_3)_2Nb(Cp)_2 \ + \ NO \ \xrightarrow{-70°C} \ Cp_2Nb(NO)(CH_3)_2 \tag{25}$$

The diamagnetic triply metal–metal-bonded alkyls of molybdenum or tungsten, M_2R_6 ($R = CH_2SiMe_3$, CH_2CMe_3), also react with NO to give diamagnetic complexes $M_2O_3(NO)_2(ONR)_2$, which in solution are most probably tetrameric species with bridging (**1b**) RNO^{2-} ligands. [89]

As has been suggested by reaction sequences (24) and (25), certain nitrosyl complexes are susceptible to nucleophilic attack by carbanions forming new N–C bonds. Another example may be the reaction of $Ni(NO)(PPh_3)_2Br$ with phenyllithium, yielding $[Ni(NO)(PPh_3)_2Ph]$ and lithium bromide. The latter decomposes in solution at temperatures above $-30°C$, but the intermediate $[Ni(PhNO)(PPh_3)_2]$ containing an η^2-coordinated $PhNO^{2-}$ ligand may exist.[162] In these reactions attack at nitrogen may be a concerted intramolecular process, although one report was published in which this reaction occured intermolecularly. Müller *et al.*[83] reported that the dimeric complex $Cp_2Cr_2(NO)_4$, which contains two bridging and two terminal nitrosyl groups, reacts with *t*-butyllithium to give, after mild hydrolysis, a complex that was assigned the following structure:

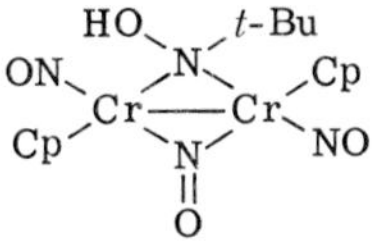

The reported spectroscopic data would also be in agreement with a structure containing an O,N-bridging $H(CH_3)N-O^-$ ligand, which is more in line with the known structural chemistry of coordinated hydroxylamines.

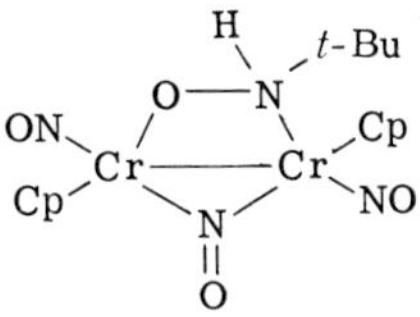

The synthetic potential of η^2-hydroxylamido (-2) complexes has not yet fully exploited, e.g., reactions at such ligands with electrophiles (CH_3^+ or protons) should give N-alkylated or N-protonated complexes with hydroxylamido (-1) ligands. UV/visible spectroscopic evidence for protonation–deprotonation equilibria has been obtained in aqueous solutions of molybdenum(II) hydroxylamido (-1) complexes.[163]

$$
\left[\begin{array}{c} O \\ \| \\ N \\ | \\ LMo \diagdown \diagup O \\ \diagdown N \diagup CH_3 \\ H \end{array}\right]^{2+} \rightleftharpoons \left[\begin{array}{c} O \\ \| \\ N \\ | \\ LMo \diagdown \diagup O \\ \diagdown N \diagdown CH_3 \end{array}\right]^{+} + H^{+} \tag{26}
$$

5. Miscellaneous reactions

Two important reactions have been reported involving the conversion of electron-rich nitrosyl complexes to hydroxylamine complexes via oxidative addition of excess hydrogen chloride or bromide.[113,158,164]

$$
Ir(NO)(PPh_3)_3 + 3HX \xrightarrow[X = Cl^-, Br^-]{} IrX_3(NH_2OH)(PPh_3)_2 + PPh_3 \tag{27}
$$

$$
Rh(NO)(PPh_3)_3 + 3HCl \longrightarrow RhCl_3(NH_2OH)(PPh_3)_2 + PPh_3 \tag{28}
$$

Interestingly, different products are obtained when the starting complexes are reacted with stoichiometric amounts of the hydrogen halide.

$$
M(NO)(PPh_3)_3 + 2HX \xrightarrow[M = Ir, Rh]{} MCl_2(NO)(PPh_3)_2 + PPh_3 + H_2 \tag{29}
$$

Hydride complexes $[IrH(NO)(PPh_3)_3]X$ and $IrHX(NO)(PPh_3)_2$ are intermediates that have been characterized. It is noteworthy that in these reactions no coordinated nitroxyl species have been identified, in contrast to the reaction of $OsCl(CO)(NO)(PPh_3)_2$ with HCl, which affords a nitrosyl derivative $OsCl_2(CO)(HNO)(PPh_3)_2$.[29,164] Similarly, $[Co(NO)(das)_2Br]^+$ reacts with protons in methanol to form $[Co(das)_2(HNO)Br]^{2+}$.[165] $Os(NO)_2(PPh_3)_2$ is thought to react with two equivalents of HCl to give $OsCl_2(NO)(HNOH)(PPh_3)_2$.[164]

The acid-catalyzed decomposition of $[Ir(NH_3)_5N_3]^{2+}$ produces two pentammine iridium(III) complexes, most probably via a common Ir(III)-nitrene intermediate.[114]

$$
[Ir(NH_3)_5N_3]^{2+} \xrightarrow{H^+} \{Ir(NH_3)_5N\}^{2+} + N_2
$$
$$
+HCl \diagdown \qquad \diagup +{}^+OH_3 \tag{30}
$$
$$
[Ir(NH_3)_5NH_2Cl]^{3+} \qquad [Ir(NH_3)_5(NH_2OH)]^{3+}
$$

Titration of $[Ir(NH_3)_5(NH_2OH)]^{3+}$ showed the presence of an ionizable proton ($pK_a = 9.4$), which is characteristic for O deprotonation of N-coordinated hydroxylamine (see Section VI).

The reduction of the pentacyanonitrosylferrate (-2) anion, $[Fe(CN)_5-NO]^{2-}$, in aqueous solution at pH 8–11 using dithionite, tetrahydroborate, or electrochemical methods yields the rather unstable $[Fe(CN)_5(H_2NOH)]^{3-}$ complex. The reduction of the nitroprusside ion has been studied by a number of groups.[166–168] An attempt has been made to study the mechanism of the reaction of H_2NOH with $[Fe(CN)_5NO]^{2-}$, which gives N_2O and $[Fe(CN)_5H_2O]^{3-}$.[168]

V. REDOX REACTIONS OF HYDROXYLAMINE WITH TRANSITION METALS

The diverse structural chemistry of coordinated hydroxylamine is complemented by a very complicated redox chemistry of this ligand with transition metals in solution. Depending on the metal, pH, additional ligands present, etc., hydroxylamine may behave as a strong reducing agent or as an oxidizing agent. The oxidation products of hydroxylamine may be N_2, N_2O, NO, NO_2, NO_2^-, or NO_3^-, excluding reactive intermediates such as nitroxyl or NO^-. On the other hand, the reduction product of H_2NOH appears to be solely ammonia.

A further complication is the generation of nitrosyl complexes from the reaction of hydroxylamine with transition metals, i.e., the nitrosylation reaction of hydroxylamine. Although this reaction has been known for a long time, the exploitation of its synthetic potential and the elucidation of its mechanism have only recently received attention. Caulton, in his review on synthetic methods in transition metal nitrosyl chemistry, commented in 1975 on the potential for further studies concerned with the mechanism of this particular reaction,[188] and since then a number of studies have appeared (see Section V,C).

A. Reduction of hydroxylamine by transition metal complexes

Since the early investigations of Kurtenacker *et al.*[169] on the reactions of H_2NOH with transition metals in aqueous solution, many kinetic studies have been conducted using reductants such as the aquo ions of Ti(III), V(II), V(III), V(IV), Cr(II), Cu(I), and U(III).

Relatively detailed studies have been reported on the reaction of titanium-(III) in acidic aqueous solution in the absence and presence of chelating ligands.[170–183] In the absence of excess chelating ligands such as oxalate or edta, the stoichiometry of the reaction is formulated as in reactions (31) and (32).

$$NH_2OH + Ti_{aq}^{3+} \xrightarrow{k_1} NH_2^{\cdot} + OH^- + Ti(IV) \tag{31}$$

$$\cdot NH_2 + Ti_{aq}^{3+} + H^+ \xrightarrow{k_2} NH_3 + Ti(IV) \tag{32}$$

Thus two equivalents of Ti^{3+} are consumed by one equivalent of H_2NOH ($k_2 \gg k_1$). In the presence of effective scavengers of the $\cdot NH_2$ radical, the stoichiometry changes to 1:1. Scavengers such as benzene, methyl methacrylate, toluene,[170] oxalic acid,[173–180] and other organic molecules[184,185] have been used to unambiguously establish the formation of $\dot{N}H_2$ or (in more acidic solutions) of $\dot{N}H_3^+$ radicals.

The reaction of Ti(III) with hydroxylamine in 0.2 M oxalic acid is obviously a favorite reaction of many electrochemists, who have studied it with every conceivable electrochemical technique. Some selected references are cited on this subject.[171–180] A second-order rate constant of 42 M^{-1} s^{-1} at 25°C for the oxidation of Ti(III) by H_2NOH measured in 0.2 M oxalic acid is a value obtained from kinetic as well as from electrochemical measurements. It is not at all clear what the actual reacting species are since Ti(III) strongly coordinates oxalate ligands.

Tomat and Rigo[172] suggested that H_2NOH (not NH_3OH^+) is the oxidant in the pH range 0–7 when Ti_{aq}^{3+} reacts with hydroxylamine in the absence of nonchelating ligands (although they used a buffer of 0.1 M H_2SO_4 and 0.1 M CH_3COOH), k_1 being the rate-determining step (1.5 × 10⁶ M^{-1} s^{-1} at 0.5°C). This interpretation does not take into account that $[Ti(H_2O)_6]^{3+}$ dissociates to $[Ti(H_2O)_5OH]^{2+}$ with a pK_a of ~3 and that the hydroxo species is known to be a much stronger reductant than the aquo ion of Ti(III). Therefore Eq. (33) may be an oversimplification of the real situation. The rate of reaction is retarded by a factor of 10^4 when Ti(edta)⁻ is the reactant.

The rate law is given by

$$v = k_1 K_a (K_a + [H^+])^{-1} [H_2NOH]_t [Ti(III)] \tag{33}$$

where K_a is the dissociation constant of NH_3OH^+ (0.3 × 10⁻⁶ M at 0.5°C).

Vanadium(II), (III), and (IV) also reduce hydroxylamine to ammonia in aqueous solution,[186–192] and in all reactions the amide radical has been detected as a reactive intermediate.[192] A detailed kinetic and polarographic study of the oxidation of $[V(OH_2)_6]^{3+}$ and $[V(III)deta]^-$ (deta, 1,6-diamino-hexane-N,N,N',N'-tetraacetic acid) has again shown H_2NOH to be the oxidant.[190,191] The stoichiometry corresponds to simple two-equivalent reduction, reaction (34).

$$2V(III) + H_2O + H_2NOH \rightarrow 2VO^{2+} + NH_3 + 2H^+ \tag{34}$$

$$NH_2OH + V(III) \xrightarrow{k_1} \cdot NH_2 + OH^- + V(IV) \tag{35}$$

$$\cdot NH_2 + V(III) + H^+ \xrightarrow{k_2} NH_3 + V(IV) \tag{36}$$

In the absence of scavengers for $\dot{N}H_2$, $k_1 \ll k_2$; $k_1 = 1.8 \times 10^3 M^{-1} \sec^{-1}$ (25°C) for $[V(OH_2)_6]^{3+}$ and $1.3 \times 10^3\ M^{-1}\ s^{-1}$ (25°C) for $[V(III)deta]^-$. It has been suggested that the substitution of NH_2OH into the coordination sphere of V(III) prior to electron transfer is the rate-determining step in both cases, although definitive evidence is lacking.

The mechanism of the reduction of H_3NOH^+ by $[Cr(OH_2)_6]^{2+}$ has been established by kinetic and isotopic labeling experiments,[194–196] and represents the best-understood example of reduction of H_2NOH using a very strong transition metal reductant. The overall reaction is represented by reaction (37).

$$2H^+ + 2Cr(H_2O)_6^{2+} + NH_3OH^+ \rightarrow NH_4^+ + 2[Cr(OH)_2]_6^{3+} + H_2O \tag{37}$$

In the acidic range ($[H^+] = 0.1–0.4\ M$) the rate was independent of $[H^+]$ (Wells and Salam[194] claim pH independence between 0 and 6), and the rate law is given in Eq. (38).

$$v = k_1[Cr^{2+}][NH_3OH^+] \tag{38}$$

where $k_1 = 1.41 \times 10^{-2}\ M^{-1} s^{-1}$ ($\mu = 1.3$; 25°C). In the presence of $Cl^{-[196]}$ or $SO_4^{2-[195]}$ the rate laws are more complicated, indicating the presence of sulfato and chloro complexes of chromium(II). In an experiment using ^{18}O-labeled hydroxylamine the inner-sphere character, i.e., O coordination of H_3NOH^+ prior to electron transfer, was demonstrated.

$$H^+ + NH_3{}^{18}OH^+ + Cr(OH_2)_6^{2+} \xrightarrow{k_1} \dot{N}H_3{}^+ + [(H_2O)_5Cr(^{18}OH_2)]^{3+} + H_2O \tag{39}$$

$$[Cr(OH_2)_6]^{2+} + \dot{N}H_3{}^+ \xrightarrow{fast} [Cr(H_2O)_5OH]^{2+} + NH_4{}^+ \tag{40}$$

Initiation of the polymerization of added methyl methacrylate again demonstrates the formation of radicals during the reaction.

The reaction of $\dot{N}H_3$ with Cr(II) is believed to occur via hydrogen atom abstraction rather than by outer-sphere electron transfer. The reaction of $[Cr(CN)_6]^{4-}$ and NH_2OH in alkaline solution could correspond to a genuine outer-sphere reduction since $[Cr(CN)_6]^{3-}$, NH_3, and water are produced.[193]

Haight *et al.*[197] investigated the reduction of H_3NOH^+ by the molybdenum(V) dimer in 3–12 *M* hydrochloric acid. Ammonia and molybdenum-(VI) are the products. Cleavage of the O–N bond with assistance from two protons to give an activated complex $Mo(V)-ONH_3$ is envisaged in the rate-determining step. Rate-determining oxo-bridge cleavage of the $Mo(V)_2$ dimer could equally well explain the experimental facts.

Halpern *et al.*[120] suggested a mechanism for the oxidation of $Co(CN)_5{}^{3-}$ by H_2NOH as is shown in reactions (41)–(43):

$$Co(CN)_5{}^{3-} + HONH_2 \xrightarrow{k} Co(CN)_5OH^{3-} + \dot{N}H_2 \qquad (41)$$

$$Co(CN)_5{}^{3-} + \dot{N}H_2 \xrightarrow{fast} Co(CN)_5NH_2{}^{3-} \qquad (42)$$

$$Co(CN)_5NH_2{}^{3-} + H_2O \rightleftharpoons Co(CN)_5NH_3{}^{2-} + OH^- \qquad (43)$$

The rate law, Eq. (44), is in agreement with reaction (41) being rate determining.

$$d[Co(CN)_5{}^{3-}]/dt = 2k[Co(CN)_5{}^{3-}][H_2NOH] \qquad (44)$$

where $k = 5.3 \times 10^{-3}\ M^{-1}\ s^{-1}$ (25°C, $\mu = 0.2\ M$).

The reduction of hydroxylamine by Cu^+ (in 0.5 *M* H_2SO_4) is first order with respect to both reagents, Cu^{2+} and ammonia (2:1) being the products.[202] $\dot{N}H_2$ radicals were detected as intermediates.

$$Cu^+ + NH_2OH \xrightarrow{k_1} Cu^{2+} + \dot{N}H_2 + OH^- \qquad (45)$$

$$Cu^+ + \dot{N}H_2 + H^+ \xrightarrow{k_2} Cu^{2+} + NH_3 \qquad k_2 \gg k_1 \qquad (46)$$

The pH dependence of the reaction has not been investigated.

Hydroxylamine oxidizes the triethanolamine complex of Fe(II) in alkaline solution.[199] Indirect electrolytic reduction of some derivatives of hydroxylamine has been studied using this iron complex.[198] Again, the reduction involves the unprotonated hydroxylamine, and ammonia and/or amines are the products. Radicals such as $\dot{N}H_2$ are generated during the reaction.[199,200]

Uranium(III) in aqueous hydrochloric or perchloric acid solution reduces hydroxylamine to ammonia while it is oxidized to U(IV).[203]

$$2U^{3+} + NH_2OH + 2H^+ \rightarrow 2U^{4+} + NH_3 + H_2O \qquad (47)$$

The reaction is first order in oxidant and reductant and the rate is strongly dependent on pH. The pH dependence of rates may be fitted to a reaction scheme in which NH_3OH^+ and NH_2OH are kinetically relevant oxidants with biomolecular rate constants k_2 and k_1.

$$v = \left(\frac{k_1 + k_2 K_p[H^+]}{1 + K_p[H^+]}\right)[U^{3+}][H_2NOH]_t \qquad (48)$$

where $K_p = 9.17 \times 10^{-9}$ (protonation constant of H_2NOH), $k_1 = 8.0 \times 10^2 \ M^{-1} s^{-1}$, and $k_2 = 1.2 \times 10^{-2} \ M^{-1} s^{-1}$ (25°C, $\mu = 2.0 \ M$). It is believed that the mechanism is inner sphere and that a $U(III)(NH_2OH)$ intermediate undergoes disproportionation to give amido radicals in the rate-determining step.

B. Oxidation of hydroxylamine by transition metal complexes

The reaction products of oxidation of hydroxylamine by transition metal oxidants are varied and depend, among other parameters, on the ratio of concentrations ([oxidant]:[H_2NOH]), pH, added ligands, and quite frequently the presence or absence of trace amounts of metal catalysts. Although dinitrogen, dinitrogen oxide, nitrite, and nitrate are the most common products, it is established that many reactions with one-electron oxidants generate the H_2NO radical in the rate-determining step. The formation of radicals in some of these systems has been demonstrated by ESR spectroscopy.[219,220] It is then the fate of this reactive intermediate that determines kind and distribution of final products. It may undergo either further oxidation to N_2O (through nitrosyl intermediates) if suitable oxidants are present, or may dimerize to decompose as N_2 and H_2O. If the metal oxidant is in large excess over hydroxylamine, oxidation to nitrite and/or nitrate is observed. The diversity and complexity of these successive reactions lead to the observed large and complex overall stoichiometries. In this section we will consider fairly detailed studies of the oxidation of NH_2OH by aquometal ions, e.g., Ce(IV), V(V), Mn(III), Fe(III), Co(III), Ag(I), and Ag(II), and by oxo ions such as VO_4^{3-}, CrO_4^{2-}, and MnO_4^{-}.

The oxidation of H_2NOH by cerium(IV) in strongly acidic or alkaline carbonate media has been studied extensively.[204–206,219,220] The experimental rate law [Eq. (49)] for the reaction in 0.5–5 M H_2SO_4 is consistent with the mechanism, reactions (50)–(55).[206]

$$-d[Ce(IV)]/dt = k[Ce(IV)][H_3NOH^+][H^+]/[HSO_4^-]^2 \qquad (49)$$

$$Ce(SO_4)^{2+} + NH_3OH^+ \rightleftharpoons Ce(SO_4)(NH_2OH)^{2+} + H_3O^+ \qquad (50)$$

$$Ce(SO_4)(NH_2OH)^{2+} \rightarrow CeSO_4^+ + NH_2O\cdot + H_3O^+ \qquad (51)$$

$$NH_2O \cdot + Ce^{4+} \rightarrow Ce^{3+} + HNO + H_3O^+ \qquad (52)$$

$$2HNO \rightarrow N_2O + H_2O \qquad (53)$$

$$HNO + 4Ce^{4+} + 2H_2O \rightarrow NO_3^- + 4Ce^{3+} + 5H_3O^+ \qquad (54)$$

The oxidation of hydroxylamine to nitrate in successive steps, with the intermediacy of NH_2O and nitroxyl radicals, has been suggested to occur with a preequilibrium as in reaction (50) and a rate-determining step as in reaction (51).

The overall stoichiometry at ratios $[Ce(IV)]:[H_2NOH] = 4:1$ is as in reaction (55):

$$16Ce(SO_4)_2 + 4NH_3OH^+ + 3H_2O \rightarrow 8Ce_2(SO_4)_3 + N_2O + 2HNO_3 + 8H_2SO_4 + 4H^+ \quad (55)$$

It is noted that there is no direct evidence for the formation of a hydroxylamine cerium(IV) complex, i.e., the inner-sphere electron transfer mechanism of reaction (51) is not proved. An outer-sphere oxidation is also consistent with the experimental data.

From spectrophotometric measurements the formation of a 1:1 complex between V(V) and H_2NOH has been established (stability constant $\beta = 12.5 \pm 0.4\ M^{-1}$) in perchloric acid media ($0.005-1.0\ M$).[208,209] The two-term rate law [Eq. (56)] for oxidation suggests two distinct inner-sphere reaction pathways, the second (k') possibly involving a bridged binuclear activated complex.

$$-d[V(V)]/dt = k[H_2NOH][V(V)] + k'[NH_2OH][V(V)]^2 \qquad (56)$$

$$V(V) + H_2NOH \overset{\beta}{\rightleftharpoons} V(V)(NH_2OH) \qquad (57)$$

$$V(NH_2OH) \overset{k}{\rightarrow} V(IV) + NH_2O + H^+ \qquad (58)$$

$$V(V) + V(\dot{N}H_2OH) \overset{k'}{\rightarrow} 2V(IV) + HNO \qquad (59)$$

The products under these conditions are nitrogen and $[V(IV)O(OH_2)_5]^{2+}$. In more acidic solutions dinitrogenoxide and nitrite have also been identified.[207] Nazer and Wells[209] showed that preequilibria between VO_2^+(aq), $VO(OH)^{2+}$ (aq), and NH_3OH^+ are probably effective at high $[H^+]$ ($1-5\ M$ $HClO_4$). As was noted by Bengtsson, with decreasing $[H^+]$ the reaction became more complicated and a "side reaction" of unspecified character was assumed. It is interesting to speculate on the nature of this reaction considering that in alkaline solutions of V(V) and H_2NOH, vanadium nitrosyl complexes are formed via an inner-sphere two-electron transfer reaction (see Section V,C).

The aquo ion of manganese(III) is also a strong oxidant in perchloric acid media. Phosphato complexes of Mn(III) are generated by this cation in phosphoric acid media. They are capable of oxidizing H_2NOH. Depending on the pH, N_2, N_2O, and NO_3^- are the products.[210] The stoichiometry and the kinetics of the reaction between $[Mn(OH_2)_6]^{3+}$ and hydroxylamine and O-methylhydroxylamine were measured by Davies and Kustin[211] in perchlorate media (0.5–3.7 M). The product distribution is a function of the ratio $R = [Mn(III)]/[NH_2OH^+]$; with $[Mn(III)]$ in excess over $[NH_2OH]$ (17 to 4) the stoichiometry is as in reactions (60) and (61).

$$6Mn(III) + NH_3OH^+ + 2H_2O \rightarrow 6Mn(II) + NO_3^- + 8H^+ \tag{60}$$

$$2Mn(III) + 2NH_3OCH_3 \rightarrow 2Mn(II) + N_2H_3(OCH_3)_2^+ + 3H^+ \tag{61}$$

The experimental rate law [Eq. (62)] for reaction (60) is in agreement with the proposed mechanism of reactions (63)–(65).

$$\text{rate} = \left(\frac{k_1[H^+] + k_1'K_H}{[H^+] + K_H}\right)[Mn(III)][NH_3OH^+] \tag{62}$$

$$Mn^{3+} + NH_3OH^+ \xrightarrow{k_1} Mn^{2+} + NH_2O\cdot + 2H^+ \tag{63}$$

$$MnOH^{2+} + NH_3OH^+ \xrightarrow{k_1'} Mn^{2+} + NH_2O\cdot + H_3O^+ \tag{64}$$

$$2H_2O + 5Mn(III) + NH_2O\cdot \xrightarrow{\text{fast}} 5Mn(II) + NO_3^- + 6H^+ \tag{65}$$

Both $[Mn(OH_2)_6]^{3+}$ and its deprotonated form $[Mn(OH_2)_5OH]^{2+}$ ($K_H = 0.93$) oxidize NH_3OH^+ in the rate-determining step to yield the radical NH_2O. No direct evidence for complex formation between reactants has been found, but cannot be ruled out.

The reaction between *trans*-cyclohexane-1,2-diamine-*N,N,N,N'*-tetraacetatomanganese(III) and hydroxylamine displays an unexpectedly complicated dependence on reductant and hydrogen ion concentrations.[212] An inner-sphere mechanism has been proposed, with the second-order dependence on reductant suggesting an activated complex containing two coordinated hydroxylamine ligands.

A reaction sequence very similar to reactions (63)–(65) has been proposed for the hydroxylamine oxidation by Ag(II) aquo ions[218] in acidic perchlorate media, whereas the oxidation with the Co(III) aquo ion in acidic perchlorate media is a paradigm for all the complications that are encountered in the redox chemistry of hydroxylamine.[216,217] The stoichiometry is again a function of the ratio $R = [Co(III)]/[H_3NOH^+]$. Excess cobalt(III) yields nitrate,

$$6Co(III) + NH_2OH + 2H_2O \rightarrow 6Co(II) + NO_3^- + 7H^+ \tag{66}$$

whereas in excess hydroxylamine ($>$fourfold) nitrogen is evolved.

$$Co(III) + NH_2OH \rightarrow Co(II) + \tfrac{1}{2}N_2 + H_2O + H^+ \tag{67}$$

A further complication is the catalysis of this reaction by cobalt(II), which may be due to the formation of an active Co(II), Co(III) dimer, which has also been postulated by Nazer and Wells.[209] In addition, the reaction sequence analogous to that of reactions (63)–(65) is operating. In contrast, the reduction of $[Co(III)(edtaH)H_2O]$ and $[Co(III)(edtaH_2)(H_2O)_2]^+$ by H_3NOH^+ in acidic media is relatively straightforward. Simple inner-sphere reductions are proposed, although no product analyses have been reported nor has the stoichiometry been established.[224]

The aquo ion of copper(II) also oxidizes hydroxylamine, but it is a much less strong oxidant compared to Ce(IV), Mn(III), and Co(III). The stoichiometry of the reaction is as in reaction (68).

$$4Cu(II) + 2NH_2OH \rightarrow 4Cu(I) + N_2O + H_2O + 4H^+ \tag{68}$$

The Cu(I) generated is rapidly complexed by addition of a substituted phenanthroline, bathocuproine disulfonate.[221] Rapid formation of a 1:1 complex of Cu(II) and NH_2OH is proposed with the rate-determining formation of $NH_2O\cdot$ radicals. Stability constants (M^{-1}) for 1:1 formation of hydroxylamine complexes of Mn(II) (3.2), Co(II) (8.0), Ni(II) (32), Cu(II) (250), Zn(II) (3.2),[223] and Ag(I) (80) have been measured in 0.5 M NaNO$_3$ at 20°C.[222] The value for Cu(II) does not contradict the proposed mechanism for an inner-sphere oxidation of hydroxylamine.

Many oxidations of hydroxylamine by metal complexes or nonmetals are catalyzed by transition metals in homogeneous solutions. The catalyzed oxidations of hydroxylamine by hydrogen peroxide[226] or peroxodisulfate[227] are strongly catalyzed by copper(II). The autoxidation of hydroxylamine was found to be catalyzed by cobalt(II) tetrasulfophthalocyanine[228] and by copper(II).[24] Cobalt(II) catalyzes the reaction of cobalt(III) with hydroxylamine.[216,217] The reduction of silver(I) salts by hydroxylamine is autocatalytic, generating nitrogen and the heterogeneous catalyst metallic silver.[229]

The reaction between iron(III) aquo ions and hydroxylamine is used for the quantitative determination of hydroxylamine.[230] This reaction is also catalyzed by copper(II) aquo ions.[213] Bengtsson[213] has studied the catalyzed and uncatalyzed reaction. An extremely complicated kinetic behavior has not allowed the elucidation of a mechanism. In contrast, the oxidation of hydroxylamine by $[Fe(CN)_6]^{3-}$ (pH 4.2–5.7) is much more straightforward.[214,215,225]

$$2NH_2OH + 2[Fe(CN)_6]^{3-} \rightarrow N_2 + 2[Fe(CN)_6]^{4-} + 2H_3O^+ \tag{69}$$

The rate law [Eq. (70)] is in agreement with the proposed mechanism of reactions (71)–(73), reaction (73) being the rate-determining step, with outer-sphere generation of $NH_2O\cdot$ radicals[214]

$$-d[Fe(CN)_6{}^{3-}]/dt = Kk_1[NH_3OH^+][Fe(CN)_6{}^{3-}][H^+]^{-1} \quad (70)$$

$$NH_3OH^+ \overset{K}{\rightleftharpoons} NH_2OH + H^+ \quad (71)$$

$$NH_2OH + [Fe(CN)_6]^{3-} \overset{k_1}{\longrightarrow} NH_2O\cdot + [Fe(CN)_6]^{4-} + H^+ \quad (72)$$

$$2NH_2O\cdot \overset{fast}{\longrightarrow} N_2 + 2H_2O \quad (73)$$

It should be noted that in these reactions unprotonated hydroxylamine is the reductant and $[Fe(CN)_6]^{3-}$ is a genuine one-electron outer-sphere oxidant, although a weak one. Reaction (69) is efficiently catalyzed by copper(II) aquo ions and by edta complexes of iron. The 1:1 complexation of Cu(II) and NH_2OH [preequilibrium (74)], followed by the rate-determining oxidation of a coordinated hydroxylamine by the oxidant [reaction (75)], with Cu^{2+} being regenerated as in reactions (76) and (77), is the proposed mechanism.[215]

$$Cu^{2+} + NH_2OH \rightleftharpoons Cu(NH_2OH)^{2+} \quad (74)$$

$$Cu(NH_2OH)^{2+} + [Fe(CN)_6]^{3-} + H_2O \overset{k}{\longrightarrow} Cu(NH_2O)^{2+} + [Fe(CN)_6]^{4-} + H_3O^+ \quad (75)$$

$$Cu(NH_2O)^{2+} + H_2NOH \overset{fast}{\longrightarrow} Cu^+ + N_2 + H_3O^+ + H_2O \quad (76)$$

$$Cu^+ + [Fe(CN)_6]^{3-} \overset{fast}{\longrightarrow} Cu^{2+} + [Fe(CN)_6]^{4-} \quad (77)$$

This is an attractive scheme and it suggests that coordinated hydroxylamine is more easily oxidized than are uncoordinated ligands, and indeed an inner-sphere pathway must be applicable in this case because the thermodynamic and kinetic aspects of reaction (74) are already known.[222] It is believed that this scheme may be valid for all copper(II)-catalyzed oxidations of hydroxylamine. When the Fe(III)–edta complex is the catalyst instead of Cu(II), a kinetic situation arises in which the formation of an [Fe(III)(edta)-$(NH_2OH)]^-$ complex is rate limiting, i.e., the substitution of H_2NOH into the coordination sphere of Fe(III) is rate-determining and subsequent steps are rapid.[215]

The redox behavior of coordinated hydroxylamine toward one-electron oxidants (or reductants) has not yet been studied in great detail.[236,237] Kinetic and thermodynamic parameters of such reactions would be of great interest in comparison with data for reactions involving uncoordinated hydroxylamine. This is a potentially very fruitful area for future research.

The oxidations of hydroxylamine with oxo anions of transition metals in high oxidation states, e.g., CrO_4^{2-} (or $Cr_2O_7^{2-}$) or MnO_4^{-}, in acidic aqueous solutions have been investigated primarily because of the analytical applications, i.e., the redox titration of H_2NOH.[230] The varying stoichiometry of these reactions has not allowed this. This is illustrated by reactions (78), (79) (obtained in acidic solution), and (80) (neutral and alkaline media),[230,235] which represent overall stoichiometries.

$$5NH_3OH^+ + 4MnO_4^- + 2H^+ \rightarrow 5NO_2^- + 4Mn^{2+} + 11H_2O \qquad (78)$$

$$5NO_2^- + 2MnO_4^- + 6H^+ \rightarrow 5NO_3^- + 2Mn^{2+} + 3H_2O \qquad (79)$$

$$10NH_2OH + 2MnO_4^- + 6H^+ \rightarrow 5N_2 + 2Mn^{2+} + 18H_2O \qquad (80)$$

Since reaction (79) is slow, reaction (81) also competes, to give N_2O.

$$H_3NOH^+ + NO_2^- \rightarrow N_2O + 2H_2O \qquad (81)$$

Dichromate gives N_2O in acidic solutions when hydroxylamine is in excess over $Cr_2O_7^{2-}$ according to reaction (82).[231]

$$2Cr_2O_7^{2-} + 6NH_3OH + 10H^+ \rightarrow 4Cr^{3+} + 3N_2O + 17H_2O \qquad (82)$$

Nitrite is produced when $[Cr_2O_7]^{2-}$ is in large excess over hydroxylamine, reaction (83).

$$2Cr_2O_7^{2-} + 3NH_3OH^+ + 10H^+ \rightarrow 4Cr^{3+} + 3NO_2^- + 11H_2O \qquad (83)$$

Two kinetic studies have been carried out on the reaction of H_3NOH^+ with CrO_4^{2-} in the pH range 0–4.7.[232,234] The rate law [Eq. (84)] of both studies is in agreement after allowing for the fact that Senent et al.[234] did not use acetate buffers[232] ($[HClO_4] = 5 \times 10^{-3}$–$5 \times 10^{-5}$ M).

$$-d[Cr(VI)]/dt = (a[H^+] + b + c[H^+]^{-1})[NH_3OH][Cr(VI)] \quad (84)$$

In acetate buffers using a wider $[H^+]$ range, a more complicated rate law is derived.[232] The two studies propose different mechanisms to account for the experimental observations. Haight et al.[232] invoke the formation of an O-bonded chromate ester that undergoes internal two-electron oxidation and reduction (involving acid catalysis). A chromium(IV) species [which disproportionates yielding Cr(III) and chromium(V), which is then rapidly reduced by excess hydroxylamine] and nitroxyl are proposed as intermediates.

$$H_2NOH + CrO_4^{2-} \rightleftharpoons H_2NOCrO_3^- \qquad (85)$$

$$H_3NOCrO_3^- + H^+ \rightarrow Cr(IV) + HNO \qquad (86)$$

$$H_3NOCrO_3 \rightarrow Cr(IV) + HNO \qquad (87)$$

$$2Cr(IV) \rightarrow Cr(V) + Cr(III) \qquad (88)$$

$$Cr(V) + H_3NOH^+ \rightarrow Cr(III) + HNO + 3H^+ \qquad (89)$$

The rate-determining steps of reactions (86) and (87) involve internal two-electron processes. Senent *et al.*[234] propose a more conventional one-electron process for the decomposition of the intermediate formed between $CrO_4{}^{2-}$ and H_2NOH of reaction (85), involving the formation of the $NH_2O\cdot$ radical and a Cr(V) species. Subsequent reactions lead to the production of N_2O and Cr(III). In the light of recent investigations of the reaction between chromate and hydroxylamine in alkaline media, in which Cr(I) nitrosyl complexes are formed, it is important to note that both of the preceding schemes produce Cr(V) species at one stage. These Cr(V) species are believed to be precursors of the nitrosyl complexes.

The reaction of molybdate(VI) oxo anions has been investigated extensively during the last 150 years. It does not always lead to simple reaction products such as N_2, N_2O, $NO_2{}^-$, or $NO_3{}^-$, but rather to nitrosyl complexes of molybdenum. These reactions will be discussed in the next section.

C. Nitrosylation reactions

1. Cyanonitrosyl complexes

It has been recognized by Hieber and co-workers[5,238] that hydroxylamine in alkaline solutions reacts with certain transition metal complexes in the presence of cyanide to give cyanonitrosylmetal complexes.

The reaction of tetracyanonickelate(II) with hydroxylamine in basic aqueous solution is a classical and often-cited reaction.

$$[Ni(CN)_4]^{2-} + 2H_2NOH \rightarrow [Ni(CN)_3NO]^{2-} + NH_3 + HCN + H_2O \qquad (90)$$

Hieber and Nast[238] proposed "disproportionation" of hydroxylamine and substitution of NO^- into the first coordination sphere of nickel(II), reactions (91)–(93), as a plausible mechanism.

$$2NH_2OH \rightarrow HNO + NH_3 + H_2O \qquad (91)$$

$$HNO \rightarrow H^+ + NO^- \qquad (92)$$

$$[Ni(CN)_4]^{2-} + NO^- \rightarrow [Ni(CN)_3NO]^{2-} + CN^- \qquad (93)$$

Before discussing this mechanism in more detail, it is appropriate to indicate the general versatility of the nitrosylation reaction. The complexes $[Cr(CN)_5NO]^{3-}$ and $[Mn(CN)_5NO]^{3-}$ have been prepared in exactly the same manner using NH_2OH as a reactant.[239–241]

$$[Cr(CN)_6]^{3-} + 2NH_2OH \rightarrow [Cr(CN)_5NO]^{3-} + NH_3 + HCN + H_2O \qquad (94)$$

$$[Mn(CN)_6]^{3-} + 2NH_2OH \rightarrow [Mn(CN)_5NO]^{3-} + NH_3 + HCN + H_2O \qquad (95)$$

Malatesta and Sacco[242] used the reaction for the preparation of iron and cobalt nitrosyl complexes, reactions (96) and (97).

$$[Fe(CNR)_4X_2] + 4NH_2OH \rightarrow [Fe(NO)_2(CNR)_2] + 2NH_4X + 2H_2O + 2CNR \quad (96)$$

$$[Co(CNR)_5X] + 2NH_2OH \rightarrow [Co(NO)(CNR)_3] + 2CNR + H_2O + NH_4X \quad (97)$$

$$CNR = p\text{-}CH_3\text{-}C_6H_4\text{-}NC, \quad \text{and} \quad X = I^-, Cl^-$$

It is also possible to use oxo complexes of the early transition metals in high oxidation states as starting materials and to react them with hydroxylamine in CN^--containing solutions. The oxygen-sensitive violet anion $[Mo(NO)(CN)_5]^{4-}$ has been prepared from molybdate(VI) and H_2NOH,[243-245] and $[Cr(NO)(CN)_5]^{3-}$ can also be prepared by this method using CrO_3.[244] $[Mo(NO)(CN)_5]^{4-}$ had already been prepared by van der Heide and Hofmann in 1896[245] by this reaction, although they did not recognize the formation of nitrosyl complexes. The crystal structure of $K_4[Mo(NO)(CN)_5]$ was reported in 1968.[248] Ammoniumvanadate(V), hydroxylamine, and cyanide ions in aqueous solutions yield $[V(NO)(CN)_5]^{3-}$,[246] the correct formulation of which was also established by X-ray crystallography.[247] A second cyanonitrosyl complex of vanadium was obtained[249] from the above reaction mixture when H_2S was bubbled through such solutions. The seven-coordinate $[V(NO)(CN)_6]^{4-}$ anion is formed.[250] A third product, $[V(NO)(H_2NO)(CN)_4]^{3-}$, has been isolated as yellow cesium salt[79] from such reaction mixtures. A series of complexes of the type $[Cr(NO)(CN)_x(en)_y]^{n\pm}$ has been prepared from solutions of KCN, CrO_3, and ethylenediamine.[251]

We now return to the important reaction (90) and its mechanism, which has been investigated in detail.[252-259] Since the time Nast *et al.*[255] introduced reaction (93) as a means of detecting the reactive NO^- intermediate, it has not been at all clear whether NO^- was really formed in reaction (90). Vepřek-Šiška and Luňák have shown in two elegant kinetic studies[252,253] that two different mechanisms are operative in the formation of $[Ni(CN)_3\text{-}NO]^{2-}$ from $[Ni(CN)_4]^{2-}$ and H_2NOH, and that most likely neither involves generation of uncoordinated NO^-. Under anaerobic alkaline conditions the reaction is slow and the products are $[Ni(CN)_3NO]^{2-}$ and ammonia; the molar ratio was always less than 1:1, with excess NH_3 produced. Small amounts of dinitrogen were invariably detected. The rate was found to be markedly retarded by added cyanide, which obviously rules out reaction (91) as the rate-determining step.

A mechanism involving the preequilibrium (98) is in agreement with the kinetic data.

$$[Ni(CN)_4]^{2-} + NH_2OH \rightleftharpoons [Ni(CN)_3(NH_2OH)]^- + CN^- \quad (98)$$

$$[Ni(CN)_3(NH_2OH)]^- + NH_2OH + OH^- \rightarrow [Ni(CN)_3NO]^{2-} + NH_3 + 2H_2O \quad (99)$$

Reaction (99) is in fact the slow oxidation of a coordinated hydroxylamine (which under the reaction conditions is probably O deprotonated) by uncoordinated hydroxylamine, which is reduced to ammonia.

Reaction (100) has been proposed to explain the formation of nitrogen with concomitant decomposition of $[Ni(CN)_3NO]^{2-}$.

$$[Ni(CN)_3NO]^{2-} + H_2NOH + CN^- \rightarrow [Ni(CN)_4]^{2-} + N_2 + H_2O + OH^- \quad (100)$$

This has been experimentally verified by Bonner and Akhtar.[254] If a suitable oxidant is present, e.g., dioxygen, the reaction of $[Ni(CN)_4]^{2-}$ with H_2NOH yielding $[Ni(CN)_3NO]^{2-}$ in alkaline solutions is much faster, the stoichiometry changes dramatically, and no ammonia is formed (101).

$$[Ni(CN)_4]^{2-} + H_2NOH + \tfrac{1}{2}O_2 + OH^- \rightarrow [Ni(CN)_3NO]^{2-} + CN^- + 2H_2O \quad (101)$$

A mechanism [reactions (102) and (103)] involving a reactive $[Ni(CN)_4-NO]^{4-}$ species that is obtained from $[Ni(CN)_3(NH_2OH)]^-$ via complete deprotonation (NO^{3-}) is in accord with the kinetic data.[252]

$$[Ni(CN)_4]^{2-} + NH_2OH + 3OH^- \rightarrow [Ni(CN)_3NO]^{4-} + 3H_2O + CN^- \quad (102)$$

$$[Ni(CN)_3NO]^{4-} + \tfrac{1}{2}O_2 + H_2O \rightarrow [Ni(CN)_3NO]^{2-} + 2OH^- \quad (103)$$

This reactive intermediate is then easily oxidized by dioxygen, reaction (103). Thus it appears that reaction (90) cannot be taken as indirect evidence for formation of a nitroxyl intermediate in the disproportionation of H_2NOH; these studies do not, however, rule out the possibility that $[Ni(CN)_4]^{2-}$ is a potential scavenger of NO^- (or HNO). Indeed, Bonner and Akhtar have convincingly shown that this is the case.[254] The decomposition of trioxodinitrate, $Na_2N_2O_3$, is known to yield HNO (or NOH),[286] and in the presence of $[Ni(CN)_4]^{2-}$ the violet $[Ni(CN)_3NO]^{2-}$ is formed. Equally convincing is the evidence for an HNO intermediate in the alkaline hydrolysis of hydroxylamine *N*-sulfonate.

2. The reaction of vanadate(V) with hydroxylamine in the presence of chelating ligands

Hartkamp[258] reported an accurate and selective method for the quantitative determination of vanadium using spectrophotometry. Beneš and Novák subsequently showed that this reaction is equally well suited for quantitative determinations of H_2NOH.[259,260] The stoichiometry of the reaction does not change under aerobic or anaerobic conditions, and no gaseous reaction products were detected (in particular, no ammonia!). Wieghardt *et al.*[68] recognized that ternary nitrosyl complexes are generated and not hydroxylamine complexes of vanadium(V).[258-260]

$$V(V)O_3^- + H_2NOH + \text{chelate} \rightarrow [V(I)(NO)\text{chelate}] + 3H_2O + 3H^+ \qquad (104)$$

An X-ray structure of $Na[V(NO)(tha)] \cdot NaClO_4 \cdot 4H_2O$, which was originally prepared by Hartkamp, is shown in Fig. 7.[261] The tris(hydroxyalkyl)amine is in this instance tris(ethanol)amine, $N(CH_2CH_2OH)_3$. The red anionic nitrosyl complex with trigonal bipyramidal environment of the vanadium center and the chelating tetradentate ligand tris(ethanol)amine exhibits a $v(NO)$ stretching frequency at 1490 cm^{-1}, an extremely low value for a linear $M-N=O$ moiety. It is conceivable that the formation of nitrosyl complexes, as in reaction (104), occurs in a series of similar reactions as reported in analytical chemistry literature, all of which have been incorrectly reported to yield highly colored hydroxylamine complexes of vanadium(V).[262-266] The reaction of pyridine-2,6-dicarboxylic acid with ammonium vanadate(V) and hydroxylamine at pH 5 leads to a colorless O,N-hydroxylamido (-1) complex of vanadium(V), the neutral complex $[VO(dipic)(H_2NO)H_2O]$.[41,42] If the same reaction is carried out in alkaline solutions with excess NH_2OH, the nitrosyl complex $Cs[V(NO)(dipic)(H_2NO)(H_2O)] \cdot 2H_2O$ can be obtained after addition of CsCl.[42] This complex was in fact one of the first complexes in which the sideways-on coordination of H_2NO^- was established by X-ray crystallography (Fig. 6). The stoichiometry of the reaction is as in reaction (105).

$$[VO_2(dipic)]^- + 2H_2NOH \rightarrow [V(NO)(dipic)(H_2NO)(H_2O)]^- + H_2O \qquad (105)$$

Again no ammonia is formed. Interestingly, the same complex is formed from the reaction of octahedral $[V(dipic)(H_2O)_3]^+$, a complex of vanadium(III), with hydroxylamine in alkaline solutions.[42] Using anaerobic conditions

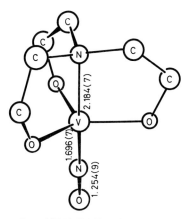

Fig. 7. The structure of the anion of $[V(tha)NO]^-$; tha, tris(ethanol)amine. (From Ref. [261].)

the stoichiometry is as in reaction (106), and then the stoichiometric amounts
of ammonia that are generated are $V(III):NH_3 = 1:1$.

$$[V(III)(dipic)(OH_2)_3]^+ + 3NH_2OH + 2OH^- \rightarrow$$

$$[V(NO)(dipic)(H_2NO)(H_2O)]^- + NH_3 + 5H_2O \quad (106)$$

A series of complexes of vanadium(V) containing one or two O,N-coordi-
nated H_2NO^- ligands, a bidentate or tridentate chelate, and a terminal oxo
group has been prepared from acidic aqueous solutions at pH 6 with hydroxylamine.[68] In alkaline solutions red-violet nitrosyl complexes are
invariably obtained without any gaseous products such as NH_3.

The stoichiometry and the product distribution of reaction (104) changes
also when the starting vanadium is in oxidation state III and not V.

$$[V(OH_2)_6]^{3+} + tha + 2H_2NOH + 4OH^- \rightarrow [V(NO)(tha)]^- + NH_3 + 11H_2O \quad (107)$$

$$tha = 1,1',1''\text{-nitrilo-tris(2-propanol)}$$

Using anaerobic conditions, ammonia is formed, reaction (107), whereas
under aerobic conditions reaction (108) prevails and no ammonia is
generated.

$$[V(OH_2)_6]^{3+} + tha + H_2NOH^+ + \tfrac{1}{2}O_2 \rightarrow [V(NO)(tha)]^- + 11H_2O + 4OH^- \quad (108)$$

Acidification of solutions containing the red nitrosyl complex $[V(NO)\text{-}
(tha)]^-$ yields a colorless complex of vanadium(V), which gives the unchanged
nitrosyl complex upon addition of base.[80]

$$[V(NO)(tha)]^- + 2H^+ \rightleftharpoons [V(tha)(H_2NO)]^+ \quad (109)$$
$$\text{red} \qquad\qquad\qquad \text{colorless}$$

The kinetics of the forward reaction (109) have been measured using the
stopped-flow technique.[80] The rate law

$$-d[V(NO)(tha)]/dt = d[V(tha)(H_2NO)]/dt$$

$$= \left(\frac{k_1 K_p'[H^+]}{1 + K_p[H^+]}\right)[V(NO)tha] \quad (110)$$

where $K_p = 40\ M^{-1}$ (25°C) and $k_1 = 33\ s^{-1}$ (25°C), shows a nonlinear
dependence on $[H^+]$ (0.005–0.75 M) and no dependence on added hydroxylamine (0.05–0.5 M). It has been suggested that K_p is the protonation
constant for the nitrosyl group, yielding an intermediate with N-coordinated
nitroxyl, reaction (111). In the rate-determining step this is converted to the
O,N-coordinated hydroxylamide (-2), reaction (112), which is rapidly
protonated in acidic solution, reaction (113),

$$[(tha)V(I)-N{=}O]^- + H \underset{\xrightarrow{\hspace{1.2cm}}}{\overset{K_p}{\xleftarrow{\hspace{1.2cm}}}} \left[(tha)V(III)-N\underset{H}{\overset{O}{\diagdown}}\right] \quad (111)$$

$$\left[(tha)V(III) - N \underset{H}{\overset{O}{\diagdown}} \right] \xrightarrow{\;k_1\;} \left[(tha)V(V) \underset{\underset{H}{\overset{\diagup}{N}}}{\overset{O}{\diagdown}} \right] \tag{112}$$

$$\left[(tha)V \underset{\underset{N}{\overset{\diagup}{\diagdown}}H}{\overset{O}{\diagdown}} \right] + H^+ \underset{\text{fast}}{\rightleftharpoons} \left[(tha)V \underset{\underset{N}{\overset{\diagup}{\diagdown}}H}{\overset{O}{\diagdown}} \right]^+ \tag{113}$$

yielding a colorless hydroxylamido (-1) complex of vanadium(V). The essence of this proposed mechanism is an intramolecular two-electron process in the rate-determining step. The principle of microscopic reversibility makes the reverse reaction an attractive scheme for the formation of nitrosyl complexes from coordinated hydroxylamine via an intramolecular redox process between nitrogen and the metal centers. This mechanism explains why the stoichiometry and the mechanism changes when vanadium(III) complexes react via reactions (106), (107), or (108) to afford nitrosyl complexes. The production of ammonia is suppressed when suitable oxidants are present, which may be dioxygen or the metal center itself.

3. The reaction of molybdate(VI) with hydroxylamine

The reaction of molybdenum(VI) oxo anions with hydroxylamine in acidic and alkaline aqueous solutions has been investigated many times in the presence and absence of chelating ligands.[245,267−282] Hofmann et al.[245,267] reported a series of compounds referred to as hydroxylamine complexes of molybdenum(VI) (d^0), although their colors varied from yellow to dark red. Subsequent investigations led to formulations such as $[Mo_{11}O_{34}NOH-(H_2O)_2]^{4-}$.[270] The structures of many of these materials remain to be elucidated. The simple question whether molybdenum(VI) is reduced in such solutions has become quite controversial. Yaguchi and Kajiwara reported a reduction to Mo(V) in the presence of edta ligands.[271] Lassner et al.[272,273] proposed an alternative model involving a ternary complex of molybdenum(VI) edta and hydroxylamine rather than the reduction of Mo(VI). A series of spectrophotometric methods for the quantitative analysis of molybdenum has been published involving the reaction of molybdate with hydroxylamine in the presence of chelates.[272−276] In each case the investigators describe their yellow compounds as hydroxylamine-containing complexes of molybdenum(VI).

Wieghardt et al.[43] and Müller et al.[277] discovered independently the nature of these reaction products as diamagnetic nitrosyl complexes of molybdenum(II) of the $\{Mo-NO\}^4$ type according to Enemark and Feltham's nomenclature.[2] Sharpless *et al.* also mention this type of complex.[13]

A variety of such complexes have been isolated from aqueous solution and characterized[30,43,57,82,87] (Table 5). It is remarkable that even in very acidic solutions molybdate(VI) reacts with hydroxylamine to form nitrosyl complexes, e.g., $Cs_2[Mo(NO)Cl_5]$ from hydrochloric acid.[277] In many instances the reaction between molybdate(VI) and hydroxylamine occurred in acidic or in alkaline solution *without* production of gaseous products, and no ammonia was detected as reaction product when yellow solutions were obtained. Since in these nitrosyl complexes the molybdenum center has a formal oxidation number II (counting NO^+ as a two-electron donor), the coordination number 7 (pentagonal bipyramid) is most common, satisfying the effective atomic number rule, but a few species with smaller coordination number are also known. Table 5 summarizes the types of complexes characterized.

Müller *et al.* have demonstrated that very complex, purely inorganic materials may be synthesized using this nitrosylation reaction. Bubbling H_2S through an aqueous alkaline solution of MoO_4^{2-} and hydroxylamine yields the tetrameric anion $[Mo_4(NO)_4(S_2)_6O]^{2-}$. Figure 8 shows its structure.[279] At specified pH and concentrations of MoO_4^{2-} and H_2NOH, an orange-yellow material precipitates, which is a useful starting material for many nitrosyl complexes.[57,272] This material contains nitrosyl and hydroxylamine in the ratio $Mo:NO:H_2NOH = 1:1:1$.[57] Reacting this material with cyanide ions yields yellow $[Mo(NO)(CN)_5]^{2-}$, which may be precipitated as the $[P(C_6H_5)_4]^+$ salt.[57] Careful oxidation of the violet $[Mo(NO)(CN)_5]^{4-}$ afforded the green, paramagnetic $[Mo(NO)(CN)_5]^{3-}$.[58,277]

There are no simple hydroxylamine complexes of molybdenum(VI) known. Only when the reducing capacity of H_2NOH is decreased, using, for

TABLE 5

$\{Mo-NO\}^4$ and $\{Mo(NO)_2\}^6$ complexes prepared from MoO_4^{2-} and H_2NOH

$[Mo(NO)(H_2NO)X_4]^{2-}$	$X^- = NCS^-,$[58] $CN^-,$[82] $N_3^-,$[57] Cl^-[57]
$[Mo(NO)(H_2NO)X_2(L-L)]^0$	$X^- = NCS^-$; (L–L) = bpy, phen[280]
$[Mo(NO)(H_2NO)(L-L)_2]^n$	(L–L) = phen,[44] 2-pic,[44] Et_2dtc[280]
$[Mo(NO)(H_2NO)(L-L-L)Y]^n$	(L–L–L) = dipic; Y = H_2O, NH_2OH[44]
	(L–L–L) = terpy; Y = H_2O, CN^-[44,43]
$[Mo(NO)(H_2NO)_2(L-L)]^+$	(L–L) = bpy,[43] phen[44]
$[Mo(NO)(L-L)_3]^{3-}$	(L–L) = $C_2O_4^{2-}$[278]
$[Mo(NO)X_5]^{2-}$	$X = Cl^-,$[277] $Br^-,$[281] CN^-[57]
$[Mo(NO)X_4]^-$	$X = Br^-$[281]
$[Mo(NO)_2X_4]^{2-}$	$X = NCS^-,$[58,280] CN^-[82]
$[Mo(NO)_2X_2(L-L)]^0$	$X = NCS^-$; (L–L) = phen, bpy[280]

example, N-methylhydroxylamine, are colorless diamagnetic complexes of the type $[MoO_2(R_1R_1NO)_2]$ obtained.[43,45]

Although no kinetic study of the nitrosylation reaction has been carried out, it is tempting to suggest a scheme [reactions (114) and (115)] that is very similar to that proposed for the reaction of VO_3^- with H_2NOH. Again, coordinated hydroxylamine complexes of Mo(VI) are formed in a rapid preequilibrium (114), with a subsequent rate-determining intramolecular two-electron process (115).

$$Mo(VI) + H_2NOH \rightleftharpoons Mo(VI)(H_2NOH) \tag{114}$$

$$Mo(VI)(NH_2OH) \rightarrow (Mo-NO)^{3+} + H_3O^+ \tag{115}$$

It is evident in alkaline solutions that deprotonated hydroxylamide (-1) ligands (possibly O,N coordinated) are more reactive than protonated $(H_2NOH)Mo(VI)$ species in acidic media. The characterization of a complex with O,N-coordinated HNO^{2-}[30,86] and the series of well-characterized η^2-H_2NO^--containing complexes show the feasibility of such a scheme from a structural point of view. A difference from the $\{V-NO\}^4$ complexes is that $\{Mo-NO\}^4$ species are generally not susceptible to electrophilic attack by protons. Therefore, the reverse reaction, i.e., formation of hydroxylamine complexes from nitrosyl complexes of molybdenum, has not yet been achieved. In contrast, Tatsumi has reported a series of complexes, e.g., $[MoF(HNO)(dpe)_2]PF_6$, where (dpe) is $Ph_2PCH_2CH_2PPh_2$.[282] However, their spectroscopic data seem to be more compatible with the formulation as hydrido complexes, $[Mo(H)(F)(NO)(dpe)_2]PF_6$.

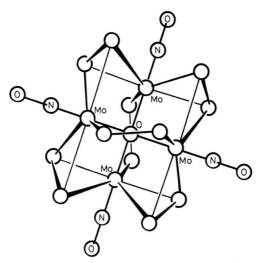

Fig. 8. The structure of the tetrameric dianion of $[Mo_4(NO)_4(S_2)_6O]^{2-}$. (From Ref. [279].)

It has also been possible to synthesize dinitrosyl complexes of molybdenum starting with molybdate(VI) and H_2NOH in aqueous solution,[58,280] e.g. $[Mo(NO)_2(NCS)_4]^{2-}$[58,280] $[Mo(NO)_2(NCS)_2(phen)]$,[280] $[Mo(NO)_2(NCS)_2(bpy)]$,[280] and $[Mo(NO)_2(CN)_4]^{2-}$.[82] Although the mechanisms have not yet been investigated in any detail, it is tempting to speculate in terms of a general scheme [reactions (116)–(118)].

$$(ON)Mo\begin{matrix}O\\|\quad H\\N\\H\end{matrix}^{2+} + OH^- \rightleftharpoons (ON)Mo\begin{matrix}O\\|\quad H\\N\end{matrix}^+ + H_2O \qquad (116)$$

$$(ON)Mo\begin{matrix}O\\|\quad H\\N\end{matrix}^+ + \tfrac{1}{2}O_2 \longrightarrow (ON)Mo(NO)^{2+} + OH^- \qquad (117)$$

$$(ON)Mo\begin{matrix}O\\|\quad H\\N\end{matrix}^+ + NH_2OH \longrightarrow (ON)Mo(NO)^{2+} + NH_3 + OH^- \qquad (118)$$

A nitrosylhydroxylamido (-1) molybdenum moiety is deprotonated in alkaline solution yielding a nitrosylhydroxylamido (-2) species that may be oxidized by dioxygen (aerobic conditions) or by excess hydroxylamine (anaerobic conditions).

It is of course gratifying that the precursors $[Mo(NO)(H_2NO)X_4]^{2-}$ $(X = NCS^-, CN^-)$ are stable and well-characterized complexes, and at least the protonation–deprotonation of reaction (116) has been observed using UV/visible spectrophotometry.[82]

4. The reaction of chromate(VI)
   with hydroxylamine

The reaction of chromate(VI) or dichromate(VI) with hydroxylamine in acidic solution has already been described. It has been known for a long time that CrO_4^{2-} (or CrO_3) reacts with H_2NOH in the presence of coordinating ligands (e.g., CN^-) to form nitrosyl chromium complexes of the type $\{Cr-NO\}^5$. A series of complexes containing ligands other than cyanide have been prepared using the nitrosylation reaction, e.g., $[Cr(NO)(acac)_2X]$ $(X = H_2O, NH_3, py)$,[283] or $[Cr(NO)(NCS)_5]^{3-}$,[277,284] $[Cr(NO)(NCS)_2(L-L)]^1$ $[(L-L) = bpy, phen]$,[284] and $[Cr(NO)(dtc)_2]$ (dtc = N,N-diethyl-

---

[1] It appears that these complexes have also been reported by Maurya, R. C.; Shukla, R. K. *J. Indian Chem. Soc.* **1982**, *59*, 340.

dithiocarbamate).[284] Using a large excess of $[NH_3OH]Cl$ over $CrO_4{}^{2-}$ in slightly acidic media (pH 5), these nitrosoyl complexes are obtained in excellent yields ($\sim 80\%$).[284] Apparently, in all cases gaseous products are evolved, the exact nature of which is not described (it is probably N_2). No investigation has been carried out concerning the mechanism that leads to $\{Cr-NO\}^5$ moieties. Note that two-electron processes as described for the reaction of molybdate(VI) with H_2NOH would afford $\{Cr-NO\}^4$ complexes. These are not observed. It is not clear at what stage of the reaction the necessary one-electron transfer is effected.

It is a plausible mechanism to assume that the actual formation of $\{Cr-NO\}^5$ complexes starts with chromium(V) species that are formed as reactive intermediates during the reduction of $CrO_4{}^{2-}$ by H_2NOH. Both mechanisms proposed for the $CrO_4{}^{2-}$ reduction have invoked Cr(V) at one stage or the other, although Haight's scheme would lead to only 50% yields of nitrosyl complexes because only one-half of Cr(VI) forms Cr(V).

$$LCr(V) + H_2NOH \rightleftharpoons LCr(V)(H_2NOH) \qquad (119)$$

$$LCr(V)(H_2NOH) \xrightarrow{\text{rds}} LCr-NO + 3H^+ \qquad (120)$$

Gould and co-workers[285,286] demonstrated in an in-depth mechanistic study that the above scheme is operative on reacting Roček's first air-stable, water-soluble carboxylato complex[287] of chromium(V) with hydroxylamine.

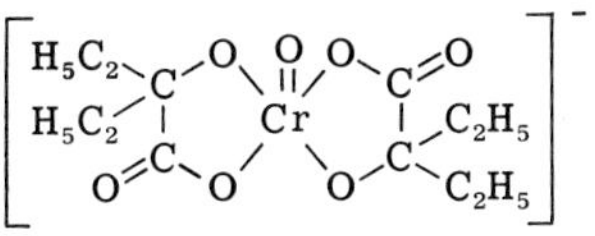

$$\left[\begin{array}{c} H_5C_2\diagdown C\diagdown^O \diagdown^O_{\,\,\parallel}{}^O\diagdown_{C}\diagup^O \\ H_5C_2\diagup C \,|\,\,\,\, Cr \,\,\,\, | \diagdown_{C_2H_5} \\ O\diagup^C\diagdown_O\diagup \diagdown_O\diagdown^C\diagdown_{C_2H_5} \end{array}\right]^{-} + H_2NOH \longrightarrow \begin{array}{l}(1-x)\,[Cr(NO)(LH)_2]^0 \\ + x[Cr(NO)(LH)(H_2O)_3]^+ \\ + xLH\end{array} \quad (121)$$

The stoichiometry was found to be $1:1$ $[Cr(V):H_2NOH]$, and no gaseous products were observed in the smooth, quantitative reaction (121).

5. Miscellaneous systems

Few other oxo complexes of transition metals have been reacted with hydroxylamine to yield nitrosyl complexes. The reaction of OsO_4 with hydroxylamine in the presence of thiocyanate ions affords red $[Os(NO)-(NCS)_5]^{2-}$.[277] Perrhenate ions, $ReO_4{}^-$, also react with H_2NOH in strongly alkaline media in the presence of NCS^- and chelating ligands, yielding nitrosyl complexes of the type $\{Re(NO)\}^6$.[288,289] In the absence of additional ligands the latter reaction has been used for spectrophotometric determination of rhenium,[290] although the nature of the yellow species formed has not been established. Formation of a nitrosyl ruthenium complex has been proposed for the reaction of RuO_2 (or $RuO_4{}^{2-}$) with hydroxylamine in alkaline aqueous solution.[291] This reaction should be an excellent synthetic route to nitrosyl ruthenium complexes.

The reaction of $(NH_4)_2[TcCl_6]$ with hydroxylamine and added ammonia yields $[Tc(NH_3)_4(NO)(H_2O)]Cl_2$, a complex that was originally formulated as hydroxylamine complex $[Tc(NH_2OH)_2(NH_3)_3(H_2O)]\ Cl_2$.[292,293]

Finally, it has been reported that a phthalocyanine complex of iron(III) reacts with hydroxylamine under carefully controlled conditions of pH (5.0) to form a nitrosyl complex.[294]

None of the above reactions has been studied in order to elucidate the mechanism.

VI. REACTIONS OF COORDINATED HYDROXYLAMINE LIGANDS

A. Acid–base properties of Pt–NH₂OH complexes

Grinberg *et al.*[295–301] reported in a series of papers the acidic properties of hydroxylamine complexes of platinum(II) in aqueous solution. Mono-, bis-, and tris(hydroxylamine) complexes behave as mono-, di-, or tribasic acids, respectively:

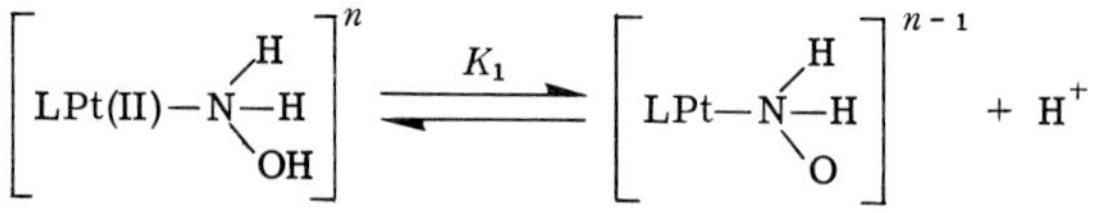

$$
\left[LPt(II){-}N\!\!\begin{array}{c}H\\ {-}H\\ OH\end{array} \right]^{n} \xrightleftharpoons{K_1} \left[LPt{-}N\!\!\begin{array}{c}H\\ {-}H\\ O\end{array} \right]^{n-1} + H^+ \tag{122}
$$

Acid dissociation constants for a series of complexes at $25°C$ ($\mu = 0.3$ M) are summarized in Table 6. The fact that O deprotonation prevails in aqueous alkaline solutions is clearly demonstrated by the differing behavior of $[Pt(NH_2OH)_4]^{2+}$, which is a relatively strong acid, and $[Pt(NH_3OCH_3)_4]^{2+}$, which exhibits no acidic properties at all.[297]

From the data in Table 6 the following conclusions may be drawn. Two hydroxylamine ligands in cis position with respect to each other are stronger acids than are their geometrical isomers with the ligands in trans position. Furthermore, the difference between successive dissociation constants is greater for the cis than for the trans isomers. This may be explained by stabilization via a strong hydrogen bond of the mono deprotonated form, if a

TABLE 6

Acid dissociation constants of coordinated hydroxylamine ligands
in platinum(II) complexes at 25°C ($\mu = 0.3$ M)

Complex	pK_1	pK_2	pK_3	Ref.
cis-[Pt(NH$_3$)$_2$(NH$_2$OH)$_2$]$^{2+}$	7.48	10.18	—	[295]
trans-[Pt(NH$_3$)$_2$(NH$_2$OH)$_2$]$^{2+}$	8.84	9.80	—	[295]
cis-[Pt(py)$_2$(NH$_2$OH)$_2$]$^{2+}$	6.92	10.16	—	[296]
trans-[Pt(py)$_2$(NH$_2$OH)$_2$]$^{2+}$	8.68	9.72	—	[296]
[Pt(NH$_2$OH)$_4$]$^{2+}$	6.46	—	—	[296]
[Pt(NH$_3$)(NH$_2$OH)$_3$]$^{2+}$	7.22	9.70	10.40	[296]
[Pt(en)(NH$_2$OH)$_2$]$^{2+}$	7.68	10.70	—	[298]
[Pt(dmso)Cl$_2$(NH$_2$OH)]	8.80	—	—	[299]
cis-[Pt(NH$_3$)(NH$_2$OH)Cl$_2$]	9.15	—	—	[300]
trans-[Pt(NH$_3$)(NH$_2$OH)Cl$_2$]	9.70	—	—	[300, 301]
trans-[Pt(NH$_2$OH)$_2$(NO$_2$)$_2$]	8.57	9.44	—	[301]
trans-[Pt(NH$_2$OH)$_2$Cl$_2$]	8.80	9.62	—	[301]
trans-[Pt(NH$_2$OH)$_2$Br$_2$]	9.07	9.77	—	[301]
trans-[Pt(NH$_2$OH)$_2$I$_2$]	9.41	9.89	—	[301]
[Ir(NH$_3$)$_5$(NH$_2$OH)]$^{3+}$	9.40	—	—	[114]

coordinated hydroxylamine is present at cis position.

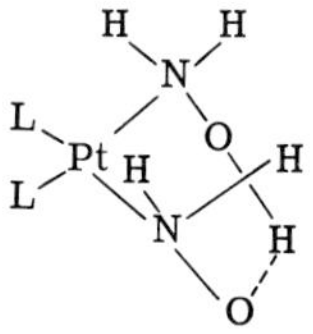

A structure similar to the one shown above, with two six-membered rings, is anticipated for Alexander's base [Pt(NH$_2$O)$_2$(NH$_2$OH)$_2$]·2H$_2$O, although intermolecular hydrogen bonding could equally well be invoked. The first dissociation constant of Pt(NH$_2$OH)$_x$ complexes is also dependent on the overall charge of the complex under investigation. Thus, positively charged cations are stronger acids than neutral compounds. The nature of the other ligands and their relative position with respect to hydroxylamine ligands influences the first dissociation constant (e.g., trans influence).

B. Redox reactions

The redox properties of coordinated hydroxylamine and of its deprotonated forms as compared to the uncoordinated ligand have not been studied in great detail, although this is a very important area of the chemistry of hydroxylamine. The proposed mechanisms of the metal-catalyzed air oxidation of H_2NOH in alkaline solutions, the interconversion of coordinated hydroxylamine to nitrosyl (oxidation) or ammonia (reduction), invoke transition metal complexes of hydroxylamine as reactive intermediates.

Hydroxylamine complexes of cobalt(III) have received some attention.[107,302] The acid hydrolysis of *cis*-chlorohydroxylamine bis(ethylenediamine)cobalt(III), i.e., substitution of coordinated chloride by an aquo ligand at $[H^+] > 0.4\ M$, is complicated at higher pH by a spontaneous redox process, yielding cobalt(II) and dinitrogen, reactions (123) and (124).[302]

$$[Co(en)_2(NH_2OH)Cl]^{2+} \rightleftharpoons [Co(en)_2(NH_2O)Cl]^+ + H^+ \tag{123}$$

$$[Co(en)_2(NH_2O)Cl]^+ \rightarrow Co(II)(en)_2^{2+} + Cl^- + \tfrac{1}{2}N_2 + H_2O \tag{124}$$

This indicates that N-coordinated hydroxylamide (-1) ligands are more easily oxidized than the protonated forms.

The reaction of hydroxylamine with nitrous acid to give N_2O has been studied carefully.[303–305] The reaction of $[Co(en)_2(NH_2OH)Cl]^{2+}$ with nitrous acid at high H^+ concentrations parallels that reaction.[107]

$$\text{rate} = k[H^+][HNO_2][Co(en)_2(NH_2OH)Cl]^{2+} \tag{125}$$

A mechanism involving electrophilic nitrosation on the oxygen of coordinated hydroxylamine has been suggested, since the O-methylated complex $[Co(en)_2(NH_2OCH_3)Cl]^{2+}$ does not react. Coordinated hydroxylamine was found to be 10% less reactive than NH_3OH^+.

$$Co(NH_2OH) + \overset{*}{N}O^+ \rightarrow Co(NH_2O\overset{*}{N}O) + H^+ \tag{126}$$

$$Co(NH_2O\overset{*}{N}O) \rightarrow Co(NH(\overset{*}{N}O)OH) \rightarrow Co(HON{=}\overset{*}{N}OH) \rightarrow \text{products} \tag{127}$$

The nitrosated intermediate undergoes intramolecular rearrangements and decomposition, yielding N_2O (and a small amount of N_2). At higher pH a more complicated kinetic behavior is found, and a deprotonated hydroxylamido (-2) cobalt(III) complex is believed to be the reactive species.

Finally, the reaction of $[Co(NH_2OH)_6]^{3+}$ with the strong oxidant cerium(IV) has been studied in sulfuric acid media.[237] A rate with a first-order dependence on the oxidant and a reaction order of 0.6 for the reductant has been reported.

$$-d[Ce(IV)]/dt = [k_0 + (k'/[H^+])][Ce(IV)][Co(NH_2OH)_6^{3+}]^{0.6} \tag{128}$$

The stoichiometry of the reaction determined by polarographic titration corresponds to the consumption of 22–24 Ce(IV) ions per $[Co(NH_2OH)_6]^{3+}$ cation. Formation of a binuclear intermediate and a coordinated $NH_2\overset{.}{O}$ radical in the rate-determining step has been proposed to account for the kinetic data.

$$[Co(H_2NOH)_6]^{3+} + Ce(IV) \overset{K}{\rightleftharpoons} [(NH_2OH)_5Co-NH_2OH-Ce(IV)] \qquad (129)$$

$$[(NH_2OH)_5CoNH_2OH-Ce(IV)] \overset{rds}{\longrightarrow} Ce(III) + H^+ + [(NH_2OH)_5Co-NH_2\dot{O}]^{3+} \qquad (130)$$

$$[(NH_2OH)_5Co(NH_2\dot{O})]^{3+} \xrightarrow[\text{1-}e \text{ transfer; fast}]{\text{intramolecular}} Co_{aq}^{2+} + 5NH_2OH + HNO + H^+ \qquad (131)$$

$$4Ce(IV) + HNO + 2H_2O \overset{fast}{\longrightarrow} NO_3^- + 4Ce^{3+} + 5H^+ \qquad (132)$$

It was found that the oxidation of coordinated hydroxylamine is energetically more demanding by 2 kcal mol^{-1}.

The electrochemical oxidation of $[(terpy)(bpy)Ru(II)NH_3]^{2+}$ in aqueous solution to give $[(terpy)(bpy)Ru(III)NO_3]^{2+}$ has been shown to occur via a series of intermediates: a hydroxylamine-containing intermediate has been detected, $[(terpy(bpy)Ru(II)NH_2OH]^{2+}$, as well as nitrosyl and nitrite complexes of ruthenium(II).[306] The reverse reaction, i.e., the reduction of coordinated nitrosyl to coordinated ammonia, occurs also for a variety of polypyridyl complexes of ruthenium and osmium.[307]

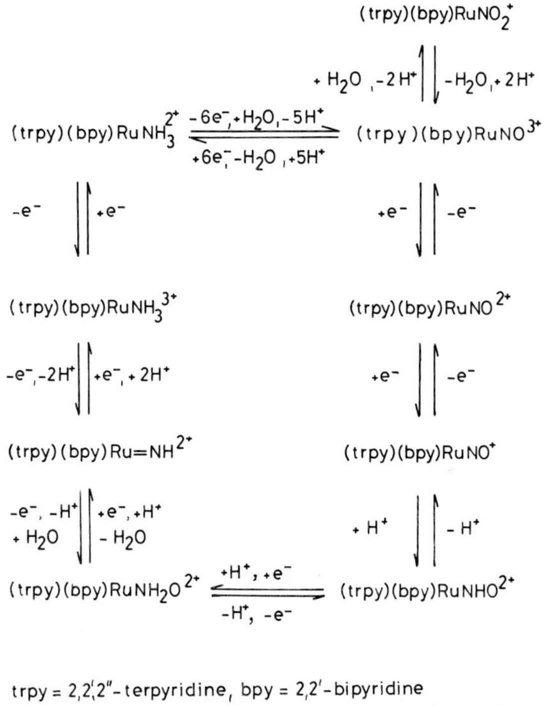

trpy = 2,2′,2″-terpyridine, bpy = 2,2′-bipyridine

Scheme 1

The mechanism involves a series of one-electron transfer steps invoking a nitroxyl intermediate and possibly the complex $[Ru(terpy)(bpy)NH_2O]^{2+}$. [306–308] It is important to note that a whole series of successive reactions occurs at one metal site without releasing free ligands into the solution. This scheme could be an important step leading to an understanding of the enzyme nitrite reductase. It certainly demonstrates that the interconversion of coordinated nitrosyl to hydroxylamine is not achieved only by intramolecular two-electron steps, as has been described in Section V,C,2, but also via a series of one-electron transfers.

C. Reactions of *O,N*-hydroxylamido (−1, −2) ligands

Evidence derived from ^{1}H NMR spectroscopy of complexes of the type $[MoX_2(R_1R_1NO)_2]$, with O,N-coordinated *N,N*-diethylhydroxylamido (−1) ligands and terminal oxo or sulfido groups, has shown that a temperature-dependent equilibrium exists in solution and that the three-membered

$\overline{\text{Mo−O−N}}$ ring is opened and re-formed.[85]

Because the N-atoms of the hydroxylamine ligands have the coordination number 4, including an Mo−N bond, simple nitrogen inversion, as in the free ligand, cannot occur. The observed configurational isomerization stems from a rapid rotation–inversion at the nitrogens, with concomitant reformation of the Mo−N bond. The rate-determining step is the dissociation of the Mo−N bond. The free energy of activation for this process is a function of X: $\Delta G^{\ddagger} = 20.6$ kcal mol^{-1} for X = O; $\Delta G^{\ddagger} = 18.4$ kcal mol^{-1} for X = S.

The same behavior has been found for the complexes $[Ti(ONEt_2)_4]$ and $[Zr(ONEt_2)_4]$ in solution.[50] Due to steric crowding (coordination number, 8), the activation energy for M−N bond breaking is lower than in the molybdenum complexes $[\Delta G^{\ddagger} = 16.0$ and 16.5 kcal mol^{-1} for the titanium(IV) and zirconium(IV) complex, respectively]. This ring cleavage process may be a necessary prerequisite for the observed reactivity of bis(*N*-methylhydroxylamido (−1) dioxomolybdenum(VI) toward heterocumulenes such as SCN$^-$, CS_2, R−C≡N, and R−N=C=O. New ligands are formed probably by the nucleophilic attack of the nitrogen of the hydroxylamido (−1) ligand (opened ring) at the electrophilic centers of the added reagents.[46-48,309]

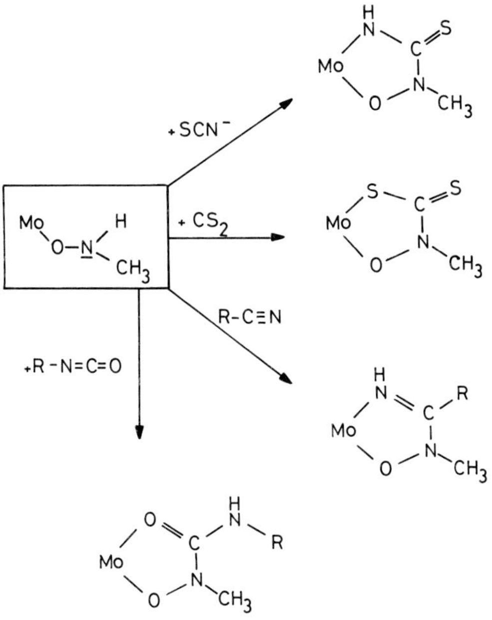

Scheme 2

It is not clear if these reactions do in fact occur at the coordinated hydroxyl-amido ligands, as in **Scheme 2**, or after dissociation. Nevertheless, co-ordinated hydroxylamines react with aldehydes to form coordinated oximes.

Both geometric isomers of $[Pt(NH_2OH)_2Cl_2]$ and $[Pt(NH_2OH)_2(NH_2O)_2] \cdot 2H_2O$ react with acetaldehyde in alkaline media to produce good yields of $[Pt[(CH_3)CHNOH]_2Cl_2]$ and $[Pt[(CH_3)HCNOH]_4]Cl_2$.[310]

$$[Pt(NH_2OH)_2Cl_2] + 2CH_3C\overset{O}{\underset{H}{\big\|}} \xrightarrow{-2H_2O} [Pt[(CH_3)CHNOH]_2Cl_2] \qquad (133)$$

The oxime ligands are N-coordinated since $[Pt[(CH_3)HCNOH]_4]^{2+}$ has been shown to be a tetrabasic acid.[311] Sideways-on coordinated hydroxyl-amido (-1) ligands react with ketones to produce O,N-coordinated oximato ligands.[56,58]

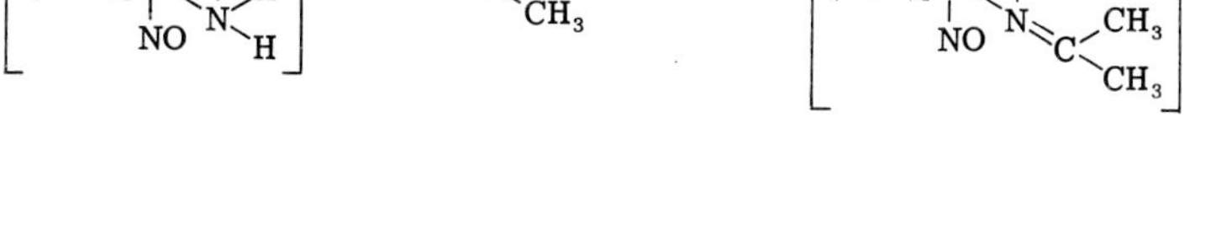

$$\left[(SCN)_4Mo\overset{O}{\underset{NO}{\big|}}\overset{H}{\underset{N}{\big|}}H\right]^{2-} + O{=}C\overset{CH_3}{\underset{CH_3}{\big<}} \xrightarrow{-H_2O} \left[(SCN)_4Mo\overset{O}{\underset{NO}{\big|}}\overset{}{\underset{N{=}C}{\big|}}\overset{CH_3}{\underset{CH_3}{\big<}}\right]^{2-} \qquad (134)$$

One of the intriguing aspects of the formation of three-membered metalla-cycles by O,N-coordinated hydroxylamido (-2) and (-1) ligands has been the possibility of a facile N–O bond cleavage, generating reactive nitrene species. The reaction of sideways-on-bound peroxo groups with olefins to produce selectively epoxides has been investigated thoroughly.[312] Since $R-NO^{2-}$ ligands are isoelectronic with the O_2^{2-} group, the hypothetical reaction (137) in analogy to (135) and (136) has been used as a preparative concept for the synthesis of aziridines.[13,45,88]

$$M\overset{O}{\underset{O}{|}} + \text{\Large)═(} \longrightarrow \triangle\overset{O}{} + M═O \tag{135}$$

$$M\overset{O}{\underset{O-R}{|}} + \text{\Large)═(} \longrightarrow \left(M\overset{C}{\underset{O-O}{\underset{R}{|}}}C \right) \longrightarrow \triangle\overset{O}{} + M-OR \tag{136}$$

$$M\overset{O}{\underset{N-R}{|}} + \text{\Large)═(} \overset{?}{\longrightarrow} \triangle\underset{\underset{R}{|}}{N} + M═O \tag{137}$$

One example has been found in which aziridines are produced most probably via nitrene transfer to an olefin.[313] The intermediate nitrene formed on thermal decomposition of $Cp_2NbMe(ONCH_3)$, in the absence of olefin, can dimerize.

$$Cp_2Nb\overset{O}{\underset{CH_3}{\underset{|}{N}}}\overset{|}{\underset{CH_3}{N}} + Ph-CH═CH_2 \longrightarrow Cp_2Nb\overset{O}{\diagdown CH_3} + \underset{H}{\overset{R}{\diagdown}}C-CH_2 \tag{138}$$

$$\Big\downarrow \Delta T$$

$$H_3CN═NCH_3$$

Azobenzene is a major product of the thermal decomposition of [(dipic)-

MoO(ONPh)(hmpt)].[88]

$$\text{(139)}$$

N–O bond cleavage and dimerization of phenylnitrene are possible reaction paths.

Paramagnetic ReO(CH$_3$)$_4$ reacts with NO to give *cis*-trimethyldioxorhenium(VII), and azomethane intermediates **I** or **II** are believed to produce nitrenes via N–O bond scission.[161]

$$\text{(140)}$$

Sharpless *et al.*[13] have also observed the transfer of an N–R group to olefins, but allylic amines rather than anticipated aziridines are formed [reaction (141)]. Again, N–O bond cleavage occurs.

$$\text{(141)}$$

The examples described involve complexes of early transition metals with rather high formal oxidation states and N-substituted hydroxylamido (-2) ligands. The thermodynamic driving force for nitrene transfer via N–O cleavage is raised by favorable metal–oxygen double-bond formation. Transition metals on the right-hand side of the periodic table have relatively low oxidation states and do not exhibit this same property, since the metal–oxygen bonds are quite weak. Therefore, the N–O bond cleavage in [Ni(PhNO)(CN–*t*-Bu)$_2$] is facilitated by concomitant oxygen atom transfer to a substrate such as triphenylphosphine. The reactions shown in the following scheme are typical oxygen atom transfer redox reactions.[119]

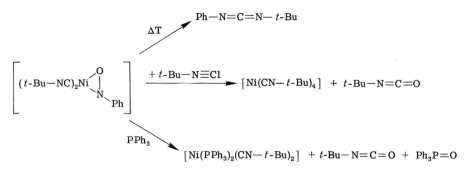

The reaction of isocyanide and nitrosobenzene in the presence of [Ni(CN–t-Bu)]$_4$ in catalytic amounts can be carried out catalytically to produce t-Bu–NCO and PhN=C=N—t-Bu and azobenzene.[119]

References

[1] Feltham, R. D.; Enemark, J. H. *Top. Stereochem.* **1981**, *12*, 155.

[2] Enemark, J. H.; Feltham, R. D. *Coord. Chem. Rev.* **1974**, *13*, 339.

[3] McCleverty, J. A. *Chem. Rev.* **1979**, *79*, 53.

[4] Bottomley, F. *In* "Reactions of Coordinated Ligands"; Braterman, P., Ed.;

[5] Nast, R.; Hieber, W.; Proeschel, E. *Z. Anorg. Allg. Chem.* **1948**, *226*, 145.

[6] Uhlenhuth, R. *Justus Liebigs Ann. Chem.* **1899**, *307*, 332.

[7] Hughes, M. N. "The Inorganic Chemistry of Biological Processes"; 2nd ed., Wiley: New York, 1981, pp. 204–211.

[8] Hooper, A. B. *In* "Microbiology"; Schlessinger, D., Ed.; Am. Soc. Microbiol.: Washington, D.C., 1978, pp. 299–304.

[9] Hollocher, T. C.; Tate, M. E.; Nicholas, D. J. D. *J. Biol. Chem.* **1981**, *256*, 10834.

[10] Lipscomb, J. D.; Hooper, A. B. *Biochemistry* **1982**, *21*, 3965.

[11] Lipscomb, J. D.; Anderson, K. K.; Münck, E.; Kent, T. A.; Hooper, A. B. *Biochemistry* **1982**, *21*, 3973.

[12] Stedman, G. *Adv. Inorg. Chem. Radiochem.* **1979**, *22*, 113.

[13] Liebeskind, L. S.; Sharpless, K. B.; Wilson, R. D.; Ibers, J. A. *J. Am. Chem. Soc.* **1978**, *100*, 7061.

[14] Padmanabhan, V. M.; Smith, H. G.; Peterson, S. W. *Acta Crystallogr.* **1967**, *22*, 928.

[15] Dicken, B. *Acta Crystallogr., Sect. B* **1969**, *B 25*, 1875.

[16] Meyers, E. A.; Lipscomb, W. N. *Acta Crystallogr.* **1955**, *8*, 583.

[17] Hughes, M. N.; Nicklin, H. G.; Shrimanker, K. *J. Chem. Soc. A* **1971**, 3485.

[18] Ebler, E.; Schott, E. *J. Prakt. Chem.* **1908**, *782*, 289.

[19] Hughes, M. N.; Nicklin, H. G. *J. Chem. Soc. A.* **1971**, 164.

[20] Hughes, M. N.; Nicklin, H. G. *J. Chem. Soc. A* **1970**, 925.

[21] Nast, R.; Föppl, I. *Z. Anorg. Allg. Chem.* **1950**, *263*, 310.

[22] Bonner, F. T.; Dzelzkalus, L. S.; Bonucci, J. A. *Inorg. Chem.* **1978**, *17*, 2487.

[23] Lunak, S.; Veprek-Siska, J. *Collect. Czech. Chem. Commun.* **1974**, *39*, 391.
[24] Moews, P. S.; Audrieth, L. F. *J. Inorg. Nucl. Chem.* **1959**, *11*, 242.
[25] Cooper, J. N.; Chilton, J. E., Jr.; Powell, R. E. *Inorg. Chem.* **1970**, *9*, 2303.
[26] Grinberg, A. A.; Stetsenko, A. I. *Russ. J. Inorg. Chem. (Engl. Transl.)* **1961**, *6*, 55.
[27] Little, R. G.; Doedens, R. J. *Inorg. Chem.* **1973**, *12*, 537.
[28] Sams, D. B.; Doedens, R. J. *Inorg. Chem.* **1979**, *18*, 153.
[29] Wilson, R. D.; Ibers, J. A. *Inorg. Chem.* **1979**, *18*, 336.
[30] Wieghardt, K.; Holzbach, W.; Weiss, J. *Z. Naturforsch., B: Anorg. Chem., Org. Chem.* **1982**, *37B*, 680.
[31] Calligaris, W.; Yoshida, T.; Otsuka, S. *Inorg. Chim. Acta* **1974**, *11*, L 15.
[32] Barrow, M. J.; Mills, O. S. *J. Chem. Soc. A* **1971**, 864.
[33] Legzdins, P.; Nurse, C. R.; Rettig, S. J. *J. Am. Chem. Soc.* **1983**, *105*, 3727.
[34] Engelhardt, L. M.; Newman, P. W. G.; Raston, C. L.; White, A. H. *Aust. J. Chem.* **1974**, *27*, 503.
[35] Adrian, H. W. W.; van Tets, A. *Acta Crystallogr., Sect. B* **1978**, *B34*, 652; **1979**, *B35*, 153.
[36] Meyers, E. A.; Lipscomb, W. N. *Acta Crystallogr.* **1955**, *8*, 583.
[37] Liebeskind, L. S.; Sharpless, K. B., Wilson, R. D.; Ibers, J. A. *J. Am. Chem. Soc.* **1978**, *100*, 7061.
[38] van Tets, A.; Adrian, H. W. W. *J. Inorg. Nucl. Chem.* **1977**, *39*, 1607; Adrian, H. W. W.; van Tets, A. *Acta Crystallogr., Sect. B* **1978**, *B34*, 88.
[39] Adrian, H. W. W.; van Tets, A. *Acta Crystallogr., Sect. B* **1977**, *B33*, 2997.
[40] Adrian, H. W. W.; van Tets, A. *Acta Crystallogr., Sect. B* **1978**, *B34*, 2632.
[41] Nuber, B.; Weiss, J. *Acta Crystallogr., Sect. B* **1981**, *B37*, 947.
[42] Wieghardt, K.; Quilitzsch, U.; Nuber, B.; Weiss, J. *Angew. Chem.* **1978**, *90*, 381; *Angew. Chem., Int. Ed. Engl.* **1978**, *17*, 351.
[43] Wieghardt, K.; Holzbach, W.; Weiss, J.; Nuber, B.; Prikner, B. *Angew. Chem.* **1979**, *91*, 582; *Angew. Chem., Int. Ed. Engl.* **1979**, *18*, 548.
[44] Wieghardt, K.; Holzbach, W.; Nuber, B.; Weiss, J. *Chem. Ber.* **1980**, *113*, 629.
[45] Saussine, L.; Mimoun, H.; Mitschler, A.; Fisher, J. *J. Nouv. Chim.* **1980**, *4*, 235.
[46] Wieghardt, K.; Hofer, E.; Holzbach, W.; Nuber, B.; Weiss, J. *Inorg. Chem.* **1980**, *19*, 2927.
[47] Wieghardt, K.; Holzbach, W.; Hofer, E.; Weiss, J. *Inorg. Chem.* **1981**, *20*, 343.
[48] Wieghardt, K.; Holzbach, W.; Weiss, J. *Z. Naturforsch., B: Anorg. Chem., Org. Chem.* **1981**, *36B*, 289.
[49] Wieghardt, K.; Holzbach, W.; Weiss, J. *Inorg. Chem.* **1981**, *20*, 3436.
[50] Wieghardt, K.; Tolksdorf, I.; Weiss, J.; Swiridoff, W. *Z. Anorg. Allg. Chem.* **1982**, *490*, 182.
[51] Wieghardt, K.; Hahn, M.; Weiss, J.; Swiridoff, W. *Z. Anorg. Allg. Chem.* **1982**, *492*, 164.
[52] Bristow, S.; Collison, D.; Garner, C. D.; Clegg, W. *J. Chem. Soc., Dalton Trans.* **1983**, 2495.
[53] Bristow, S.; Garner, C. D.; Clegg, W. *Inorg. Chim. Acta* **1983**, *76*, L-261.
[54] Gheller, S. F.; Hambley, T. W.; Traill, P. R.; Brownlee, R. T. C.; O'Connor, M. J.; Snown, M. R.; Wedd, A. G. *Austr. J. Chem.* **1982**, *35*, 2183.
[55] Dickman, M. H.; Doedens, R. J. *Inorg. Chem.* **1982**, *21*, 682.
[56] Müller, A.; Mohan, N.; Sarkar, S.; Eltzner, W. *Inorg. Chim. Acta* **1981**, *55*, L-33; Müller, A.; Mohan, N. *Z. Anorg. Allg. Chem.* **1981**, *480*, 157.
[57] Wieghardt, K.; Backes-Dahmann, G.; Swiridoff, W.; Weiss, J. *Inorg. Chem.* **1983**, *22*, 1221.
[58] Müller, A.; Eltzner, W.; Sarkar, S.; Bögge, H.; Aymonino, P. J.; Mohan, N.; Seyer, U.; Subramanian, P. *Z. Anorg. Allg. Chem.* **1983**, *503*, 22.

[59] Jaitner, P.; Huber, W.; Scheidsteger, O.; Huttner, G. *Angew. Chem. J. Organomet. Chem.* **1983**, *259*, C1.

[60] Stevens, R. E.; Gladfelter, W. L. *J. Am. Chem. Soc.* **1982**, *104*, 6454.

[61] Bertrand, J. A.; Smith, J. H.; Eller, P. G. *Inorg. Chem.* **1974**, *13*, 1649.

[62] Aime, S.; Gervasio, G.; Milone, L.; Rosetti, R.; Staughellini, P. L. *J. Chem. Soc., Chem. Commun.* **1976**, 370.

[63] Khare, G. P.; Doedens, R. J. *Inorg. Chem.* **1976**, *15*, 86.

[64] Khare, G. P.; Doedens, R. J. *Inorg. Chem.* **1977**, *16*, 907.

[65] Connor, J. A.; Ebsworth, E. A. V. *Adv. Inorg. Chem. Radiochem.* **1964**, *6*, 280.

[66] Hartkamp, H. *Angew. Chem.* **1959**, *71*, 651.

[67] Drew, R. E.; Einstein, F. B. W. *Inorg. Chem.* **1973**, *12*, 829.

[68] Wieghardt, K.; Quilitzsch, U. *Z. Anorg. Allg. Chem.* **1979**, *457*, 75.

[69] Svensson, I. B.; Stomberg, R. *Acta Chem. Scand.* **1971**, *25*, 898.

[70] Anderson, O. P.; Kuechler, T. C. *Inorg. Chem.* **1980**, *19*, 1417.

[71] Dickman, M. H.; Doedens, R. J. *Inorg. Chem.* **1981**, *20*, 2677.

[72] Porter, C. C.; Dickmann, M. H.; Doedens, R. J. *Inorg. Chem.* **1983**, *22*, 1964.

[73] Kharitonov, Yu. Ya.; Sarukhanov, M. A. "The Chemistry of Complexes of Metals with Hydroxylamine"; Nauka: Moscow, 1977.

[74] Kharitonov, Yu. Ya.; Sarukhanov, M. A. *Koord. Khim.* **1978**, *4*, 323.

[75] Kharitonov, Yu. Ya.; Sarukhanov, M. A. "Vibrational Spectra of Hydroxylamine and Its Coordination Compounds"; Tashkent: Uzb. SSR, 1971.

[76] Scheuermann, W.; Van Tets, A. *J. Raman Spectrosc.* **1977**, *6*, 100.

[77] Singh, A.; Sharma, C. K.; Rai, A. K.; Gupta, V. D.; Mehrotra, R. C. *J. Chem. Soc. A* **1971**, 2440.

[78] Wieghardt, K.; Quilitzsch, U. *Z. Anorg. Allg. Chem.* **1979**, *457*, 75.

[79] Quilitzsch, U.; Wieghardt, K. *Z. Naturforsch., B: Anorg. Chem., Org. Chem.* **1979**, *34B*, 640.

[80] Wieghardt, K.; Quilitzsch, U. *Z. Naturforsch., B: Anorg. Chem., Org. Chem.* **1981**, *36B*, 683.

[81] Drago, R. S.; Donoghue, J. T.; Herlocker, D. W. *Inorg. Chem.* **1965**, *4*, 836.

[82] Wieghardt, K. Unpublished results.

[83] Müller, J.; Schmitt, S. *J. Organomet. Chem.* **1978**, *160*, 109.

[84] Ivashkovich, E. M.; Zhelekhovskaya, L. M. *Russ. J. Inorg. Chem. (Engl. Transl.)* **1979**, *24*, 33.

[85] Hofer, E.; Holzbach, W.; Wieghardt, K. *Angew. Chem.* **1981**, *93*, 303; *Angew. Chem., Int. Ed. Engl.* **1981**, *20*, 282.

[86] Wieghardt, K.; Holzbach, W. *Angew. Chem.* **1979**, *91*, 583; *Angew. Chem., Int. Ed. Engl.* **1979**, *18*, 549.

[87] Wieghardt, K.; Holzbach, W.; Nuber, B.; Weiss, J. *Chem. Ber.* **1980**, *113*, 629.

[88] Muccigrosso, D. A.; Jacobsen, S. E.; Apgar, P. A.; Mares, F. *J. Am. Chem. Soc.* **1978**, *100*, 7063.

[89] Middleton, A. R., Wilkinson, G. *J. Chem. Soc., Dalton Trans.* **1981**, 1898.

[90] Ghosh, P.; Bandyopadhyay, P.; Chakrarorty, A. *J. Chem. Soc., Dalton Trans.* **1983**, 401.

[91] Feldt, W. *Ber. Dtsch. Chem. Ges.* **1894**, *27*, 401.

[92] Kharitonov, Yu. Ya.; Sarukhanov, M. A.; Baranovskii, I. B. *Zh. Prikl. Spektrosk.* **1971**, *15*, 739.

[93] Ivashkovich, E. M.; Lazorchak, N. I.; Gumanitskaya, O. D.; Shpontak, M. Y. *Russ. J. Inorg. Chem. (Engl. Transl.)* **1976**, *21*, 1170.

[94] Simmons, E. L.; Wendlandt, W. W. *J. Inorg. Nucl. Chem.* **1971**, *33*, 3955.

[95] Ivashkovich, E. M.; Skoblei, M. I. *Russ. J. Inorg. Chem.* **1974**, *19*, 411.

[96] Ivashkovich, E. M. *Russ. J. Inorg. Chem.* **1982**, *27*, 395.

[97] Issleib, K.; Kreibich, A. *Z. Anorg. Allg. Chem.* **1961**, *313*, 338.
[98] Salyn, Ya. V.; Nefedov, V. I.; Sarukhanov, M. A.; Kharitonov, Yu. Ya. *Koord. Khim.* **1975**, *1*, 945.
[99] Koerner von Gustorf, E.; Jun, M. J. *Z. Naturforsch., B: Anorg. Chem., Org. Chem., Biochem., Biophys., Biol.* **1965**, *20B*, 521.
[100] Koerner von Gustorf, E.; Henry, M. C.; Sacher, R. E.; Di Pietro, C. *Z. Naturforsch., B: Anorg. Chem., Org. Chem., Biochem., Biophys., Biol.* **1966**, *21b*, 1152.
[101] Koerner von Gustorf, E. *Angew. Chem.* **1966**, *78*, 780.
[102] Middleton, A. R.; Thornback, J. R.; Wilkinson, G. *J. Chem. Soc., Dalton Trans.* **1980**, 174.
[103] Golub, A. M.; Ivashkovich, E. M. *Zh. Ukr. Khim.* **1978**, *44*, 14.
[104] Werner, A.; Berl, E. *Ber. Dtsch. Chem. Ges.* **1905**, *38*, 893.
[105] White, R. C.; Finley, S. W. *Inorg. Chem.* **1964**, *3*, 1329.
[106] Kharitonov, Yu. Ya.; Sarukhanov, M. A.; Baranovskii, I. B. *Russ. J. Inorg. Chem. (Engl. Transl.)* **1967**, *12*, 82.
[107] Hughes, M. N.; Shrimanker, K.; Wimbledon, P. E. *J. Chem. Soc., Dalton Trans.*, **1978**, 1634.
[108] Hughes, M. N.; Shrimanker, K. *Inorg. Chim. Acta* **1967**, *18*, 69.
[109] Klein, H. F.; Karsch, H. H. *Chem. Ber.* **1976**, *109*, 1453.
[110] Herlocker, D. W.; Drago, R. S. *Inorg. Chem.* **1968**, *7*, 1479.
[111] Herlocker, D. W. *J. Inorg. Nucl. Chem.* **1968**, *30*, 2197.
[112] Sarukhanov, M. A.; Valdman, S. S.; Parpiev, N. A. *Russ. J. Inorg. Chem. (Engl. Transl.)* **1975**, *20*, 1327.
[113] Reed, C. A.; Roper, W. R. *J. Chem. Soc. A* **1970**, 3054.
[114] Lane, B. C.; McDonald, J. W.; Myers, V. G.; Basolo, F.; Pearson, R. G. *J. Am. Chem. Soc.* **1971**, *93*, 4934.
[115] Uhlenhuth, R. *Justus Liebigs Ann. Chem.* **1900**, *311*, 120.
[116] Falqui, M. T.; Ponticelli, G.; Sotgui, F. *Ann. Chim. (Rome)* **1966**, *56*, 464.
[117] Yoneda, H. *Bull. Chem. Soc. Jpn.* **1957**, *30*, 132.
[118] Babaeva, A. V.; Bukalov, I. E. *Izv. Sekt. Platiny Drugikh Blagorodn. Met., Inst. Obshch. Neorg. Khim., Akad. Nauk SSSR* **1955**, *31*, 67.
[119] Otsuka, S.; Aotani, Y.; Tatsuno, Y.; Yoshida, T. *Inorg. Chem.* **1976**, *15*, 656.
[120] Berman, R. S.; Kochi, J. K. *Inorg. Chem.* **1980**, *19*, 248.
[121] Beck, W.; Schmidtner, K. *Chem. Ber.* **1967**, *100*, 3363.
[122] Okunaka, M.; Matsubayashi, G.; Tanaka, T. *Bull. Chem. Soc. Jpn.* **1975**, *48*, 1826.
[123] Okunaka, M.; Matsubayashi, G.; Tanaka, T. *Inorg. Nucl. Chem. Lett.* **1976**, *12*, 813.
[124] Okunaka, M.; Matsubayashi, G.; Tanaka, T. *Bull. Chem. Soc. Jpn.* **1977**, *50*, 1070.
[125] Okunaka, M.; Matsubayashi, G.; Tanaka, T. *Bull. Chem. Soc. Jpn.* **1977**, *50*, 907.
[126] Lossen, W. *Justus Liebigs. Ann. Chem.* **1871**, *160*, 242.
[127] Alexander, H. *Justus Liebigs. Ann. Chem.* **1888**, *246*, 239.
[128] Stetsenko, A. I. *Russ. J. Inorg. Chem. (Engl. Transl.)* **1961**, *6*, 903.
[129] Kharitonov, Yu. Ya.; Sarukhanov, M. A. *Russ. J. Inorg. Chem. (Engl. Transl.)* **1966**, *11*, 1359.
[130] Babaeva, A. V.; Mosgagina, M. A. *Dokl. Akad. Nauk SSSR* **1953**, *89*, 293.
[131] Stetsenko, A. I.; Strelin, S. G. *Zh. Neorg. Khim.* **1973**, *18*, 2642.
[132] Chugaev, L. A.; Chernyaev, I. I. *Izv. Inst. Izuch. Platiny Drugich Blagorodn. Met., Akad. Nauk SSSR* **1920**, *1*, 3.
[133] Grinberg, A. A.; Stetsenko, A. I. *Russ. J. Inorg. Chem. (Engl. Transl.)* **1961**, *6*, 55.
[134] Chernayev, I. I. *Izv. Inst. Izuch. Platiny Drugich Blagorodn. Met., Akad. Nauk SSSR* **1926**, *4*, 243.
[135] Pinkard, F. W.; Saenger, H.; Wardlow, W. *J. Chem. Soc.* **1933**, 1056.

[136] Chernayev, I. I. *Izv. Inst. Izuch. Platiny Drugich Blagorodn. Met., Akad. Nauk SSSR* **1928,** *6,* 55; Chernayev, I. I.; Chugaev, L. A. *J. Russ. Phys. Chem. (Engl. Transl.)* **1923,** *51,* 211 and 731.

[137] Grinberg, A. A.; Stetsenko, A. I.; Mitkinova, I. D. *Zh. Neorg. Khim.* **1966,** *11,* 2075.

[138] Sarukhanov, M. A.; Stetsenko, A. I. *Russ. J. Inorg. Chem. (Engl. Transl.)* **1971,** *16,* 220.

[139] Sarukhanov, M. A.; Stetsenko, A. I. *Russ. J. Inorg. Chem. (Engl. Transl.)* **1975,** *20,* 118.

[140] Sarukhanov, M. A.; Mridkha, M. S.; Kharitonov, Yu. Ya.; Stetsenko, A. I.; Konstantinova, K. K. *Koord. Khim.* **1981,** *7,* 934.

[141] Stetsenko, A. I.; Gelfman, M. I.; Mitkinova, N. D.; Strelin, S. G. *Russ. J. Inorg. Chem. (Engl. Transl.)* **1968,** *13,* 575.

[142] Kukushkin, Yu. N.; Stetsenko, A. I.; Duibanova, V. G.; Strelin, S. G. *Russ. J. Inorg. Chem. (Engl. Transl.)* **1974,** *19,* 1530.

[143] Kukushkin, Yu. N.; Stetsenko, A. I.; Strelin, S. G.; Duibanova, V. G. *Russ. J. Inorg. Chem. (Engl. Transl.)* **1972,** *17,* 561.

[144] Stetsenko, A. I.; Nefedov, V. I.; Abzaeva, T. G.; Salyn, Ya. V. *Izv. Akad. Nauk, SSSR, Ser. Khim.* **1974,** *3,* 530.

[145] Sarukhanov, M. A.; Kharitonov, Yu. Ya.; Stetsenko, A. I.; Abzaeva, I. G. *Koord. Khim.* **1977,** *3,* 739.

[146] Stetsenko, A. I.; Abzaeva, T. G.; Mitkinova, N. D.; Pogoreva, V. G.; Konovalov, L. V.; Barsukov, A. V. *Russ. J. Inorg. Chem. (Engl. Transl.)* **1976,** *21,* 97.

[147] Zarli, B.; Volponi, L.; Sindellari, L. *J. Inorg. Nucl. Chem.* **1973,** *35,* 231.

[148] Kohlschütter, V.; Hofmann, K. A. *Justus Liebigs Ann. Chem.* **1899,** *307,* 314.

[149] Hofmann, K. A. *Z. Anorg. Allg. Chem.* **1897,** *15,* 75.

[150] Zvyagintsev, O. E.; Kuznetsov, V. A. *Russ. J. Inorg. Chem. (Engl. Transl.)* **1959,** *4,* 393.

[151] Kozlov, A. G. *Russ. J. Inorg. Chem. (Engl. Transl.)* **1961,** *6,* 668.

[152] Kharitonov, Yu. Ya.; Sarukhanov, M. A. *Russ. J. Inorg. Chem. (Engl. Transl.)* **1968,** *13,* 186.

[153] Crismer, B. *Ber. Dtsch. Chem. Ges.* **1890,** *23,* 223.

[154] Sarukhanov, M. A.; Valdman, S. S.; Parpiev, N. A. *Uzb. Khim. Zh.* **1975,** 11.

[155] Valdman, S. S.; Musaev, Z. M.; Sharipov, Kh. T. *Russ. J. Inorg. Chem. (Engl. Transl.)* **1981,** *26,* 1674.

[156] Sarukhanov, M. A.; Valdman, S. S.; Parpiev, N. A. *Zh. Neorg. Khim.* **1973,** *18,* 838.

[157] Goremykin, V. I. *Izv. Akad. Nauk SSSR, Ser. Khim.* **1943,** 248.

[158] Dolcetti, G.; Hoffmann, N. W.; Collman, J. P. *Inorg. Chim. Acta* **1972,** *6,* 531.

[159] Carlin, R. L.; Baker, M. J. *J. Chem. Soc.* **1964,** 5008.

[160] Maatta, E. A.; Wentworth, R. A. D. *Inorg. Chem.* **1980,** *19,* 2597.

[161] Mertis, K.; Wilkinson, G. *J. Chem. Soc., Dalton Trans.* **1976,** 1488.

[162] Seidel, W.; Geinitz, D. *Z. Chem.* **1975,** *15,* 71.

[163] Wieghardt, K. Unpublished results.

[164] Grundy, K. R.; Reed, C. A.; Roper, W. R. *J. Chem. Soc., Chem. Commun.* **1970,** 1501.

[165] Enemark, J. H.; Feltham, R. D.; Riker-Nappier, J.; Bizot, K. F. *Inorg. Chem.* **1975,** *14,* 624.

[166] Mulvey, D.; Waters, W. A. *J. Chem. Soc., Dalton Trans.* **1975,** 951.

[167] Mašek, J.; Bapat, M. G.; Čosovic, B.; Dempir, J. *Collect. Czech. Chem. Commun.* **1969,** *34,* 485; Mašek, J.; Dempir, J. *ibid.,* 727; Mašek, J.; Mášlová, E. *ibid.* **1974,** *39,* 2141.

[168] Luňák, S.; Vepřek-Šiška, J. *Collect. Czech. Chem. Commun.* **1974,** *39,* 2719.

[169] Kurtenacker, A.; Wengefeld, F. *Z. Anorg. Allg. Chem.* **1923,** *131,* 310; **1924,** *140,* 301; Kurtenacker, A.; Werner, F. *ibid.* **1927,** *160,* 333.

[170] Davis, P.; Evans, M. G.; Higginson, W. C. *J. Chem. Soc.* **1951,** 2563.

[171] Blažek, A.; Koryta, J. *Collect. Czech. Chem. Commun.* **1953**, *18*, 326.
[172] Tomat, R.; Rigo, A. *J. Electroanal. Chem.* **1972**, *35*, 21.
[173] Petek, M.; Neal, T. E.; McNeely, R. L.; Murray, R. W. *Anal. Chem.* **1973**, *45*, 32.
[174] Clauss, H. *Ber. Bunsenges. Phys. Chem.* **1974**, *78*, 703.
[175] Fischer, O.; Dračka, O.; Fischerova, E. *Collect. Czech. Chem. Commun.* **1961**, *26*, 1505; **1962**, *27*, 1119.
[176] Delahay, P.; Mattax, C. C.; Berzius, T. *J. Am. Chem. Soc.* **1954**, *76*, 5319.
[177] Herman, H. B.; Bard, A. J. *Anal. Chem.* **1964**, *36*, 510.
[178] Christie, J. H.; Laner, G. *Anal. Chem.* **1964**, *36*, 2037.
[179] Saveant, J. M.; Vianello, E. *Electrochim. Acta* **1965**, *10*, 905.
[180] Lingane, P. J.; Christie, J. H. *J. Electroanal. Chem. Interfacial Electrochem.* **1967**, *13*, 227.
[181] Albisetti, C. J.; Coffman, D. D.; Hoover, F. W.; Jenner, E. L.; Mochel, W. E. *J. Am. Chem. Soc.* **1959**, *81*, 1489.
[183] Edge, D. J.; Norman, R. O. C. *J. Chem. Soc.* B **1969**, 182.
[184] Tomat, R.; Rigo, A. *J. Electroanal. Chem.* **1975**, *63*, 329.
[185] Gilbert, B. C.; Marriott, P. R. *J. Chem. Soc., Perkin Trans.* 2 **1979**, 1425.
[186] Ellis, C. M.; Vogel, A. I. *Analyst* **1956**, *81*, 693.
[187] Beneš, R.; Novák, J. *Collect. Czech. Chem. Commun.* **1970**, *35*, 3788.
[188] Caulton, K. G. *Coord. Chem. Rev.* **1975**, *14*, 317.
[189] Beneš, R.; Novák, J. *Collect. Czech. Chem. Commun.* **1971**, *36*, 293.
[190] Tomat, R.; Rigo, A. *J. Electroanal. Chem. Interfacial Electrochem.* **1974**, *50*, 345.
[191] Tomat, R.; Rigo, A. *J. Inorg. Nucl. Chem.* **1974**, *36*, 611.
[192] Vara Prasad, D. V. P. R.; Mahadevan, V. *Eur. Polym. J.* **1981**, *17*, 1185.
[193] Fischerová, E.; Fischer, O. *Collect. Czech. Chem. Commun.* **1965**, *30*, 675.
[194] Wells, C. F.; Salam, M. A. *J. Chem. Soc.* A **1968**, 1568.
[195] Schmidt, W.; Swinehart, J. H.; Taube, H. *Inorg. Chem.* **1968**, *7*, 1984.
[196] Ševčik, P.; Treindl, E. *Chem. Zvesti* **1969**, *23*, 822.
[197] Haight, G. P.; Swift, A. C. *J. Phys. Chem.* **1961**, *65*, 1921.
[198] Feroci, G.; Lund, H. *Acta Chem. Scand., Ser.* B **1976**, *B30*, 651.
[199] Koryta, J. *Collect. Czech. Chem. Commun.* **1954**, *19*, 666.
[200] Ferris, J. P.; Gerne, R. *Tetrahedron Lett.* **1964**, *24*, 1613.
[201] Chock, P. B.; Dewar, R. B. K.; Halpern, J.; Wong, L. Y. *J. Am. Chem. Soc.* **1969**, *91*, 82.
[202] Tomat, R.; Rigo, A. *J. Electroanal. Chem. Interfacial Electrochem.* **1975**, *59*, 191.
[203] Adamčiková, L.; Treindl, L. *Collect. Czech. Chem. Commun.* **1978**, *43*, 1844.
[204] Hlasivcová, N.; Novák, J.; Zýka, J. *Collect. Czech. Chem. Commun.* **1967**, *32*, 4403.
[205] Waters, W. A.; Wilson, I. R. *J. Chem. Soc.* A **1966**, 534.
[206] Treindl, L.; Viludová, A. *Collect. Czech. Chem. Commun.* **1974**, *39*, 3456.
[207] Beneš, R.; Hlasivcová, N.; Novák, J. *Collect. Czech. Chem. Commun.* **1971**, *36*, 1654.
[208] Bengtsson, G. *Acta Chem. Scand.* **1972**, *26*, 2494; **1973**, *27*, 2554 and 3053.
[209] Nazer, A. F.; Wells, C. F. *J. Chem. Soc., Dalton Trans.* **1980**, 1532.
[210] Hlasivcová, N.; Novák, J. *Collect. Czech. Chem. Commun.* **1969**, *34*, 3995.
[211] Kustin, K.; Davies, G. *Inorg. Chem.* **1969**, *8*, 484.
[212] Arselli, P.; Mentashi, E. *J. Chem. Soc., Dalton Trans.* **1983**, 689.
[213] Bengtsson, G. *Acta Chem. Scand.* **1973**, *27*, 1717.
[214] Jindal, V. K.; Agrawal, M. C.; Mushran, S. P. *J. Chem. Soc.* A **1970**, 2060.
[215] Bridgart, G. J.; Waters, W. A.; Wilson, I. R. *J. Chem. Soc., Dalton Trans.* **1973**, 1582.
[216] Šrámková, B.; Zýka, J.; Doležal, J. *J. Electroanal. Chem.* **1971**, *30*, 169.
[217] Šrámková, B.; Šrámek, J.; Zýka, J. *Anal. Chim. Acta* **1972**, *62*, 113.

[218] Honig, D. S.; Kustin, K.; Martin, J. F. *Inorg. Chem.* **1972,** *11,* 1895.
[219] Adams, J. Q.; Thomas, J. R. *J. Chem. Phys.* **1963,** *39,* 1904; Adams, J. Q.; Nicksic, S. W.; Thomas, J. R. *ibid.* **1965,** *45,* 654.
[220] Gutch, C. J.; Waters, W. A. *J. Chem. Soc.* **1965,** 751; Stone, R.; Waters, W. A. *Proc. Chem. Soc., London* **1962,** 253.
[221] Imamura, T.; Fujimoto, M. *Bull. Chem. Soc. Jpn.* **1975,** *48,* 2971.
[222] Szilard, I. *Acta Chem. Scand.* **1963,** *17,* 2674.
[223] Nyman, C. J. *J. Am. Chem. Soc.* **1955,** *77,* 1371.
[224] Banerjea, A. K.; Basak, A. K.; Banerjea, D. *Indian J. Chem.* **1979,** *18A,* 332.
[225] Anderson, J. H. *Analyst* **1964,** *89,* 357; **1966,** *91,* 532.
[226] Erlenmeyer, H.; Flierl, C.; Sigel, H. *J. Am. Chem. Soc.* **1969,** *91,* 1065.
[227] Swaroop, R.; Gupta, Y. K. *J. Inorg. Nucl. Chem.* **1974,** *36,* 169.
[228] Wagnerová, D. M.; Schwertnerová, E.; Veprek-Šiška, J. *Collect. Czech. Chem. Commun.* **1974,** *39,* 3036.
[229] Lloyd, C. P.; Pickering, W. F. *J. Inorg. Nucl. Chem.* **1967,** *29,* 1907.
[230] Bray, W. C.; Simpson, M. E.; McKenzie, A. A. *J. Am. Chem. Soc.* **1919,** *41,* 1363.
[231] Hlasivcová, N.; Novák, J. *Collect. Czech. Chem. Commun.* **1971,** *36,* 2027.
[232] Scott, R. A.; Haight, G. P., Jr.; Cooper, J. N. *J. Am. Chem. Soc.* **1974,** *96,* 4136.
[233] Sen Gupta, K. K.; Sen Gupta, S.; Sen, P. K.; Chatterjee, H. R. *Indian J. Chem., Sect. A* **1977,** *15,* 506.
[234] Senent, S.; Ferrari, L.; Arranz, A. *Rev. Roum. Chim.* **1981,** *26,* 377.
[235] Hlasivcová, N.; Novák, J. *Collect. Czech. Chem. Commun.* **1971,** *36,* 186.
[236] Hughes, M. N.; Shrimanker, K.; Wimbledon, P. E. *J. Chem. Soc., Dalton Trans.* **1978,** 1634.
[237] Olexová, A.; Treindle, L. *Chem. Zvesti* **1981,** *35,* 605.
[238] Hieber, W.; Nast, R. *Z. Anorg. Chem.* **1940,** *244,* 23.
[239] Vannerberg, N. G. *Acta Chem. Scand.* **1966,** *20,* 1571.
[240] Hieber, W.; Nast, R.; Proeschel, E.; Schuk, R. *Z. Anorg. Chem.* **1948,** *256,* 159.
[241] Cotton, F. A.; Monchamp, R. R.; Henry, R. J. M.; Young, R. C. *J. Inorg. Nucl. Chem.* **1959,** *10,* 28.
[242] Malatesta, L.; Sacco, A. *Z. Anorg. Allg. Chem.* **1953,** *274,* 341.
[243] Nast, R.; Gehring, G. *Z. Anorg. Chem.* **1948,** *256,* 169.
[244] Griffith, W. P.; Lewis, J.; Wilkinson, G. *J. Chem. Soc.* **1959,** 872.
[245] van der Heide, K.; Hofmann, K. *Z. Anorg. Chem.* **1896,** *12,* 282.
[246] Griffith, W. P.; Lewis, J.; Wilkinson, G. *J. Chem. Soc.* **1959,** 1632.
[247] Jagner, S.; Vannerberg, N. G. *Acta Chem. Scand.* **1968,** *22,* 3330.
[248] Svedung, D. H.; Vannerberg, N. G. *Acta Chem. Scand.* **1968,** *22,* 1551.
[249] Müller, A.; Werle, P.; Diemann, E.; Aymonino, P. J. *Chem. Ber.* **1972,** *105,* 2419.
[250] Drew, M. G. B.; Pygall, C. F. *Acta Crystallogr., Sect. B* **1977,** *B33,* 2838.
[251] Keller, A.; Jezowska-Trzebiatowska, B. *Inorg. Chim. Acta* **1981,** *51,* 123.
[252] Veprek-Šiška, J.; Luňák, S. *Collect. Czech. Chem. Commun.* **1972,** *37,* 3846.
[253] Veprek-Šiška, J.; Luňák, S. *Collect. Czech. Chem. Commun.* **1974,** *39,* 41.
[254] Bonner, F. T.; Akhtar, M. J. *Inorg. Chem.* **1981,** *20,* 3155.
[255] Nast, R.; Nyul, K.; Grziwok, E. *Z. Anorg. Allg. Chem.* **1952,** *267,* 306.
[256] Bonner, F. T.; Ravid, B. *Inorg. Chem.* **1975,** *14,* 558.
[257] Ackermann, M. N.; Powell, R. E. *Inorg. Chem.* **1966,** *5,* 1335.
[258] Hartkamp, H. *Z. Anal. Chem.* **1964,** *202,* 13.
[259] Benes, R.; Novak, J. *Collect. Czech. Chem. Commun.* **1971,** *36,* 1800.
[260] Benes, R.; Novak, J.; Sulcek, Z. *Collect. Czech. Chem. Commun.* **1972,** *37,* 1118.
[261] Wieghardt, K.; Kleine-Boymann, M.; Swiridoff, W.; Weiss, J. Unpublished results.

[262] Lukachina, V. V.; Pilipenko, A. T.; Karpova, O. I. *Zh. Anal. Khim.* **1973**, *28*, 86.
[263] Maltseva, L. S.; Shalamova, G. G.; Gusev, S. I. *Zh. Anal. Khim.* **1974**, *29*, 2053.
[264] Pilipenko, A. T.; Karpova, O. I.; Lukachina, V. V.; Trachevskii, V. V. *Zh. Anal. Khim.* **1972**, *27*, 78.
[265] Savvin, S. B.; Mineeva, V. A.; Marov, I. N.; Kalinichenko, N. B. *Zh. Anal. Khim.* **1972**, *27*, 1972.
[266] Savvin, S. B.; Mineeva, V. A.; Okhanova, L. A.; Pachadzhanov, D. N. *Zh. Anal. Khim.* **1971**, *26*, 2364.
[267] Kohschütter, V.; Hofmann, K. A. *Justus Liebigs Ann. Chem.* **1899**, *307*, 314.
[268] Canneri, G. *Gazz. Chim. Ital.* **1927**, *57*, 872.
[269] Jakob, W. F.; Jezowska, B. *Rocz. Chem.* **1931**, *11*, 229.
[270] Jakob, W. F.; Dyrek, M. *Rocz. Chem.* **1968**, *42*, 1393.
[271] Yaguchi, H.; Kajiwara, T. *Jpn. Anal.* **1965**, *14*, 785.
[272] Lassner, E. *J. Less-Common. Met.* **1968**, *15*, 143; Lassner, E.; Püschel, R.; Schedle, H. *ibid.*, 151; Lassner, E.; Schedle, H. *Talanta* **1968**, *15*, 623; Lassner, E. *ibid.* **1972**, *19*, 1121.
[273] Lassner, E.; Püschel, R.; Katzengruber, K.; Schedle, H. *Mikrochim. Acta* **1969**, *1*, 134.
[274] Savvin, S. B.; Mineeva, V. A.; Okhanova, C. A.; Pachadzhanov, D. N. *Zh. Anal. Khim.* **1971**, *26*, 532.
[275] Savvin, S. B.; Mineeva, V. A.; Okhanova, L. A. *Zh. Anal. Khim.* **1972**, *27*, 2198.
[276] Popa, G.; Dumitrescu, V. *Rev. Chim. (Bucharest)* **1975**, *26*, 761.
[277] Sarkar, S.; Müller, A. *Z. Naturforsch., B: Anorg. Chem., Org. Chem.* **1978**, *33b*, 1053.
[278] Müller, A.; Sarkar, S.; Mohan, N.; Bhattacharyya, R. *Inorg. Chem. Acta.* **1980**, *45*, L245.
[279] Müller, A.; Eltzner, W.; Bögge, H.; Sarkar, S. *Angew. Chem.* **1982**, *94*, 555; *Angew. Chem., Int. Ed. Engl.* **1982**, *21*, 535.
[280] Bhattacharyya, R.; Bhattacharjee, G. P. *J. Chem. Soc., Dalton Trans.* **1983**, 1593.
[281] Sarkar, S.; Müller, A. *Angew. Chem.* **1977**, *89*, 189; *Angew. Chem., Int. Ed. Engl.* **1977**, *16*, 183.
[282] Tatsumi, T.; Sekizawa, K.; Tominaga, H. *Bull. Chem. Soc. Jpn.* **1980**, *53*, 2297.
[283] Keller, A.; Jezowska-Trzebiatowska, B. *Bull. Acad. Pol. Sci.* **1980**, *28*, 73.
[284] Bhattacharyya, R. G.; Bhattacharjee, G. P.; Roy, P. S. *Inorg. Chim. Acta* **1981**, *54*, L63.
[285] Rajasekar, N.; Subramaniam, R.; Gould, E. S. *Inorg. Chem.* **1982**, *21*, 4111.
[286] Rajasekar, N.; Subramaniam, R.; Gould, E. S. *Inorg. Chem.* **1983**, *22*, 971.
[287] Krumpolc, M.; Roček, J. *J. Am. Chem. Soc.* **1979**, *101*, 3206.
[288] Bhattacharyya, R. G.; Roy, P. S. *J. Coord. Chem.* **1982**, *12*, 129.
[289] Bhattacharyya, R. G.; Roy, P. S. *Transition Met. Chem. (Weinhem, Ger.)* **1982**, *7*, 285.
[290] Borisova, L. V.; Ermakov, A. N.; Ismagulova, A. B. *Analyst* **1982**, *107*, 495.
[291] Norkus, P. K.; Yankauskas, Yu. Yu. *Zh. Anal. Khim.* **1972**, *27*, 2424.
[292] Eakins, J. D.; Humphreys, D. G.; Mellish, C. E. *J. Chem. Soc.* **1963**, 6012.
[293] Armstrong, R. A.; Taube, H. *Inorg. Chem.* **1976**, *15*, 1904.
[294] Smith, T. D.; Tan, C. H.; Cookson, D. J.; Pilbrow, J. R. *J. Chem. Soc., Dalton Trans.* **1980**, 1297.
[295] Grinberg, A. A.; Stetsenko, A. I. *Russ. J. Inorg. Chem. (Engl. Transl.)* **1962**, *7*, 1396.
[296] Grinberg, A. A.; Stetsenko, A. I.; Guryanova, G. P. *Russ. J. Inorg. Chem. (Engl. Transl.)* **1966**, *11*, 1008.
[297] Grinberg, A. A.; Stetsenko, A. I.; Mitkinova, N. D. *Russ. J. Inorg. Chem. (Engl. Transl.)* **1966**, *11*, 1110.
[298] Grinberg, A. A.; Stetsenko, A. I.; Strelin, S. G. *Russ. J. Inorg. Chem. (Engl. Transl.)* **1968**, *13*, 427.
[299] Kukushkin, Yu. N.; Stetsenko, A. I.; Duibanova, V. G.; Strelin, S. G. *Russ. J. Inorg. Chem. (Engl. Transl.)* **1971**, *16*, 1632.

[300] Kukushkin, Yu. N.; Stetsenko, A. I.; Duibanova, V. G.; Strelin, S. G. *Russ. J. Inorg. Chem. (Engl. Transl.)* **1974,** *19*, 1364.

[301] Mitkinova, N. D.; Konstantinova, K. K.; Duibanova, V. G.; Stetsenko, A. I. *Russ. J. Inorg. Chem. (Engl. Transl.)* **1977,** *22*, 1192.

[302] Chan, S. C.; Leh, F. *J. Chem. Soc. A* **1967,** 574.

[303] Hughes, M. N.; Stedman, G. *J. Chem. Soc.* **1963,** 2824.

[304] Hughes, M. N.; Morgan, T. D. B.; Stedman, G. *J. Chem. Soc. B* **1968,** 344.

[305] Hussain, M. A.; Stedman, G.; Hughes, M. N. *J. Chem. Soc. B* **1968,** 597.

[306] Thompson, M. S.; Meyer, T. J. *J. Am. Chem. Soc.* **1981,** *103*, 5577.

[307] Murphy, W. R.; Takeuchi, K. J.; Meyer, T. J. *J. Am. Chem. Soc.* **1982,** *104*, 5817.

[308] Armor, J. N. *Inorg. Chem.* **1973,** *12*, 1959.

[309] Wieghardt, K.; Holzbach, W.; Hofer, E.; Weiss, J. *Chem. Ber.* **1981,** *114*, 2700.

[310] Kukushkin, Yu. N.; Stetsenko, A. I.; Strelin, S. G.; Reshetnikova, Z. V. *Russ. J. Inorg. Chem. (Engl. Transl.)* **1971,** *16*, 1790.

[311] Grinberg, A. A.; Stetsenko, A. I.; Strelin, S. G. *Russ. J. Inorg. Chem. (Engl. Transl.)* **1968,** *13*, 569.

[312] Chaumette, P.; Mimoun, H.; Saussine, L.; Fischer, J.; Mischler, A. *J. Organomet. Chem.* **1983,** *250*, 291.

[313] Middleton, A. R.; Wilkinson, G. *J. Chem. Soc., Dalton. Trans.* **1980,** 1888.

ADVANCES IN INORGANIC AND BIOINORGANIC MECHANISMS, VOL. 3

Reactions Involving Coordinated Ligands

Rudi van Eldik

Institute for Physical and Theoretical Chemistry
University of Frankfurt
Frankfurt, Federal Republic of Germany

I. INTRODUCTION

This article concerns the kinetics and mechanisms of reactions involving, for the most part, coordinated oxyanions. Hipp and Busch[1] reviewed the reactions of a number of ligands and emphasized the effect the metal ion has on reactivity. The properties of a ligand are usually strongly modified by coordination to a metal center due to changes in the electron density. In the review by Hipp and Busch, many examples were given in which the metal–ligand bond is broken or formed during the reaction; in contrast, we concentrate on reactions of coordinated ligands in which the metal–ligand bond remains intact.

The main emphasis will be on ligands that are bonded via an oxygen atom to the metal center. A systematic discussion of the formation and reactivity of such coordinated ligand species will be presented. These processes involve only the formation or breakage of "secondary" bonds on the coordinated ligand. In such a treatment the reactivity of the metal–ligand bond plays a minor role, in contrast to the usual emphasis placed on such bonds in treating the reactions of metal complexes. It is therefore hoped that this contribution will underline the importance of processes involving the reactions of coordinated ligands as well as their general applicability. No attempt will be made to present a complete literature coverage in all cases, but representative examples will be selected to illustrate the significance of the reactions and the corresponding mechanisms.

The simplest form of reaction involving coordinated ligands is protonation of a hydroxo, or deprotonation of an aquo, ligand. Such processes often form an integral part of various reaction mechanisms. For instance, during base hydrolysis reactions of metal amine complexes, deprotonation of the amine ligand to produce the conjugate base species is the key to the entire reaction sequence. Coordinated ligands are in general much more acidic than are the free ligands. A number of examples and the corresponding pK_a values illustrate this point.

Many aquo complexes[1] have pK_a values close to 7, as compared to the value of 15.7 for uncoordinated (solvent) water: $Co(NH_3)_5OH_2^{3+}$ (6.3,[2] 6.6,[3]), $trans$-$Co(NH_3)_4(CN)OH_2^{2+}$ (7.6),[4] cis-$Co(tren)(OH)OH_2^{2+}$ (7.9),[5] cis-$Rh(en)_2(OH_2)_2^{3+}$ (6.1),[6] cis-$Rh(en)_2(OH)OH_2^{2+}$ (8.1),[6] $trans$-$Rh(en)_2$-$(OH)OH_2^{2+}$ (7.7),[7] $trans$-$Rh(en)_2(X)OH_2^{2+}$ (X = Cl, 6.1; X = Br, 6.4; X = I, 6.8),[8] and $Ir(NH_3)_5OH_2^{3+}$ (6.7).[9] Some metal aquo complexes are even more acidic: cis-$Co(cyclam)(OH_2)_2^{3+}$ (4.9),[10] $trans$-$Rh(en)_2(OH_2)_2^{3+}$ (4.3),[7] $trans$-$Rh(NH_3)_4(OH_2)_2^{3+}$ (4.9),[11] $Fe(H_2O)_6^{3+}$ (2.7,[12] 2.9,[13,14]), and $Pt(NH_3)_5OH_2^{4+}$ (4.1).[15] The increase of between 8 and 13 orders of magnitude in the value of K_a for free and coordinated water results from a corresponding increase in the deprotonation rate constant, which is 2.5×10^{-5} s^{-1} at 25°C for uncoordinated water and, for example, 3.6×10^3 s^{-1} at 25°C for $Co(NH_3)_5OH_2^{3+}$.[16] The reverse protonation reactions are all diffusion controlled with rate constants of the order of 10^{10} M^{-1} s^{-1}.[16] The decrease in nucleophilicity of coordinated hydroxo ligands as compared to free hydroxide ion creates the interesting possibility of allowing rapid deprotonation of coordinated water molecules, which is an essential aspect of many catalytic processes.[1]

---

[1] Abbreviations used in complexes here and throughout: acac, acetylacetone; bpy, bipyridine; cyclam, 1,4,8,11-tetraazacyclotetradecane; en, ethylenediamine; nta, nitrilotriacetic acid; phen, phenanthroline; tetren, tetraethylenepentamine; tn, trimethylenediamine; tren, N,N',N''-triaminotriethylamine.

The extent to which a metal center can influence the acidity of a coordinated ligand strongly depends on the nature of the bonding atoms and the distance that the acidic group is removed from the metal center. In the case of monodentate carbonato complexes the acidic oxygen atom is separated from the metal center by $-O-C-$ and the decrease in the pK_a value is significantly smaller than for coordinated water. Typical values are $Co(NH_3)_5$-OCO_2H^{2+} (6.4,[17] 6.7[9]), $Rh(NH_3)_5OCO_2H^{2+}$ (7.0),[9] *trans*-$Rh(en)_2$-$(OH_2)OCO_2H^{2+}$ (5.8),[7] and *cis*-$Rh(en)_2(OH_2)OCO_2H^{2+}$ (4.7).[6] A more complete list of values is given in Section II. These pK_a values are approximately four units smaller than the pK_a of HCO_3^- (i.e., 10.3 at 25°C),[18] and the difference is significantly less than in the case of coordinated water. Very similarly, the pK_a values of $Co(NH_3)_5O_2CCO_2H^{2+}$ and $Co(NH_3)_5O_2$-$CCH_2CO_2H^{2+}$ are 2.2[19] (2.05[20]) and 3.45,[21] respectively, compared to 3.7[22,23] and 5.15[21] for $C_2O_4H^-$ and $CH_2C_2O_4H^-$, respectively. Thus there is an increase in K_a on coordination of less than two orders of magnitude.

The quoted examples also illustrate the effects of the central metal atom, the overall charge on the complex, and the other surrounding ligands on the acidity of coordinated ligands. These are, however, small in comparison to the effect of coordination itself.

II. FORMATION AND REACTIVITY OF METAL CARBONATO COMPLEXES

Earlier reviews on the chemistry of metal carbonato complexes[24,25] were recently updated by a review on the chemistry of metal carbonato and carbon dioxide complexes.[26] Because much of the kinetic and mechanistic detail of such formation and aquation reactions is covered in the latter review, only the most important tendencies will be summarized here to form a basis for the forthcoming sections.

When $CO_2(g)$ is dissolved in aqueous solution, the following reactions/equilibria occur[26]:

$$CO_2(aq) + H_2O \underset{k_{-1}}{\overset{k_1}{\rightleftharpoons}} H_2CO_3 \overset{K_0}{\rightleftharpoons} HCO_3^- + H^+ \qquad K_1 \qquad (1)$$

$$HCO_3^- \rightleftharpoons CO_3^{2-} + H^+ \qquad\qquad K_2 \qquad (2)$$

Typical values at 25°C are $k_1 = 3.7 \times 10^{-2}$ s^{-1},[27] $k_{-1} = 18.3$ s^{-1},[28] $pK_0 = 3.7$,[29] $pK_1 = p(K_0k_1/k_{-1}) = 6.4$, and $pK_2 = 10.3$.[18] Of the different species, CO_2 is the most reactive and can bind to a metal hydroxo species according to the general mechanism[26]

$$ML_n(OH_2)^{p+} \rightleftharpoons ML_n(OH)^{(p-1)+} + H^+ \qquad K_3 \qquad (3)$$

$$ML_n(OH)^{(p-1)+} + CO_2 \underset{k_{-2}}{\overset{k_2}{\rightleftharpoons}} ML_n(OCO_2H)^{(p-1)+} \qquad (4)$$

$$ML_n(OCO_2H)^{(p-1)+} \rightleftharpoons ML_n(OCO_2)^{(p-2)+} + H^+ \qquad K_4 \qquad (5)$$

The produced monodentate carbonato species is stable and may, depending on the nature of L, undergo a ring-closing reaction to produce a bidentate carbonato species

$$ML_n(OCO_2)^{(p-2)+} \overset{k_3}{\rightarrow} ML_{n-1}CO_3^{(p-2)+} + L \qquad (6)$$

The kinetics of CO_2 uptake by metal aquo species has been studied for numerous systems, from which the general conclusion was drawn that only the metal hydroxo species are capable of taking up CO_2. The reaction rates are such that no M–O bond breakage occurs during the uptake process. A summary of the available rate and activation parameters for CO_2 uptake reactions along with the pK_a values of the corresponding aquo complexes is given in Table 1. It was shown[8,26] that $\log k_2$ increases linearly with pK_3, i.e., with increasing nucleophilicity of the bound hydroxide ion. The activation parameters in Table 1 present a good isokinetic relationship,[26] from which it follows that $\Delta G_0^{\ddagger} = 59.8 \pm 0.5$ kJ mol^{-1} at the isokinetic temperature of 318 ± 6 K.

On treating monodentate carbonato complexes of the type ML_n-$(OCO_2)^{(p-2)+}$ with acid, CO_2 is released and the corresponding aquo complex is produced, i.e., the reverse step of reaction (4) becomes rate determining. Isotope labeling experiments[35,36] have proved conclusively that acid-catalyzed aquation takes place without metal–oxygen bond breakage. Detailed kinetic studies[26] have demonstrated that only the protonated bicarbonato complex, i.e., $ML_n(OCO_2H)^{(p-1)+}$, undergoes rapid decarboxylation (O–C bond cleavage). In addition, the kinetic data enable the estimation of pK_4, which is summarized along with rate and activation parameters for the decarboxylation reaction in Table 2. The constant K_4 reflects the strength of the M–O bond and, indirectly, that of the O–C bond such that a decrease in $\log k_{-2}$ with increasing pK_4 is observed.[8,26] The activation parameters in Table 2 exhibit a good isokinetic relationship, from which it follows that $\Delta G_0^{\ddagger} = 73.2 \pm 0.4$ kJ mol^{-1} at the isokinetic temperature of 318 ± 13 K, in close agreement with that reported for the reverse CO_2 uptake process. The greatest number of values of pK_4 are close to 6, i.e., close to the first acid dissociation constant of carbonic acid (pK_1), illustrating

TABLE 1

Rate and activation parameters for CO_2 uptake reactions[a]

Complex ion	pK_3[b]	k_2 $(M^{-1}s^{-1})$	$\Delta H_2^{\ddagger}$ (kJ mol^{-1})	$\Delta S_2^{\ddagger}$ (J K^{-1} mol^{-1})	Ref.
$Co(NH_3)_5OH^{2+}$	6.6	220 ± 40	64 ± 4	15 ± 12	[3]
$\alpha,\beta S$-Co(tetren)OH^{2+}	6.3	166 ± 15	64 ± 5	14 ± 17	[30]
cis-Co(tren)(OH$_2$)OH^{2+}	5.3	44 ± 2	61 ± 1	-8 ± 1	[5]
cis-Co(tren)(OH)$_2$$^+$	7.9	170 ± 10	134 ± 25	222 ± 92	[5]
cis-Co(en)$_2$(OH$_2$)OH^{2+}	6.1	225 ± 15	64 ± 4	14 ± 13	[31]
$trans$-Co(NH$_3$)$_4$(CN)OH$^+$	7.6	338 ± 32	66 ± 6	21 ± 21	[32]
cis-Co(cyclam)(OH$_2$)OH^{2+}	4.9	57 ± 4	62 ± 4	3 ± 15	[10]
cis-Co(cyclam)(OH)$_2$$^+$	8.0	196 ± 24	64 ± 2	12 ± 8	[10]
$trans$-Co(cyclam)(OH$_2$)OH^{2+}	2.9	37 ± 0.1	121 ± 1	193 ± 2	[10]
$trans$-Co(cyclam)(OH)$_2$$^+$	7.2	70 ± 0.2	118 ± 3	186 ± 8	[10]
cis-Cr(C$_2$O$_4$)$_2$(OH$_2$)OH^{2-}	7.0	170 ± 10	51 ± 1	-36 ± 5	[33]
$Cu(gly)_2OH$	9.4	590 ± 30	—	—	[10]
$Ir(NH_3)_5OH^{2+}$	6.7	590 ± 30	—	—	[9]
$Rh(NH_3)_5OH^{2+}$	6.8	490 ± 120	71 ± 4	50 ± 13	[9]
cis-Rh(en)$_2$(OH$_2$)OH^{2+}	6.1	69 ± 18	67 ± 2	16 ± 7	[6]
cis-Rh(en)$_2$(OH)$_2$$^+$	8.1	215 ± 29	61 ± 1	5 ± 3	[6]
$trans$-Rh(en)$_2$(OH$_2$)OH^{2+}	4.3	81 ± 3	54 ± 4	-23 ± 13	[7]
$trans$-Rh(en)$_2$(OH)$_2$$^{+c}$	7.7	140 ± 20	68 ± 4	30 ± 12	[7]
$trans$-Rh(en)$_2$(OCO$_2$)OH	7.4	330 ± 80	64 ± 3	19 ± 10	[7]
$trans$-Rh(en)$_2$(Cl)OH$^+$	6.1	260 ± 12	52 ± 5	-22 ± 18	[8]
$trans$-Rh(en)$_2$(Br)OH$^+$	6.4	395 ± 33	63 ± 5	16 ± 16	[8]
$trans$-Rh(en)$_2$(I)OH$^+$	6.8	422 ± 43	74 ± 3	50 ± 11	[8]
$Zn(CR)OH^+$	8.7	225 ± 23	—	—	[34]

[a] Temperature, 25°C; ionic strength, 0.5 M.

[b] Acid dissociation constant of corresponding aquo complex ion [see equilibrium (3)].

[c] The rate constant for CO_2 uptake for this species is third order. For the purpose of comparison it is represented here as $k[CO_2]$, with $[CO_2] = 0.01$ M, a typical value in a large proportion of the data reported.

the similarity between the nature of the CO_2/H_2O and CO_2/MOH interactions.

The kinetic data in Tables 1 and 2 illustrate that k_2 and k_{-2} vary little with pK_3 and pK_4, respectively. In addition, labilizing effects of the ligand trans to the bicarbonate group is not significant. These tendencies can all be ascribed to the secondary nature of the bond formation/breakage processes occurring during CO_2 uptake/decarboxylation. The pressure dependence of these reactions[40] demonstrates the validity of these conclusions and suggests that the transition state has a volume approximately halfway between that of the reactant and product species. The intimate mechanism of CO_2 uptake

Rudi van Eldik

TABLE 2

Rate and activation parameters for decarboxylation reactions of
monodentate carbonato complexes[a]

Complex ion	pK_4	k_{-2} (s^{-1})	$\Delta H^{\ddagger}_{-2}$ (kJ mol^{-1})	$\Delta S^{\ddagger}_{-2}$ (J K^{-1} mol^{-1})	Ref.
$Co(NH_3)_5OCO_2H^{2+}$	6.4	1.25 ± 0.06	71 ± 2	-2 ± 4	[17]
	6.7	1.10 ± 0.05	70 ± 1	-8 ± 4	[9]
$\alpha,\beta S$-Co(tetren)OCO$_2$H^{2+}	6.4	0.28 ± 0.03	65 ± 8	-36 ± 18	[30]
cis-Co(tren)(OH$_2$)OCO$_2$H^{2+}	5.9	1.19 ± 0.06	60 ± 2	-43 ± 5	[37]
cis-Co(en)$_2$(OH$_2$)OCO$_2$H^{2+}	5.8	0.81 ± 0.04	60 ± 3	-46 ± 12	[31]
trans-Co(en)$_2$(OH$_2$)OCO$_2$H^{2+}	5.6	2.1 ± 0.1	58 ± 7	-42 ± 25	[31]
cis-Co(en)$_2$(NH$_3$)OCO$_2$H^{2+}	6.7	0.60 ± 0.02	68 ± 2	-17 ± 7	[38]
trans-Co(en)$_2$(NH$_3$)OCO$_2$H^{2+}	6.3	0.66 ± 0.02	70 ± 4	-13 ± 12	[38]
trans-Co(en)$_2$(Cl)OCO$_2$H$^+$	6.5	1.02 ± 0.05	73 ± 5	0 ± 17	[39]
trans-Co(NH$_3$)$_4$(CN)OCO$_2$H$^+$	6.9	0.38 ± 0.01	90 ± 4	46 ± 12	[32]
Ir(NH$_3$)$_5$OCO$_2$H^{2+}	6.8	1.45 ± 0.07	79 ± 2	25 ± 4	[9]
Rh(NH$_3$)$_5$OCO$_2$H^{2+}	7.0	1.13 ± 0.06	71 ± 2	-4 ± 4	[9]
cis-Rh(en)$_2$(OH$_2$)OCO$_2$H^{2+}	4.7	0.72 ± 0.02	81 ± 1	24 ± 4	[6]
trans-Rh(en)$_2$(OH$_2$)OCO$_2$H^{2+}	5.8	2.92 ± 0.08	46 ± 1	-82 ± 2	[7]
trans-Rh(en)$_2$(OCO$_2$H)$_2$$^+$	5.8	2.26 ± 0.05	68 ± 1	-9 ± 3	[7]
trans-Rh(en)$_2$(OCO$_2$)OCO$_2$H	6.4	1.3 ± 0.1	—	—	[7]
trans-Rh(en)$_2$(Cl)OCO$_2$H$^+$	6.4	1.26 ± 0.04	73 ± 2	-1 ± 7	[8]
trans-Rh(en)$_2$(Br)OCO$_2$H$^+$	6.4	1.12 ± 0.03	76 ± 1	11 ± 3	[8]
trans-Rh(en)$_2$(I)OCO$_2$H$^+$	6.6	0.55 ± 0.04	72 ± 1	-10 ± 5	[8]

[a] Temperature, 25°C; ionic strength, 0.5 M.

(MO–H bond breakage) and decarboxylation (MO–C bond breakage) can
be visualized in terms of the transition state[40,41]

$$\tag{7}$$

The acid-catalyzed and spontaneous aquation reactions of bidentate and
bridged carbonato complexes involve an initial ring-opening reaction,
during which the M–O bond is broken, followed by rapid decarboxylation
of the monodentate carbonato intermediate.[26] Since a metal–ligand bond
is broken in the initial step of the reaction, the overall process does not
involve reactions only on the coordinated ligand and is therefore not discussed in any further detail. The subsequent decarboxylation process was
treated previously. The ring-opening rate constant for such processes
strongly depends on the nature of the central metal atom, the surrounding

nonparticipating ligands, and the overall charge on the complex. This is within expectation for such primary bond-breakage processes.

Decarboxylation does not occur only on carbonato ligands. It was reported[42,43] that the 2-methyl-2-amino-malonato ligand bound to a Co(III) center can decarboxylate to produce α-alanine, during which no metal–ligand bond breakage occurs.

Finally, it should be mentioned that CO_2 uptake and decarboxylation processes form an integral part of many isotope exchange, substitution, and isomerization reactions of metal carbonato complexes.[24–26] Many systems have been reported in which CO_2 exhibits catalytic properties in a variety of reactions, and the formation of intermediate carbonato complexes was suggested (for example, see Refs. [44] and [45]). Similarly, the process of CO_2 fixation and insertion into M–H, M–C, M–N, and M–O bonds has received significant attention in recent years.[26]

III. FORMATION AND REACTIVITY OF METAL SULFITO COMPLEXES

Until recently, work in this area mainly concerned the chemistry of S-bonded sulfito complexes, with special emphasis on the strong trans-labilizing influence of S-bonded sulfito ligands in octahedral Co(III) species.[46–54] The reactions of S-bonded sulfito complexes mainly involve metal–ligand bond breakage processes and will only be referred to in a comparative way. By analogy with the uptake of CO_2 to give carbonato complexes (where O bonding is the only possibility), a similar process was expected to lead to the formation of O-bonded sulfito complexes. It is only in recent years that the chemistry of O-bonded sulfito complexes has received significant attention, although various early studies had suggested the existence of such species (see further discussion for more detail).

Sulfur dioxide dissolves readily in water to produce "sulfurous acid" [the left-hand side of equilibrium (8); "H_2SO_3"], which consists mainly of dissolved and hydrated SO_2.[55,56] The aqueous chemistry of dissolved SO_2 can be summarized as follows:

$$H_2O + SO_2 \underset{k_{-4}}{\overset{k_4}{\rightleftharpoons}} HSO_3^- + H^+ \qquad K_5 \qquad (8)$$

$$HSO_3^- \rightleftharpoons SO_3^{2-} + H^+ \qquad K_6 \qquad (9)$$

$$2HSO_3^- \underset{k_{-5}}{\overset{k_5}{\rightleftharpoons}} S_2O_5^{2-} + H_2O \qquad K_7 \qquad (10)$$

Typical values for the rate and overall equilibrium constants at 25°C are $k_4 = 1 \times 10^{8}$[57] and 3.4×10^{6} s^{-1},[58] $k_{-4} = 2.5 \times 10^{9}$[57] and 2×10^{8} M^{-1} s^{-1},[58] $pK_5 = 1.9$,[2,59] $pK_6 = 6.3$,[2,59] $k_5 = 7 \times 10^{2}$ M^{-1} s^{-1},[57] $k_{-5} = 1 \times 10^{4}$ s^{-1},[57] and $K_7 = 0.07$[60] and 0.088 M.[61] Furthermore, the constant K_5 is slightly temperature dependent and an average value of $\Delta H_5 = -18$ kJ mol^{-1} was determined[2] from literature data.[59,62] A disulfite species is formed at high concentrations of bisulfite according to equilibrium (10). The magnitude of K_7 is such that solutions of $Na_2S_2O_5$ exist as $\geq 99\%$ HSO_3^- at $[Na_2S_2O_5] \leq 0.05$ M, conditions usually adopted in kinetic experiments.

In constast to S-bonded sulfito complexes, very little was reported on the formation and reactivity of O-bonded sulfito complexes before 1980. It was suggested by Stranks and co-workers[63] that the reaction of sulfite with *trans*-Co(en)$_2$(OH)OH$_2$$^{2+}$ involves addition of SO_2 to the aquo/hydroxo ligand, followed by rapid intramolecular rearrangement from O- to S-bonded sulfite. They[63] were unable to detect the O-bonded intermediate. In a later study[64] the participation of an O-bonded intermediate was also suggested for the reaction of *trans*-Co(en)$_2$(OH$_2$)SO$_3$$^+$ with SO_3^{2-}. A similar intermediate was suggested in the reduction of Co(nta)(OH$_2$)$_2$ by sulfite.[65] The rapid formation of Cr(OH$_2$)$_5$OSO$_2$$^+$ during the reaction of Cr(OH$_2$)$_6$$^{3+}$ and HSO_3^- was reported in two independent studies.[66,67] Finally, the oxidation of Fe(CN)$_5$SO$_3$$^{4-}$ by IrCl$_6$$^{2-}$ was suggested to involve the formation of the Fe(CN)$_5$OSO$_2$$^{4-}$ species.[68] In all of these examples, only indirect evidence for the existence of O-bonded sulfito complexes was offered. Recently, however, the situation has changed drastically, and more direct evidence has become available.

The first study that reported a detailed analysis of kinetic data on the formation and stability of O-bonded sulfito complexes was performed on the Co(NH$_3$)$_5$OH$_2$$^{3+}$/SO$_2$ system.[2] Bubbling SO_2 gas or adding $Na_2S_2O_5$, NaHSO$_3$, or Na$_2$SO$_3$ to buffered solutions of Co(NH$_3$)$_5$OH$_2$$^{3+}$ at a pH between 4 and 7 resulted in an instantaneous color change from light to dark red. The spectral changes are in good agreement with that expected for the formation of an O-bonded sulfito complex. Immediate acidification of such solutions caused an instantaneous color change and the resulting spectra were in agreement with that of Co(NH$_3$)$_5$OH$_2$$^{3+}$. Thus the process involves reversible SO_2 uptake to produce Co(NH$_3$)$_5$OSO$_2$$^+$, which slowly decomposes to Co^{2+} and SO_4^{2-} ($t_{1/2} \sim 50$ s). An ^{17}O-tracer NMR study[69] revealed that no Co–O bond is broken during the SO_2 uptake and reverse acid-catalyzed aquation reactions. The suggested mechanism[2] is similar to that outlined in equilibria (3)–(5) for CO_2 uptake/decarboxylation and can be summarized as follows:

$$
\begin{array}{ccccc}
Co(NH_3)_5OH^{2+} & + & SO_2 & \xrightleftharpoons[k_{-6}]{k_6} & Co(NH_3)_5OSO_2H^{2+} \\
\Big\updownarrow K_4 & & \Big\updownarrow K_5 & & \Big\updownarrow K_8 \\
Co(NH_3)_5OH_2^{3+} & & HSO_3^- & & Co(NH_3)_5OSO_2^+ \\
& & \Big\updownarrow K_6 & & \Big\downarrow k_7 \\
& & SO_3^{2-} & & Co^{2+} + 5\,NH_3 + SO_4^{2-}
\end{array}
\tag{11}
$$

All participating hydrogen ions are omitted from scheme (11) for the sake of simplicity and the K values are defined as acid dissociation constants. Typical values for the rate constants are[2] $k_6 = 2.6 \pm 0.6 \times 10^8$ and $k_{-6}/K_8 = 2.2 \pm 0.4 \times 10^6 \; M^{-1} \; s^{-1}$ at 10°C; k_7 (intramolecular redox reaction) $= 2.5 \pm 0.4 \times 10^{-3} \; s^{-1}$ at 15°C. On the assumption that $K_8 \sim K_5$, as for the corresponding carbonate system, then $k_{-6} \sim 4 \times 10^4 \; s^{-1}$ at 10°C.

The redox reaction (k_7) is much slower than SO_2 uptake or elimination, but is significantly faster than that usually found for S-bonded sulfito complexes.[47,48] The formation of $Co(NH_3)_5OSO_2^+$ becomes an equilibrium process during the study of the redox reaction. Kinetic data were obtained as a function of temperature, pH, and the concentration of total sulfite, and are in agreement with the expression

$$
k_{obs} = \frac{k_7 K_4 K_9 [SO_2]}{K_4[H^+] + K_4 K_9 [SO_2] + [H^+]^2}
\tag{12}
$$

which is the rate law for the following suggested redox mechanism.[2]

$$
\begin{array}{l}
Co(NH_3)_5OH^{2+} + SO_2 \xrightleftharpoons{K_9} Co(NH_3)_5OSO_2^+ + H^+ \\[1em]
Co(NH_3)_5OSO_2^+ \xrightarrow{k_7} Co^{2+}(aq) + 5\,NH_3 + SO_3^- \\[1em]
Co(NH_3)_5OSO_2^+ + SO_3^- \xrightarrow{fast} (NH_3)_5Co\text{---}O\text{-}S \\[2em]
\Big\downarrow fast \\[1em]
Co^{2+}(aq) + 5\,NH_3 + SO_2 + SO_4^{2-}
\end{array}
\tag{13}
$$

A fit of the experimental data results in $k_7 = 1.4 \pm 0.3 \times 10^{-2}$ s^{-1} at 25°C, $\Delta H^{\ddagger} = 112 \pm 5$ kJ mol^{-1}, and $\Delta S^{\ddagger} = 96 \pm 17$ J K^{-1} mol^{-1}. The mechanism outlined in sheme (13) was selected on the basis of numerous experiments performed to study its nature.[2] The rate-determining step is direct reduction by the sulfito ligand, with formation of Co^{2+} and the radical ion SO$_3^-$. However, the activation parameters for k_7 are similar in magnitude to those frequently observed for various substitution reactions of pentamminecobalt(III) complexes. This could suggest that the rate-determining step is loss of NH$_3$, probably trans to the sulfito ligand, which would fit the data equally well. A recent pressure-dependence study of the redox reaction[70] resulted in a volume of activation of $+34 \pm 3$ cm^3 mol^{-1} at 25°C. This can either be interpreted as evidence for considerable bond breakage and charge neutralization during the rate-determining electron transfer step, or as evidence for the dissociative release of NH$_3$ as suggested above.[70] It is most likely that both processes, i.e., electron transfer and release of NH$_3$, contribute to the overall observed activation parameters.

Sykes and co-workers[71] observed that sulfite complexes with Cr(NH$_3$)$_5$-OH$_2^{3+}$ with retention of the metal–oxygen bond. In a later detailed kinetic study[72] it was found that both Cr(NH$_3$)$_5$OH$_2^{3+}$ and Rh(NH$_3$)$_5$OH$_2^{3+}$ react with sulfite to produce O-bonded sulfito species, which on acidification decompose to the aquo complexes. The kinetic data confirm an SO$_2$ uptake mechanism similar to that outlined in scheme (11) for the corresponding Co(III) complex. The values of k_6 are $1.8 \pm 0.3 \times 10^8$ [Rh(III) complex] and $2.9 \pm 0.6 \times 10^8$ M^{-1} s^{-1} [Cr(III) complex] at 25°C, which are slightly smaller values than $4.7 \pm 0.3 \times 10^8$ M^{-1} s^{-1} found for the Co(III) complex at 25°C.[2] Both complexes [Cr(NH$_3$)$_5$OSO$_2^+$ and Rh(NH$_3$)$_5$OSO$_2^+$] undergo subsequent reactions[73] during which O- to S-bonded isomerization and/or substitution occurs to produce the corresponding *trans*-disulfito species.

In subsequent studies El-Awady and Harris[74,75] investigated SO$_2$ uptake by Co(tren)(OH$_2$)$_2^{3+}$ and the corresponding aquo/hydroxo and dihydroxo complexes. They found that not only the latter two species react with SO$_2$ to produce O-bonded sulfito complexes as expected, but that the diaquo complex can do the same, although at an appreciably lower rate. The rate and activation parameters for these reactions are summarized in Table 3 along with those for other Co(III) complexes. The intermediate Co(tren)(OH$_2$)-OSO$_2^+$ species undergoes a rapid elimination of SO$_2$ upon acidification, for which the rate data are $k = 4.7 \times 10^5$ M^{-1} s^{-1} at 10°C, $\Delta H^{\ddagger} = 46.4 \pm 3.3$ kJ mol^{-1}, and $\Delta S^{\ddagger} = 29 \pm 2$ J K^{-1} mol^{-1}. The magnitude of k is in close agreement with that reported for the corresponding pentammine complex.[2] On the assumption that the pK_8 value for Co(tren)(OH$_2$)(OSO$_2$H)$^{2+}$ is close to that for *cis*-Co(en)$_2$(OH$_2$)(OSeO$_2$H)$^{2+}$,[78] the magnitude of the SO$_2$

TABLE 3

Rate and activation parameters for SO_2 uptake reactions
of Co(III) complexes at 25°C

Complex ion	k_6 ($M^{-1} s^{-1}$)	$\Delta H_6^{\ddagger}$ (kJ mol^{-1})	$\Delta S_6^{\ddagger}$ (J K^{-1} mol^{-1})	Ref.
$Co(NH_3)_5OH^{2+}$	4.7×10^8	41 ± 1	59 ± 2	[2]
$\alpha,\beta S\text{-Co(tetren)}OH^{2+}$	3.3×10^8	-2.1 ± 4.2	-88 ± 15	[76]
$Co(tren)(OH_2)_2^{3+}$	9×10^6	150 ± 4	393 ± 12	[74]
$Co(tren)(OH_2)OH^{2+}$	5.3×10^7	18.8 ± 0.4	-33 ± 1	[74]
$Co(tren)(OH)_2^+$	2.4×10^9	48 ± 1	96 ± 13	[74]
$cis\text{-Co(en)}_2(OH_2)OH^{2+}$	1.0×10^8	25.1 ± 0.4	-6.7 ± 0.8	[77]

elimination step could be estimated, and this is given in Table 4. The subsequent reactions of the $Co(tren)(OH_2)(OSO_2)^+$ species strongly depend on the pH of the solution. At low pH (between 2.9 and 5.4) an internal redox reaction produces Co(II) and sulfate as for the pentammine complex[2]: $k = 1 \times 10^{-3}$ s^{-1} at 25°C, $\Delta H^{\ddagger} = 101 \pm 8$ kJ mol^{-1}, and $\Delta S^{\ddagger} = 34 \pm 27$ J K^{-1} mol^{-1}. El-Awady and Harris[75] suggest that both $Co(tren)(OH_2)$-$(OSO_2)^+$ and $Co(tren)(OH_2)(OSO_2H)^{2+}$ undergo ring closure accompanied by electron transfer to produce the redox products. The feasibility of such a doubly O-bonded intermediate was justified by reports on the preparation of closely related complexes.[79,80] At high pH (7.2 to 8.9) internal redox no longer takes place and a subsequent substitution process occurs during which $Co(tren)(SO_3)_2^-$ is produced. The mechanism consistent with the experimental rate law involves either SO_3^{2-} addition to Co-$(tren)(OH_2)(OSO_2)^+$ or HSO_3^- addition to the deprotonated congener $Co(tren)(OH)(OSO_2)$ as the rate-determining step. IR spectra of the final

TABLE 4

Rate and activation parameters for acid-catalyzed decomposition
of Co(III) sulfito complexes at 25°C

Complex ion	$k_{-6}{}^a$ (s^{-1})	$\Delta H_{-6}^{\ddagger}$ (kJ mol^{-1})	$\Delta S_{-6}^{\ddagger}$ (J K^{-1} mol^{-1})	Ref.
$Co(NH_3)_5OSO_2H^{2+}$	1.0×10^3	—	—	[2]
$Co(tetren)OSO_2H^{2+}$	1.1×10^3	30 ± 10	-85 ± 36	[76]
$Co(tren)(OH_2)OSO_2H^{2+}$	3.8×10^3	46 ± 3	29 ± 2	[74]
$cis\text{-Co(en)}_2(OH_2)OSO_2H^{2+}$	6.0×10^3	—	—	[77]

^{a} Estimated on the assumption that the pK_a value of the bisulfito complex is 3.9 at 25°C.[74]

product suggest that it contains one O-bonded sulfito ligand, Co(tren)-$(OSO_2)(SO_3)^-$. Baldwin and Schiavon *et al.* argue against the possibility of a rapid substitution reaction since such processes are usually controlled by the water-exchange rate of such species. However, another study[54] clearly demonstrated the very significant kinetic trans effect of sulfite ligands, such that a five-coordinate intermediate could be formed that rapidly reacts with sulfite to produce the *trans*-disulfito species. Very similarly, the O-bonded sulfito ligand could also labilize the cis position to produce the Co(tren)-$(OSO_2)SO_3^-$ species at a rate significantly faster than the normal water exchange of Co(III) amine complexes.[81]

Harris and co-workers[82] studied SO_2 uptake by $\alpha,\beta S$-Co(tetren)OH_2^{3+} to produce $\alpha,\beta S$-Co(tetren)OSO_2^+, which undergoes acid-catalyzed elimination of SO_2, or an intramolecular isomerization reaction to produce the S-bonded isomer. The uptake of SO_2 is fast, as expected (Table 3), and the rate parameters for SO_2 elimination are close to those for Co(tren)(OH_2)-OSO_2^+ (Table 4). An interesting aspect of this system is the absence of a redox reaction and the presence of a linkage isomerization step. This is probably due to the fact that the ligand trans to the sulfito group is part of the tetren chelate, which prohibits further substitution by sulfite or water in the trans position and which may be a prerequisite for an internal electron transfer process to occur.[70] The linkage isomerization reaction is characterized by the parameters[82] $k = 2.8 \times 10^{-4}$ s^{-1} at 25°C, $\Delta H^{\ddagger} = 56 \pm 4$ kJ mol^{-1}, and $\Delta S^{\ddagger} = -124 \pm 12$ J K^{-1} mol^{-1}. The low activation parameters may be indicative of an internal S_N2 mechanism involving concerted Co–O and Co–S bond breakage and formation, respectively. Further information on the nature of the isomerization mechanism is expected to be obtained from high-pressure kinetic and ^{17}O-tracer experiments.[83]

Very similarly it was found[73,84] that *trans*-Co(NH$_3$)$_4$(CN)OH$^+$ can take up SO_2 to produce *trans*-Co(NH$_3$)$_4$(CN)OSO$_2$, which isomerizes to *trans*-Co(NH$_3$)$_4$(CN)SO$_3$. At present the reactions of Co(phen)$_2$(OH$_2$)$_2^{3+}$ and Co(bpy)$_2$(OH$_2$)$_2^{3+}$ with SO_2/HSO$_3^-$/SO$_3^{2-}$ are under investigation.[85] Preliminary results have demonstrated an extremely rapid association of SO_2 with the diaquo species, followed by formation of the S-bonded sulfito complex.

In a recent study[15] it was reported that Pt(NH$_3$)$_5$OH^{3+} reacts in aqueous sulfite solution to produce Pt(NH$_3$)$_5$OSO$_2^{2+}$. The rate constant for SO_2 addition was found to be 6.8×10^6 M^{-1} s^{-1} at 25°C, which is consistent with a mechanism that does not involve metal–oxygen bond rupture. In this study the workers succeeded in isolating the O-bonded species as [Pt(NH$_3$)$_5$-OSO$_2$]SO$_3 \cdot 2H_2O$, which, in the presence of sulfite, undergoes a slow rearrangement to form *cis*-Pt(NH$_3$)$_4$(SO$_3$)$_2$. It was suggested that there is a rapid *cis*-NH$_3$ replacement by water followed by a rate-determining O- to S-bonded sulfite isomerization step ($k = 4 \times 10^{-5}$ s^{-1} at 25°C), followed by the rapid substitution of the *cis*-aquo ligand by sulfite. In the case of the Pt(NH$_3$)$_4$(OH)$_2$$^{2+}$ species,[86] it was found that Pt(NH$_3$)$_4$(OH)OSO$_2$$^+$ undergoes a fairly fast rearrangement process in the presence of sulfite to produce *trans*-Pt(NH$_3$)$_4$(SO$_3$)$_2$. In this case the rate constant for SO$_2$ uptake is 2×10^7 M^{-1} s^{-1} at 25°C.

A comparison of the available rate parameters for SO$_2$ uptake by Co(III) complexes is given in Table 3. These are on the average $\sim 10^6$ times faster than the corresponding CO$_2$ uptake rate constants (see Table 1) and are accounted for by a combination of lower $\Delta H^{\ddagger}$ and higher $\Delta S^{\ddagger}$ values.[2] A similar reactivity difference is found in the hydration rates of SO$_2$ and CO$_2$,[58] and in their solubility data (3.6 and 0.08 M at 0°C, respectively).[87] The activation parameters in Table 3 and those reported for the Pt(IV) complexes[15,86] show a good isokinetic correlation (Fig. 1), from which it follows that $\Delta G_0^{\ddagger} = 26.4 \pm 1.9$ kJ mol^{-1} at the isokinetic temperature of 314 ± 12 K. The good correlation indicates that a common mechanism is operative. The value of $\Delta G_0^{\ddagger}$ is more than 30 kJ mol^{-1} smaller than the value for CO$_2$ uptake, and further accounts for the observed reactivity difference referred to above.

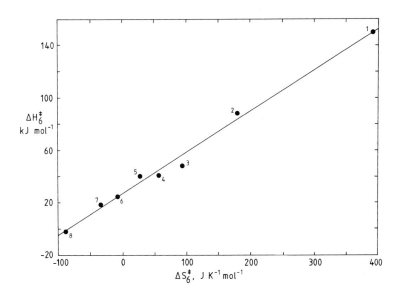

Fig. 1. Isokinetic plot for various SO$_2$ uptake reactions: (1) Co(tren)(OH$_2$)$_2$$^{3+}$, Ref. [74]; (2) Pt(NH$_3$)$_5$OH^{3+}, Ref. [15]; (3) Co(tren)(OH)$_2$$^+$, Ref. [74]; (4) Co(NH$_3$)$_5$OH^{2+}, Ref. [2]; (5) Pt(NH$_3$)$_4$(OH)$_2$$^{2+}$, Ref. [86]; (6) *cis*-Co(en)$_2$(OH$_2$)OH^{2+}, Ref. [77]; (7) Co(tren)(OH$_2$)-OH^{2+}, Ref. [74]; (8) $\alpha,\beta S$-Co(tetren)OH^{2+}, Ref. [76].

A direct comparison of the elimination rates of SO_2 and CO_2 is not possible since the pK_a values of the protonated sulfito complexes could not be obtained from the pH dependence of the elimination rate constant, as in the case of the carbonato complexes (Table 2). However, a reasonable estimate[74] demonstrates that the elimination of SO_2 is at least 10^3 times faster than the elimination of CO_2 (i.e., decarboxylation). These tendencies confirm the general conclusion that uptake and elimination of SO_2 do not involve metal–oxygen bond breakage.

IV. REACTIONS OF METAL SULFINATO AND RELATED COMPLEXES

A good correlation exists between the kinetic and structural (ground state) trans effects of S-bonded sulfito complexes of Co(III).[88] The structural trans effect of sulfinato complexes is significantly smaller than that of sulfito complexes,[88] and structural and other analyses have demonstrated that the *p*-toluenesulfinato and benzenesulfinato complexes of pentamminecobalt(III) are S bonded. It follows that the complexes are prepared via simple substitution on the Co(III) center and are not of further interest in the context of the present discussion.

On the other hand, sulfinato complexes of pentamminecobalt(III) are generally O bonded, and both M–O and O–S bond cleavage may occur during aquation and base-hydrolysis reactions.[89] ^{18}O-Tracer experiments have demonstrated that no O–S bond cleavage occurs during the acid- and base-hydrolysis reactions of $Co(NH_3)_5OSO_2CH_3{}^{2+}$ and $Co(NH_3)_5OSO_2$-$CF_3{}^{2+}$. Competition experiments indicate that the acid-independent aquation reaction is leaving-group dependent, while the base-hydrolysis process follows the trends as expected for a S_N1CB mechanism.

In the case of the $Co(NH_3)_5O_3SF^{2+}$ complex,[90] solvolysis proceeds via Co–O cleavage to produce $Co(NH_3)_5(solvent)^{3+}$ and concurrently via S–F cleavage to produce $Co(NH_3)_5F^{2+}$. The compound FSO_3H is a known fluorination agent,[91] such that the $Co(NH_3)_5O_3SF^{2+}$ complex can transfer F^- intramolecularly to the metal. The latter reaction was suggested[90] to involve an isomerization step followed by S–F bond cleavage, i.e., rapid elimination of SO_3.

$$(NH_3)_5Co-O\overset{\displaystyle\left.\begin{array}{c}F\\ \diagdown\\ S\end{array}\right.^{2+}}{\diagup\diagdown}\!\!\!\begin{array}{c}O\\ \\ O\end{array}\quad\xrightarrow{\text{slow}}\quad (NH_3)_5Co-FSO_3^{2+}\quad\xrightarrow{\text{fast}}\quad (NH_3)_5CoF^{2+}\ +\ SO_3$$

$$(14)$$

It follows that the Co–O bond is broken during direct solvolysis and linkage isomerization of the FSO_3^- ligand, but that the Co–F bond remains intact during the subsequent elimination of SO_3. Similar isomerization reactions have been observed for O- to N-bonded nitrite (see Section VI), N- to O-bonded $NH_2SO_3^-$ [92], S- to N-bonded SCN^- [93,94], N- to O-bonded NH_2COO^- [95], N- to O-bonded $OC(NH_2)N(CH_3)_2$ [96], and N- to O-bonded amide complexes of pentamminecobalt(III).[97] It has furthermore been suggested that the base hydrolysis of $Co(NH_3)_5O_3SF^{2+}$ also involves a base-catalyzed linkage isomerization process, followed by dissociative loss of FSO_3^- (via Co–F breakage) to produce a five-coordinate intermediate, as is usually postulated for such mechanisms.

The ability of coordinated hydroxide to react with various electrophiles has led to interesting synthetic routes. Along these lines it was recently reported[98] that $Co(NH_3)_5OH^{2+}$ reacts with C_6H_5NCS to produce

$$(NH_3)_5CoS-C\underset{NHC_6H_5}{\overset{O}{\diagup\!\!\!\diagdown}}$$

which probably occurs via the intermediate

$$(NH_3)_5CoO-C\underset{NHC_6H_5}{\overset{S}{\diagup\!\!\!\diagdown}}$$

The Co–O bond presumably remains intact during the formation of the latter species.

V. REACTIONS OF METAL SELENITO COMPLEXES

Fowless and Stranks[78,99,100] reported the synthesis and characterization of a series of selenito complexes in which the ligand is oxygen bonded to the metal center. The SeO_3^{2-} ion is very similar in its behavior to the SO_3^{2-} ion. It is dibasic, producing $HSeO_3^-$ and H_2SeO_3 in acidic medium; at higher concentrations species such as[101] $H(SeO_3)_2^{3-}$, $H_2(SeO_3)_2^{2-}$, $H_3(SeO_3)_2^-$, and $H_4(SeO_3)_2$ have been reported. In addition, the following equilibrium is established rapidly in solution.[102]

$$2HSeO_3^- \rightleftharpoons Se_2O_5^{2-} + H_2O \tag{15}$$

Typically, pK_a values for H_2SeO_3 are 2.4 and 8.1 at 20°C and 1 M ionic strength.[78] From this information it is expected that the formation of O-bonded selenito complexes should closely parallel that of sulfito complexes. Fowless and Stranks[78] succeeded in isolating the following species as complex salts: $Co(NH_3)_5OSeO_2^+$, *cis*- and *trans*-$Co(en)_2(OH_2)OSeO_2H^{2+}$, *cis*-$Co(en)_2(OH_2)OSeO_2^+$, $Co(en)_2O_2SeO^+$, and $Co(tn)_2O_2SeO^+$. In the latter two complexes selenite acts as a bidentate ligand and produces a

ring structure. A remarkable feature of these species is the stability of the protonated selenito complexes, in contrast to the situation for the protonated carbonato and sulfito species. The pK_a values of the biselenito complexes are 4.35 and 4.55 for the cis and trans species, respectively.

Detailed kinetic studies[99] of the formation and decomposition reactions of selenito complexes have revealed some interesting tendencies. The formation rate constant increases with increasing $[HSeO_3^-]$ and a limiting rate is reached at high $[HSeO_3^-]$. This was interpreted as evidence for an interchange mechanism, which can be formulated for a particular complex ion as follows:

$$cis\text{-}Co(tn)_2(OH_2)_2^{3+} + HSeO_3^- \overset{K_{10}}{\rightleftharpoons} cis\text{-}Co(tn)_2(OH_2)_2^{3+} \cdot HSeO_3^-$$

$$\downarrow k_8$$

$$cis\text{-}Co(tn)_2(OH_2)OSeO_2H^{2+} + H_2O$$
(16)

For this specific complex, $K_{10} = 6.7\ M^{-1}$ and $k_8 = 63\ s^{-1}$ at 30°C, pH 3.3, and ionic strength 1.0 M. Similar tendencies were observed for the other investigated complexes. Typical rate and activation parameters for such

TABLE 5

Rate and activation parameters for the formation
of Co(III) selenito complexes[a]

Complex	k_8 (s^{-1})	$\Delta H_8^{\ddagger}$ (kJ mol^{-1})	$\Delta S_8^{\ddagger}$ (J K^{-1} mol^{-1})
cis-$Co(tn)_2(OH_2)_2^{3+}$	17.4	54 ± 1	−42 ± 4
trans-$Co(tn)_2(OH_2)_2^{3+}$	66	51 ± 3	−42 ± 5
cis-$Co(en)_2(OH_2)_2^{3+}$	16.8	54 ± 2	−40 ± 4
trans-$Co(en)_2(OH_2)_2^{3+}$	47	48 ± 2	−53 ± 5
$Co(NH_3)_5OH_2^{3+}$	8.8	57 ± 2	−38 ± 4

[a] Data taken from Ref. [99]; temperature, 25°C; ionic strength, 1.0 M; [Se(IV)] = 0.1 M, pH 3.3.

formation reactions are summarized in Table 5. The reverse aquation reaction

$$R\text{-}OSe_2OH^{2+} + H_2O \xrightarrow{k_9} R\text{-}OH_2{}^{3+} + HSeO_3{}^{-} \qquad (17)$$

was found to be independent of $[H^+]$ in the range $3 > pH > 1$; the corresponding rate and activation parameters are summarized in Table 6.

The values of k_8 and k_9 and their activation enthalpies clearly demonstrate that these processes differ significantly from normal substitution reactions that involve M–O bond breakage. For example, k_8 for *cis*-Co(en)$_2$(OH$_2$)$_2{}^{3+}$ is $\sim 10^6$ times greater than the value for the water-exchange rate constant.[99] It follows that these reactions do not involve Co–O bond breakage, and a direct interaction between the aquo ligand and $HSeO_3{}^{-}$, or between the hydroxy ligand and H_2SeO_3, must account for the rapid formation of the selenito complexes. The following scheme was suggested to account for the observed effects.[99]

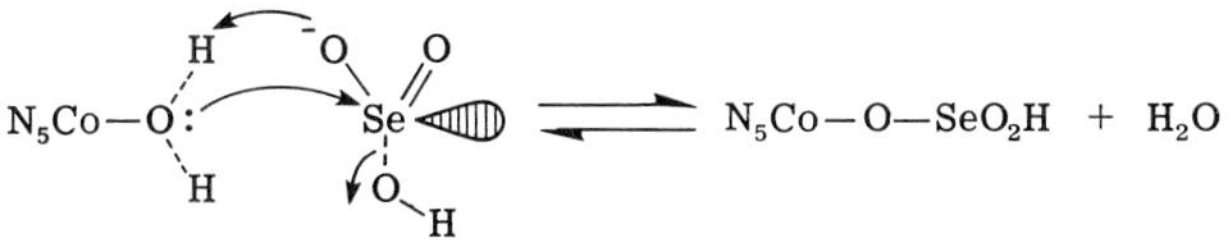

$$N_5Co\text{-}O\overset{H}{\underset{H}{\cdots}}\overset{O}{\underset{O}{\underset{H}{Se}}} \rightleftharpoons N_5Co\text{-}O\text{-}SeO_2H + H_2O \qquad (18)$$

A similar suggestion was made to account for the reaction of the corresponding hydroxo complexes at high pH,[100] for which the rate law [Eq. (19)] was obtained.

$$k_{anation} = k_{10}[HSeO_3{}^{-}] + k_{11}[Se(IV) \text{ dimer}] \qquad (19)$$

Typical values of k_{10} and k_{11} for *cis*-Co(en)$_2$(OH$_2$)OH^{2+} are 4 and 110 M^{-1} s^{-1} at 25°C, respectively. The first term of Eq. (19) involves the reaction of the hydroxo complex with $HSeO_3{}^{-}$, whereas the second term involves the reaction with the dimeric $H_2(SeO_3)_2{}^{2-}$ and $H(SeO_3)_2{}^{3-}$ species. The rate data

TABLE 6

Rate and activation parameters for the aquation of a series
of Co(III) selenito complexes[a]

Complex	k_9 (s^{-1})	$\Delta H_9^{\ddagger}$ (kJ mol^{-1})	$\Delta S_9^{\ddagger}$ (J K^{-1} mol^{-1})
cis-Co(tn)$_2$(OH$_2$)OSeO$_2$H^{2+}	0.20	53 ± 3	−82 ± 11
trans-Co(tn)$_2$(OH$_2$)OSeO$_2$H^{2+}	1.0	54 ± 3	−64 ± 8
cis-Co(en)$_2$(OH$_2$)OSeO$_2$H^{2+}	0.18	52 ± 3	−86 ± 11
trans-Co(en)$_2$(OH$_2$)OSeO$_2$H^{2+}	1.1	49 ± 2	−81 ± 11
Co(NH$_3$)$_5$OSeO$_2$H^{2+}	0.28	51 ± 3	−78 ± 10

[a] Data taken from Ref. [100]; temperature, 25°C; ionic strength, 1 M; pH 1.

for the complex, as well as for $Co(NH_3)_5OH^{2+}$, $Rh(NH_3)_5OH^{2+}$, *cis*-$Co(tn)_2(OH_2)OH^{2+}$, *trans*-$Co(en)_2(OH_2)OH^{2+}$, and *trans*-$Co(tn)_2(OH_2)$-OH^{2+}, are such that substitution at the Se(IV) center is believed to account for the observed tendencies.[100] The complexes all exhibit $\Delta H^{\ddagger}$ values in the range 50–60 kJ mol^{-1} at pH 10, which further underlines the suggested mechanism. The corresponding rate constants are six times faster than those for oxygen exchange on the Se(IV) center, and the intimate nature of the k_{10} and k_{11} steps can be visualized as follows:

$$M-\overset{\displaystyle H}{\underset{\displaystyle H}{O}}\cdots\overset{O}{\underset{O}{Se}}=O \qquad\qquad M-\overset{\displaystyle H}{\underset{\displaystyle H}{O}}\cdots\overset{O}{\underset{O}{Se}}-O\overset{H-O}{\underset{O\quad O^-}{Se}} \tag{20}$$

In reference to the previous discussion of reactions involving the uptake and loss of CO_2 and SO_2 (Sections II and III), the question arises whether the observed tendencies described earlier could be interpreted in terms of the participation of a hydrated SeO_2 species, i.e., H_2SeO_3, similar to H_2CO_3 and H_2SO_3. Even if this species is present only in very low concentrations, it does create the possibility of SeO_2 being the reactive species, as in the case of CO_2 and SO_2. If this is the case, the acidification of selenito complexes will proceed via the loss of SeO_2, i.e., O–Se bond breakage, on the protonated $M-OSeO_2H$ species. Similarly, the formation of selenito complexes will involve the uptake of SeO_2 by metal hydroxy species. Whether this possibility is feasible is under investigation in our laboratory.

Furthermore, the general scheme for the formation of sulfito complexes (scheme 11) would further suggest the possibility of forming an Se-bonded selenito complex by treating hydroxo complexes with SeO_3^{2-} at high pH. Such a reaction will proceed via a rate-determining substitution process during which metal–oxygen bond breakage occurs. Evidence for the formation of Se-bonded selenito complexes under such conditions was indeed reported[103] for the $Co(NH_3)_5SeO_3^+$ species. It was suggested[103] that the Se-bonded species can isomerize to the O-bonded species in solution to produce an equilibrium mixture of both species. Similar isomerization reactions were reported for sulfito complexes (see Section III).

VI. FORMATION AND REACTIVITY OF NITRITO COMPLEXES

Nitrosation reactions of metal aquo/hydroxo complexes have received much attention from different research groups. Almost 30 years ago Pearson and co-workers[104] suggested that $Co(NH_3)_5OH_2^{3+}$ and NO_2^- react in acidic medium, without breakage of the metal–oxygen bond, to produce $Co(NH_3)_5ONO^{2+}$, which subsequently isomerizes to $Co(NH_3)_5NO_2^{2+}$. This suggestion was later confirmed by Murmann and Taube,[105] who illustrated with ^{18}O-tracer experiments that the nitrito complex formation reaction does not involve metal–oxygen bond breakage. The identity of the nitrosating agent was not clear and different opinions have been reported.[106] Similar reactions were later reported[107] for Rh(III), Ir(III), and Pt(IV) complexes. Throughout, the rate of the nitrosation process is given by

or

$$\text{Rate} \propto [\text{aquo complex}][NO_2^-][HNO_2]$$
$$\text{Rate} \propto [\text{hydroxo complex}][HNO_2]^2 \tag{21}$$

Protonation of nitrite ion results in the formation of HNO_2, which partially decomposes into HNO_3 and NO. Typical pK_a values for HNO_2 are 2.8 (1 M ionic strength)[108] and 3.3 (dilute solutions)[109] at 25°C. In principle, two nitrosation agents have been suggested to exist in acidic nitrite solutions, viz. NO^+ and N_2O_3,[110] and different groups have favored one or the other of these.[108]

$$H^+ + HNO_2 \overset{K_{11}}{\rightleftharpoons} H_2NO_2^+ \overset{k_{12}}{\longrightarrow} NO^+ + H_2O \tag{22}$$

$$2HNO_2 \overset{K_{12}}{\rightleftharpoons} N_2O_3 + H_2O \quad (N_2O_3 \equiv NO^+NO_2^-) \tag{23}$$

For reaction (22) $K_{11}k_{12} = 617 \pm 80^{[111]}$ or $230^{[112]}$ M^{-1} s^{-1} at 0°C. For reaction (23) $K_{12} = 0.2^{[113,114]}$ or $0.16^{[115]}$ M^{-1} at 20°C.

A detailed kinetic study[108] has revealed that the nitrosation of $Co(NH_3)_5$-OH_2^{3+} can be summarized by the following mechanism:

$$Co(NH_3)_5OH^{2+} + N_2O_3 \overset{k_{13}}{\longrightarrow} Co(NH_3)_5ONO^{2+} + HNO_2$$

$$Co(NH_3)_5OH_2^{3+} \qquad 2HNO_2 \tag{24}$$

$$H^+ + NO_2^-$$

According to the rate data,[108] only the hydroxo complex can "take up" NO^+ from N_2O_3 (i.e., $NO^+NO_2^-$) to produce the nitrito species, and k_{13} has a value of $7.9 \pm 1.0 \times 10^3$ M^{-1} s^{-1} at 25°C and ionic strength 1 M. The activation parameters for k_{13} ($\Delta H^\ddagger = 50 \pm 3$ kJ mol^{-1} and $\Delta S^\ddagger = -3 \pm 8$ J K^{-1} mol^{-1}) underline the secondary bond formation character of the nitrosation step. If the $Co(NH_3)_5ONO^{2+}$ species is acidified before significant isomerization to the nitro complex occurs, the $Co(NH_3)_5OH_2^{3+}$

species is regenerated rapidly, which demonstrates the reversibility of the NO^+ uptake process.[108] Along these lines, other nitrito complexes have been characterized in the nitrosation of species such as *cis-* and *trans-* $Co(en)_2(OH_2)_2{}^{3+}$ and -$Co(en)_2(Cl)OH_2{}^{2+}$. In the case of the diaquo complexes, dinitrito, nitronitrito, and dinitro species have been identified.[116,117]

The nature of the subsequent isomerization reaction has been investigated by various groups. This step occurs in the solid[118] as well as the liquid phase, and is generally catalyzed by metal[119] and hydroxy ions[116,120] in aqueous solution. The isomerization process is intramolecular and various intermediates or transition state species have been postulated.

$$
\begin{array}{ccc}
\overset{O}{\underset{N\diagdown O}{M\diagup}} & M-N{\overset{O}{\underset{O}{\diagup\diagup}}}\,\ominus & M{\overset{N}{\underset{O}{-\!\!\diagup\diagup\diagdown O}}}
\end{array}
\tag{25}
$$

The effect of pressure on this process[117,121] indicates a decrease in volume in going from the linear nitrito to the nonlinear nitro complex, and typical volumes of activation of between -3 and $-7\ cm^3\ mol^{-1}$ have been reported.

It was reported[122] that oxygen scrambling occurs during base-catalyzed isomerization[116] and acid-catalyzed elimination of NO^+. The latter observation is rather puzzling[123] and would suggest metal–oxygen bond breakage during the elimination process, possibly via intermediates such as

$$
(NH_3)_5Co{\overset{N}{\underset{O\text{----}H}{\diagdown}}}{\overset{O}{\diagdown}} \quad \text{or} \quad (NH_3)_5Co{\overset{O}{\underset{O}{\diagup\diagdown}}}N
\tag{26}
$$

This situation contrasts with that discussed previously for the elimination reactions involving CO_2 and SO_2, and can possibly be related to the nature of the eliminated NO^+ species during the acidification procedure.

VII. FORMATION AND REACTIVITY OF CARBOXYLATO AND RELATED COMPLEXES

Carboxylato complexes of transition metals, in particular of pentamminecobalt(III), have in general been prepared via anation of the corresponding aquo or DMF complexes [124], for which the first reports go back as far as 1857.[125] Such processes are usually slow in the case of Co(III) because they are controlled by metal–solvent bond breakage. An alternative route that involves the direct acylation of the hydroxo ligand was suggested[126,127] and was further developed.[124] In general, carboxylato complexes can be synthesized from the reaction of the acid anhydride or related acylating reagent with the hydroxo ligand if the latter is reasonably nucleophilic. For example, $Co(NH_3)_5OH^{2+}$ reacts rapidly with acetic anhydride to form $Co(NH_3)_5OOCCH_3^{2+}$,[124,126] and with propionic anhydride to produce $Co(NH_3)_5OOCC_2H_5^{2+}$.[128] The rate of these processes increases significantly with increasing nucleophilicity of the hydroxo ligand, i.e., pK_a value of the corresponding aquo ligand.

For the reaction of propionic anhydride with metal hydroxo complexes, Eq. (27) holds,

$$\log k = 0.18pK_a + \text{constant} \qquad (27)$$

demonstrating the linear dependence on the nucleophilicity of the hydroxy ligand. The slope of the $\log k$ versus pK_a plot is in close agreement with that reported[8] for the corresponding relationship between k_2 and K_3 (see Section II), for which the slope is 0.16 ± 0.04.[26] The relative insensitivity of k on the value of pK_a is consistent with the secondary nature of the bond formation process. ^{18}O-Tracer experiments have clearly demonstrated[129] that the metal–oxygen bond remains intact during the formation of $Co(NH_3)_5$-$OOCCH_3^{2+}$ in the reaction of the hydroxo complex with acetyl anhydride in solvents of low polarity. In more polar solvents, it was found[129] that linkage isomerization of the acetato ligand occurs probably via an intermediate of the type

$$(NH_3)_5C\overset{O}{\underset{O}{\diamond}}\!\!\diamond CCH_3^{2+} \qquad (28)$$

The mechanism of such an acylation process can be formulated as follows[124,127]:

$$(NH_3)_5Co^*OH^{2+} \; + \; (CH_3CO)_2O \longrightarrow (NH_3)_5Co\overset{CH_3}{\underset{\underset{\underset{CH_3}{|}}{\underset{C=O}{|}}}{\overset{|}{\underset{|}{-{}^*O-C-O}}}} \qquad (29)$$

$$\Big\downarrow \text{fast}$$

$$(NH_3)_5Co\overset{O}{\overset{\|}{-{}^*O-C-CH_3}} \; + \; CH_3COO^- \; + \; H^+$$

The ability of coordinated hydroxide to add rapidly to carbonyl substrates has also been observed in the reaction of $cis\text{-}Co(en)_2(OH_2)OH^{2+}$ with acetylacetone to give $Co(en)_2acac^{2+}$.[130] A whole series of carboxylate complexes were synthesized using this mechanistic property.[127,129]

During base hydrolysis of $Co(NH_3)_5OOCCH_3^{2+}$, ^{18}O-tracer experiments demonstrated that bond cleavage is almost entirely at the Co–O center.[131] In the case of halogenated acetato complexes the extent of O–C bond cleavage increases with the acidity of the parent acetic acid.[131,132] The latter process was visualized as follows[132]:

$$Co-O-\underset{\underset{O}{\|}}{C}-R \longrightarrow Co-O-\underset{\underset{O^-}{|}}{\overset{\overset{OH}{|}}{C}}-R \rightleftharpoons Co-\overset{\overset{O^-}{|}}{\underset{\underset{O^-}{|}}{O-C}}-R \longrightarrow Co-O^- + RCO_2^- \tag{30}$$

Interestingly enough, base hydrolysis of a bidentate carboxylate ligand such as in $Co(en)_2C_2O_4^+$ proceeds via C–O and not via Co–O bond cleavage,[133,134] once again emphasizing the nature of the coordinated ligand and its role in such processes. A recent study[134] of the cyclization of *cis*-$Co(en)_2(OH_2)C_2O_4H^{2+}$ revealed that in the pH range 0 to 4 the reaction is intramolecular, with attack on the protonated carboxylate group by coordinated water. This process is preferred over displacement of water from the coordination sphere and presumably results from the facile loss of water from carbon in the intermediate species. The cyclization of *trans*-$Co(en)_2$-$(OH_2)C_2O_4H^{2+}$ and *trans*-$Co(en)_2(OH_2)C_2O_4^+$ proceeds quantitatively via *cis*-$Co(en)_2(OH_2)C_2O_4^+$,[135] which chelates intramolecularly as just discussed.

VIII. FORMATION AND REACTIVITY OF IODATO AND PERCHLORATO COMPLEXES

It has in general been observed that oxy anions such as NO_2^-, SO_3^{2-}, and IO_3^-, which exchange oxygen rapidly with solvent water, undergo fast complexation with substitution-inert complexes having aquo ligands.[136,137] It was suggested that oxygen exchange occurs according to

$$-M^*OH_2^{3+} + OXO_x^{n-} \rightarrow -M^*OXO_x^{(3-n)+} + H_2O \tag{31}$$

where $^*O$ indicates a labeled oxygen.

Two groups[67,138] investigated the very rapid complexation of $Cr(NH_3)_5$-OH_2^{3+}, *cis*-$Co(en)_2(OH_2)_2^{3+}$, and $Cr(H_2O)_6^{3+}$ by iodate, which is according to the above-outlined model, and in fact substitution occurs at the iodine(V) center. These reactions are of T-jump rate and were investigated as a function of various kinetic parameters. In the case of the $Cr(H_2O)_6^{3+}$ species the interpretation of the kinetic data is complicated by the formation of predominantly a diiodato complex.[67,139] The reciprocal relaxation time shows a linear dependence on $[H^+]$ for all three studied complexes,[67,138] and the suggested reaction mechanism includes the following steps:

$$M^{3+} + H^+ + IO_3^- \overset{K_{14}}{\rightleftharpoons} M^{3+} \cdot IO_3^- + H^+ \rightleftharpoons MOIO_2^{2+} + H_2O + H^+$$

$$\Big\updownarrow K_{13} \tag{32}$$

$$M^{3+} + HIO_3 \overset{K_{15}}{\rightleftharpoons} M^{3+} \cdot HIO_3 \overset{k_{14}}{\longrightarrow} MOIO_2H^{3+} + H_2O$$

The linear acid dependence was ascribed to the participation of HIO_3. Both IO_3^- and HIO_3 produce ion pairs with the complexes (M^{3+}), followed by reversible substitution at the iodine(V) center. These rates are much faster than those expected for substitution at the Cr(III) center.[140] With $K_{13} = 0.47\ M^{-1}$, $K_{14} = 7\ M^{-1}$, and $K_{15} = 0.3\ M^{-1}$,[67] k_{14} turns out be 7×10^3 s^{-1} for $M^{3+} = Cr(H_2O)_6^{3+}$. Very similar results were reported for the other complexes mentioned, and the following generalized transition state was suggested[138]:

$$(33)$$

Further proof for the suggested mechanism comes from the observation that the values of $k_{14}K_{13}$ are very similar to those reported for oxygen exchange between iodate and solvent water. The nature of the substitution process on the iodine center remains unknown, although arguments in favor of an associative mechanism including hydrogen bonding [structure (33)] have been presented.[138]

The iodato complex $Cr(NH_3)_5IO_3^{2+}$ rapidly equilibrates to produce $Cr(NH_3)_5OH_2^{3+}$ and iodate without metal–oxygen bond rupture.[138,141] This process is the reverse of the formation reaction and is once again controlled by substitution at the iodine center.

Sargeson and co-workers[142] succeeded in isolating a perchlorato complex, $[Co(NH_3)_5OClO_3](ClO_4)_2$. This aquates within a few seconds at room temperature, suggesting that the metal–oxygen bond remains intact. Tracer experiments demonstrated that Cl–O bond cleavage occurs during base hydrolysis, but that Co–O bond cleavage occurs during acid hydrolysis.[89]

IX. FORMATION AND REACTIVITY OF CARBAMATO AND RELATED COMPLEXES

In the hope that coordinated NCO^- would be susceptible to nitrosation by NO^+, as is free NCO^-, to produce coordinated CO_2 and N_2, Sargeson and co-workers[95] investigated the acid hydrolysis and nitrosation of $Co(NH_3)_5NCO^{2+}$.[143] The products of acid hydrolysis were $Co(NH_3)_6^{3+}$, $Co(NH_3)_5O_2CNH_2^{2+}$, and $Co(NH_3)_5OH_2^{3+}$. Of these species the formation of the O-bonded carbamato complex is of interest at present, and detailed kinetic and ^{18}O-tracer studies[95] revealed the following overall reaction mechanism:

$$(NH_3)_5CoNCO^{2+} + H^+ \rightleftharpoons (NH_3)_5CoNCOH^{3+}$$

$$(NH_3)_5CoNCOH^{3+} + H_2O \rightarrow (NH_3)_5CoNH_2COOH^{3+}$$

$$(NH_3)_5CoNH_2COOH^{3+} \rightleftharpoons (NH_3)_5CoNHCOOH^{2+} + H^+$$

$$(NH_3)_5CoNH_2COOH^{3+} \rightarrow (NH_3)_5CoOOCONH_2^{3+} + H^+ \tag{34}$$

$$(NH_3)_5CoNH_2COOH^{3+} + H_2O \rightarrow (NH_3)_5CoOH_2^{3+} + NH_3 + CO_2$$

$$(NH_3)_5CoNHCOOH^{2+} \rightarrow (NH_3)_5CoNH_2^{2+} + CO_2$$

$$(NH_3)_5CoNH_2^{2+} + H^+ \rightarrow (NH_3)_6Co^{3+}$$

The ^{18}O-tracer studies showed that only one oxygen atom of CO_2 was labeled and that it originated from the solvent. Of particular interest is the ability of coordinated NCO^- to react with the solvent to produce $-NH_2COOH$ and $-NHCOOH$ complexes, where the latter species can decarboxylate to produce $Co(NH_3)_6^{3+}$ and CO_2. The stability of the metal–nitrogen bond is responsible for the increased reactivity of the coordinated NCO^- as compared to free NCO^-. During nitrosation the metal–nitrogen bond is broken and $Co(NH_3)_5OH_2^{3+}$ is the main product.

In a similar manner it was found that metal ions can significantly promote the base hydrolysis of nitriles.[144,145] The base hydrolysis of $Co(NH_3)_5(N{\equiv}CCH_3)^{3+}$ was investigated[145] to determine whether the complex would react by direct attack of OH^- at the nitrile group to give the N-bonded acetamido complex, or by attack of a deprotonated amine on the nitrile group to produce coordinated acetamide. The kinetic results suggest a direct attack of OH^- at the carbon atom of the nitrile group:

$$(NH_3)_5CoN\equiv CCH_3^{3+} + OH^- \rightarrow (NH_3)_5CoNH-\overset{\overset{\displaystyle O}{\|}}{C}-CH_3^{2+} \tag{35}$$

The alternative mechanism

$$(NH_3)_5CoN\equiv CCH_3^{3+} + OH^- \longrightarrow (NH_3)_5CoN\equiv CCH_2^{-/2+} + H_2O \tag{36}$$

$$(NH_3)_5CoNH-\overset{\overset{\displaystyle O}{\|}}{C}-CH_3^{2+} \underset{H_2O}{\longleftarrow} (NH_3)_5Co-\bar{N}=C=CH_2^{2+}$$

is unlikely since similar results were obtained for the hydrolysis of the corresponding benzonitrile complex, which cannot undergo the outlined deprotonation. The rate constants for the base hydrolysis of the bound aceto- and benzonitriles are 10^6 times faster than for the free nitriles.[145]

X. FORMATION AND REACTIVITY OF ARSENATO COMPLEXES

Lincoln and co-workers[146] investigated the formation and hydrolysis reactions of $Co(NH_3)_5HAsO_4^+$ and $Co(NH_3)_5H_2AsO_4^{2+}$. They found that both processes proceed via arsenic–oxygen bond formation and breakage. The observed rate constants are significantly faster than those usually observed for substitution reactions of $Co(NH_3)_5OH_2^{3+}$.[147] Furthermore, ^{18}O-tracer experiments also underline the validity of the above suggestions, and the data are interpreted in terms of substitution at the As(V) center via an associative transition state of the type[146]

$$\tag{37}$$

These results strongly contrast those reported for the corresponding phosphato complexes in which all reactions (anation and hydrolysis) proceed via Co–O bond cleavage.[148–150] This is ascribed to the fact that the P(V) center does not easily expand its coordination number to five as in the case of As(V), such that coordinated water is ineffective as a nucleophile toward P(V).

XI. FORMATION AND REACTIVITY OF MOLYBDATE, TUNGSTATE, AND CHROMATE COMPLEXES

Reactions of substitutionally inert aquo complexes with oxy anions that exhibit fast oxygen exchange characteristics may be extremely fast since no metal–oxygen bond has to be broken on the aquo complex. The oxy anions of Mo(VI), W(VI), and Cr(VI) exhibit such behavior. Typical mechanistic information on such reactions will therefore be presented in a comparative way.

Early work[151] demonstrated that Mo(VI) reacts with $Co(NH_3)_5OH_2^{3+}$ to form $Co(NH_3)_5OMoO_3^+$, in which Mo(VI) is tetrahedrally coordinated. A kinetic study[152] revealed that for the reaction

$$Co(NH_3)_5OH_2^{3+} + MoO_4^{2-} \xrightarrow{k_{15}} Co(NH_3)_5OMoO_3^+ + H_2O \qquad (38)$$

$$k_{15} = k_a + k_b[H^+] + k_c[MoO_4^{2-}] + k_d[H^+][MoO_4^{2-}] \qquad (39)$$

where $k_a = 96 \pm 7 \ M^{-1} \ s^{-1}$, $k_b = 1.1 \pm 0.2 \times 10^9 \ M^{-2} \ s^{-1}$, $k_c = 2.2 \pm 0.2 \times 10^3 \ M^{-2} \ s^{-1}$, and $k_d = 1.02 \pm 0.05 \times 10^{11} \ M^{-3} \ s^{-1}$ at 25°C and ionic strength 1 M. This rate law and the data are consistent with substitution at the Mo(VI) center. Two paths are considered to account for the kinetic data, i.e., one in which $Co(NH_3)_5OH_2^{3+}$ reacts with MoO_4^{2-} as in reaction (38), and another in which $Co(NH_3)_5OH^{2+}$ is the reactive species. The corresponding rate constants were found[152] to be 6.6×10^4 and 3.2×10^5 $M^{-1} \ s^{-1}$ at 25°C, once again demonstrating that substitution at the Co(III) center cannot be rate determining. Furthermore, it has also been reported that Mo(VI) can increase its coordination number on protonation, even to six.[153,154] The quoted rate constants are in good agreement with those usually found for substitution at Mo(VI) centers.[155]

In a similar study involving $Co(NH_3)_5OH_2^{3+}$ and WO_4^{2-}, it was found that precipitates were produced.[156] The cis-$Co(en)_2(OH_2)_2^{3+}$ species, however, reacted with WO_4^{2-} in the pH range 8 to 9 without any such complications occurring. Since the pK_a values of the diaquo complex are 5.8 and 7.9, respectively,[156] the dominant species in this pH range are $Co(en)_2(OH_2)OH^{2+}$ and $Co(en)_2(OH)_2^+$. The complexation with WO_4^{2-} was found to be rapid and must involve substitution at the W(VI) center.[157] The WO_4^{2-} ion is the dominant form under the mentioned conditions since the pK_a of HWO_4^- is 10.5.[158]

$$Co(en)_2(OH_2)OH^{2+} + HWO_4^- \xrightarrow{k_{16}} Co(en)_2(OH)OWO_3 + H_2O + H^+$$

$$Co(en)_2(OH)_2^+ + HWO_4^- \xrightarrow{k_{17}} Co(en)_2(OH)OWO_3 + H_2O$$

$$Co(en)_2(OH_2)_2^{3+} + WO_4^{2-} \xrightarrow{k_{18}} Co(en)_2(OH)OWO_3 + H_2O + H^+$$

$$Co(en)_2(OH_2)OH^{2+} + WO_4^{2-} \xrightarrow{k_{19}} Co(en)_2(OH)OWO_3 + H_2O$$

(40)

The $[H^+]$ dependence of the observed rate constant is ascribed to the first two reactions in the overall reaction (40), for which $k_{16} = 1 \times 10^7$ and $k_{17} = 3.2 \times 10^7\ M^{-1}\,s^{-1}$ at 25°C. Due to the existence of proton ambiguities, contributions from the last two reactions were also considered, and maximum values of $k_{18} = 660$ and $k_{19} = 2.8 \times 10^4\ M^{-1}\,s^{-1}$ could be estimated.

The values of k_{16} and k_{17} are significantly larger than those for the reactions of WO_4^{2-} with H_2O and OH^-, respectively, for which the rate constants are $0.33\ s^{-1}$ and $2.2\ M^{-1}\,s^{-1}$.[159] This is probably due to the protonation of WO_4^{2-}, during which an increase in coordination number from four to six is expected to account for the observations. Furthermore, the substitution reactions of MoO_4^{2-} and WO_4^{2-} have been shown to give fairly satisfactory free-energy relationships, which is in line with the common occurrence of an increase in coordination number during substitution[160,161] A subsequent step to the reactions outlined in reaction (40) is most probably chelation of the WO_4 moiety.

Finally, some results have been reported for the reactions of $Co(NH_3)_5$-OH_2^{3+} and cis-$Co(NH_3)_4(OH_2)_2^{3+}$ with CrO_4^{2-} or $HCrO_4^-$. Following an earlier study[162] in which some evidence for the formation of a $Co(NH_3)_5$-$OCrO_3^+$ species via substitution at the Cr(VI) center was reported, a detailed kinetic study[163,164] revealed that the hydroxo complex coordinates to $HCrO_4^-$ according to the mechanism

$$(NH_3)_5Co-OH^{2+} + HOCrO_3^- \rightleftharpoons (NH_3)_5Co-O-\overset{H}{\underset{OH}{\overset{|}{Cr}}}{\overset{O}{\underset{O}{\diagdown}}}O^+$$

$$\Big\downarrow H^+$$

$$(NH_3)_5Co-O-Cr{\overset{O}{\underset{O}{\diagdown}}}O^+ + H_2O$$

(41)

The second step includes proton-assisted elimination of water. The pH dependence of the data[163] suggests that $Co(NH_3)_5OH_2^{3+}$ and $Co(NH_3)_5OH^{2+}$ react with $HCrO_4^-$ with rate constants of 0.8 and 80 $M^{-1}\,s^{-1}$ at 25°C, respectively. This reactivity difference is in agreement with that found for SO_2 uptake reactions of Co(III) complexes (see Section IV).

XII. ADDITION TO AND ABSTRACTION FROM COORDINATED LIGANDS

In the previous sections the focus has been mainly on coordinated ligands bonded through oxygen to the central metal atom. In this section we would like to report some of the general results found for other types of ligands not discussed in the preceding sections. In this respect it is important to refer to the review by Hipp and Busch[1] mentioned in Section I. In general, the ligand electron density is usually modified by the metal ion in such a way that the species experiences enhancement of either electrophilic or nucleophilic attack.

The central carbon atom in acetylacetonato complexes, for example, can undergo substitution with a large number of nucelophiles[1,165]:

$$L_nM\diagdown\!\!\!\diagup\!\!-H + XY \longrightarrow L_nM\diagdown\!\!\!\diagup\!\!-X + HY \quad (42)$$

These include Cl, Br, I, SCN, NO_2, CHO, and COR. The hexafluoroacetylacetonato complex of bis(ethylenediamine)cobalt(III) was found[166] to show large reversible spectral changes on the addition of OH^-, which was ascribed to the equilibrium

$$(en)_2Co\diagdown\!\!\!\diagup\!\!CH + OH^- \rightleftharpoons (en)_2Co\diagdown\!\!\!\diagup\!\!C\colon + H_2O \quad (43)$$

An NMR investigation[167] revealed that the process should rather be formulated as follows:

$$(en)_2Co\diagdown\!\!\!\diagup\!\!CH + OH^- \rightleftharpoons (en)_2Co\diagdown\!\!\!\diagup\!\!CH \quad (44)$$

A recent kinetic study of the system[168] indicated that the process involves measurable (stopped-flow and T-jump) addition and abstraction of OH^-, and the data underline the validity of the formulation in equilibrium (44).

Another type of addition reaction involves the complexes of α-amino acid esters. In basic media these species ring-close according to reaction (45).[1]

$$(NH_3)_4Co\overset{NH_2-CH_2COOR}{\underset{NH_3}{}} + OH^- \longrightarrow (NH_3)_4Co\overset{H_2N-CH_2}{\underset{H_2N-C}{}}\Big| + OR^- + H_2O$$

$$\tag{45}$$

This probably involves the participation of a conjugate base species. Similarly, addition of base to $Co(en)_2NH_2CH_2CONHR$ results in the formation of $Co(en)_2NH_2CH_2COO$ and NH_2R, during which the cobalt–oxygen bond remains intact.

Similarly, it was reported[169,170] that lactone formation can be enhanced by factors up to 10^5 by employing the entropic advantages of an intra-molecular process. ^{18}O studies show that the cyclization of *cis*-$Co(en)_2$-$(OH_2)(glyOH)^{3+}$ and *cis*-$Co(en)_2(OH_2)(glyO)^{2+}$, where glyOH and glyO are N-bonded glycine and glycinate, respectively, occurs intramolecularly, with displacement of coordinated water.[171] The rate of this process strongly depends on pH: $t_{1/2} \sim 40$ s at pH $< 1, t_{1/2} \sim 400$ s at pH ~ 5, and $t_{1/2} \sim 10$ h at pH > 9. The suggested mechanism follows that adopted for lactonization reactions of purely organic hydroxy acids. The results are best interpreted in terms of a mechanism that includes the rate-determining general acid-catalyzed decay of structure (46) in acidic and neutral media.

$$\tag{46}$$

In basic solution the rate-determining step is cyclization to form

$$\tag{47}$$

Comparison with oxygen exchange data for glycine shows large accelerations for the metal-based system, and the rates compare favorably with those reported for intramolecular lactone formation in purely organic molecules. In a subsequent study[172] it was shown that chelation of *trans*-$Co(en)_2$-$(NH_2CH_2COO)OH_2^{2+}$ does not proceed via the cis species as intermediate, and it was proposed to occur directly via backside expulsion of water by the carboxylate group. Tracer experiments demonstrated the retention of all the oxygen atoms on the carboxylate group under all conditions, requiring loss of bound water or hydroxide in forming the chelate.

XIII. CONCLUDING REMARKS

This article has been devoted to a discussion of different reactions involving coordinated ligands. It can in general be concluded that such metal-bonded species exhibit a rather unexpected reactivity in comparison to the uncoordinated species. Such processes should therefore not only be considered in reactions including the participation of metal species, but also in the formulation and development of catalytic and biologically orientated processes. Such reactions are of relevance in metal-catalyzed enzyme processes in which hydration, phosphorylation, and hydrolysis are involved, and may introduce new routes for metal-promoted reactions in general.

Probably the most well-known example in the area of bioinorganic reactions is the catalytic action of carbonic anhydrase in the hydration of CO_2 and the dehydration of HCO_3^-.[173] It is quite certain that CO_2 does not coordinate directly to the Zn metal ion in carbonic anhydrase, but rather to functional groups such as imidazoles of histidine and water close to the metal center.[174] The pH dependence of the catalytic activity is characterized by a pK_a near 7, and this can be ascribed to the deprotonation of a histidine side chain or a coordinated water molecule. Pocker *et al.*[174] described a mechanism in which the Zn-bound water molecule provides both donor and acceptor functions. The rates of the processes are such that a simple model based on the reactivity of a Zn–OH group cannot account for the tremendous effect,[26] and other aspects must play a significant role.

Processes similar to those discussed for SO_2 uptake by metal aquo and hydroxo species, followed by rapid intramolecular redox reactions (Section III), may account for some aspects of the atmospheric oxidation of SO_2 to SO_4^{2-}. Such processes are known to be catalyzed by the presence of aquated Fe(III) and Mn(II) species,[175] and further investigations are being undertaken in order to elucidate the intimate mechanisms.

Finally, metal ions play a significant role in the redox reactions of coordinated ligands. The metal ion can act either as a source or as a sink of electrons to reduce or oxidize a coordinated ligand, or as a transmitter of electrons, as in the case of oxidative addition or reductive elimination reactions. The latter have been treated in detail elsewhere,[176] and the importance of the role played by coordinated ligands as bridging groups has been emphasized frequently.[177] Such processes often form an integral part of catalytic reactions in which ligand–metal interactions form the basis of the overall reaction sequence.

Acknowledgments

The author gratefully acknowledges financial support from the Deutsche Forschungsgemeinschaft, the Fonds der Chemischen Industrie, and the Scientific Affairs Division of NATO under Grant RG 114.81.

References

[1] Hipp, C. J.; Busch, D. H. *ACS Monogr.* **1978,** *174,* 221.

[2] van Eldik, R.; Harris, G. M. *Inorg. Chem.* **1980,** *19,* 880.

[3] Chaffee, E.; Dasgupta, T. P.; Harris, G. M. *J. Am. Chem. Soc.* **1973,** *95,* 4169.

[4] Krishnamoorthy, C. R.; Palmer, D. A.; van Eldik, R.; Harris, G. M. *Inorg. Chim. Acta* **1979,** *35,* L361.

[5] Dasgupta, T. P.; Harris, G. M. *J. Am. Chem. Soc.* **1975,** *97,* 1733.

[6] Palmer, D. A.; van Eldik, R.; Kelm, H.; Harris, G. M. *Inorg. Chem.* **1980,** *19,* 1009.

[7] van Eldik, R.; Palmer, D. A.; Harris, G. M. *Inorg. Chem.* **1980,** *19,* 3673.

[8] van Eldik, R.; Palmer, D. A.; Kelm, H.; Harris, G. M. *Inorg. Chem.* **1980,** *19,* 3679.

[9] Palmer, D. A.; Harris, G. M. *Inorg. Chem.* **1974,** *13,* 965.

[10] Dasgupta, T. P.; Harris, G. M. *J. Am. Chem. Soc.* **1977,** *99,* 2490.

[11] Skibsted, L. H.; Ford, P. C. *Acta Chem. Scand.* **1980,** *A34,* 109.

[12] Jost, A. *Ber. Bunsenges. Phys. Chem.* **1976,** *80,* 316.

[13] Swaddle, T. W.; Merbach, A. E. *Inorg. Chem.* **1981,** *20,* 4212.

[14] Doss, R.; van Eldik, R.; Kelm, H. *Ber. Bunsenges. Phys. Chem.* **1982,** *86,* 925.

[15] Koshy, K. C.; Harris, G. M. *Inorg. Chem.,* in press. Recent measurements in this laboratory indicated this value to be in error.

[16] Eigen, M.; Kruse, W.; Maass, G.; de Maeyer, L. *Prog. React. Kinet.* **1964,** *2,* 285.

[17] Dasgupta, T. P.; Harris, G. M. *J. Am. Chem. Soc.* **1968,** *90,* 6360.

[18] Paabo, M.; Bates, R. G. *J. Phys. Chem.* **1969,** *73,* 3014.

[19] Ting, S.; Kelm, H.; Harris, G. M. *Inorg. Chem.* **1966,** *5,* 696.

[20] Andrade, C.; Taube, H. *Inorg. Chem.* **1966,** *5,* 1087.

[21] Joubert, P. R.; van Eldik, R. *Int. J. Chem. Kinet.* **1976,** *8,* 411.

[22] van Eldik, R.; Harris, G. M. *Inorg. Chem.* **1975,** *14,* 10.

[23] Brown, P. M.; Harris, G. M. *Inorg. Chem.* **1968,** *7,* 1872.

[24] Krishnamurty, K. V.; Harris, G. M.; Sastri, V. S. *Chem. Rev.* **1970,** *70,* 171.

[25] Piriz Mac-Coll, C. R. *Coord. Chem. Rev.* **1969,** *4,* 147.

[26] Palmer, D. A.; van Eldik, R. *Chem. Rev.* **1983,** *83,* 651.

[27] Johnson, K. S. *Limnol. Oceanogr.* **1982,** *27,* 849.

[28] van Eldik, R.; Palmer, D. A. *J. Solution Chem.* **1982,** *11,* 339.

[29] Wissbrun, K. F.; French, D. M.; Patterson, A., Jr. *J. Phys. Chem.* **1954,** *58,* 693.

[30] Dasgupta, T. P.; Harris, G. M. *Inorg. Chem.* **1978,** *17,* 3304.

[31] Wan, W. K.; Ph.D. Thesis, State University of New York at Buffalo, New York, **1978.**

[32] Krishnamoorthy, C. R.; Palmer, D. A.; van Eldik, R.; Harris, G. M. *Inorg. Chim. Acta* **1979,** *35,* L361.

[33] Palmer, D. A.; Dasgupta, T. P.; Kelm, H. *Inorg. Chem.* **1978**, *17*, 1173.
[34] Woolley, P. *Nature* (*London*) **1975**, *258*, 677.
[35] Hunt, J. P.; Rutenberg, A. C.; Taube, H. *J. Am. Chem. Soc.* **1952**, *74*, 268.
[36] Bunton, C. A.; Llewellyn, D. R. *J. Chem. Soc.* **1953**, 1692.
[37] Dasgupta, T. P.; Harris, G. M. *Inorg. Chem.* **1978**, *17*, 3123.
[38] Ficner, S. A.; Ph.D. Thesis, State University of New York at Buffalo, New York, **1980.**
[39] Inoue, T.; Harris, G. M. *Inorg. Chem.* **1980**, *19*, 1091.
[40] Spitzer, U.; van Eldik, R.; Kelm, H. *Inorg. Chem.* **1982**, *21*, 2821.
[41] Harris, G. M.; Dasgupta, T. P. *J. Indian Chem. Soc.* **1977**, *54*, 62.
[42] Job, R. C.; Bruice, T. C. *J. Am. Chem. Soc.* **1974**, *96*, 809.
[43] Glusker, J. P.; Carrell, H. L.; Job, R.; Bruice, T. C. *J. Am. Chem. Soc.* **1974**, *96*, 5741.
[44] Early, J. E.; Alexander, W. *J. Am. Chem. Soc.* **1970**, *92*, 2294.
[45] Sastri, M. N.; Raman, V. A. *Z. Phys. Chem.* (*Leipzig*) **1975**, *256*, 993.
[46] Siebert, H.; Wittke, G. *Z. Anorg. Allg. Chem.* **1973**, *399*, 43.
[47] Thacker, M. A.; Scott, K. L.; Simpson, M. E.; Murray, R. S.; Higginson, C. E. *J. Chem. Soc., Dalton Trans.* **1974**, 647.
[48] Scott, K. L. *J. Chem. Soc., Dalton Trans.* **1974**, 1486.
[49] Richards, L.; Halpern, J. *Inorg. Chem.* **1976**, *15*, 2571.
[50] Yandell, J. K.; Tomlins, L. A. *Aust. J. Chem.* **1978**, *31*, 561.
[51] Raston, C. L.; White, A. H.; Yandell, J. K. *Aust. J. Chem.* **1978**, *31*, 993, 999.
[52] Palmer, J. M.; Deutsch, E. *Inorg. Chem.* **1975**, *14*, 17.
[53] Seibles, L.; Deutsch, E. *Inorg. Chem.* **1977**, *16*, 2273.
[54] Spitzer, U.; van Eldik, R. *Inorg. Chem.* **1982**, *21*, 4008.
[55] Schmidt, M.; Siebert, W. *In* "Comprehensive Chemical Kinetics"; Vol. 6; Barnford, C. H.; Tipper, C. F. H., Eds.; Elsevier: Amsterdam, 1972, p. 878.
[56] Schmidt, M. *In* "Sulfur in Organic and Inorganic Chemistry"; Vol. 2; Senning, A. (Ed.); Dekker: New York, 1972; p. 84.
[57] Betts, R. H.; Voss, R. H. *Can. J. Chem.* **1970**, *48*, 2035.
[58] Eigen, M.; Kustin, K.; Maass, G. *Z. Phys. Chem.* (*Wiesbaden*) **1961**, *30*, 130.
[59] Gmelin's "Handbuch der Anorganischen Chemie"; Verlag Chemie: Weinheim, 1960, Syst. No. 9, Part B2, p. 466.
[60] Bourne, D. W. A.; Higuchi, T.; Pitman, I. H. *J. Pharm. Sci.* **1974**, *63*, 865.
[61] Connick, R. E.; Tam, T. M.; von Deuster, E. *Inorg. Chem.* **1982**, *21*, 103.
[62] "Stability Constants" *Spec. Publ.—Chem. Soc.* **1970**, *17*, 229.
[63] Murray, R. S.; Stranks, D. R.; Yandell, J. K. *J. Chem. Soc., Chem. Commun.* **1969**, 604.
[64] Farrel, S. M.; Murray, R. S. *J. Chem. Soc., Dalton Trans.* **1977**, 322.
[65] Thacker, M. H.; Higginson, W. C. E. *J. Chem. Soc., Dalton Trans.* **1975**, 704.
[66] Caryle, D. W.; King, E. L. *Inorg. Chem.* **1970**, *9*, 2333.
[67] Bazsa, Gy.; Diebler, H. *React. Kinet. Catal. Lett.* **1975**, *2*, 217.
[68] James, A. D.; Murray, R. S. *J. Chem. Soc., Dalton Trans.* **1977**, 319.
[69] van Eldik, R.; von Jouanne, J.; Kelm, H. *Inorg. Chem.* **1982**, *21*, 2818.
[70] van Eldik, R. *Inorg. Chem.* **1983**, *22*, 353.
[71] Ramasami, T.; Wharton, R. K.; Sykes, A. G. *Inorg. Chem.* **1975**, *14*, 359.
[72] van Eldik, R. *Inorg. Chim. Acta* **1980**, *42*, 49.
[73] van Eldik, R. Unpublished results.
[74] El-Awady, A. A.; Harris, G. M. *Inorg. Chem.* **1981**, *20*, 1660.
[75] El-Awady, A. A.; Harris, G. M. *Inorg. Chem.* **1981**, *20*, 4251.
[76] Dash, A. C.; El-Awady, A. A.; Harris, G. M. *Inorg. Chem.* **1981**, *20*, 3160.
[77] Dasgupta, T. P.; Harris, G. M. Unpublished results.
[78] Fowless, A. D.; Stranks, D. R. *Inorg. Chem.* **1977**, *16*, 1271.

[79] Baldwin, M. E. *J. Chem. Soc.* **1961**, 3123.

[80] Schiavon, G.; Marchetti, F.; Paradisi, C. *Inorg. Chem. Acta* **1979**, *33*, L101.

[81] Kruse, W.; Taube, H. *J. Am. Chem. Soc.* **1961**, *83*, 1280.

[82] Dash, A. C.; El-Awady, A. A.; Harris, G. M. *Inorg. Chem.* **1981**, *20*, 3160.

[83] Schneider, K.; Aygen, S.; van Eldik, R. Prepared for publication.

[84] Kraft, J.; van Eldik, R. Prepared for publication.

[85] Joshi, V.; van Eldik, R., Harris, G. M. Prepared for publication.

[86] Joshy, K. C.; Harris, G. M. Prepared for publication.

[87] "Handbook of Chemistry and Physics"; 49th ed.; CRC Press: Cleveland, 1968, pp. B189, B252.

[88] Elder, R. C.; Heeg, M. J.; Payne, M. D.; Trkula, M.; Deutsch, E. *Inorg. Chem.* **1978**, *17*, 431.

[89] Buckingham, D. A.; Cresswell, P. J.; Sargeson, A. M.; Jackson, W. G. *Inorg. Chem.* **1981**, *20*, 1647.

[90] Jackson, W. G.; Begbie, C. M. *Inorg. Chem.* **1981**, *20*, 1654.

[91] Jache, A. W. *Adv. Inorg. Chem. Radiochem.* **1974**, *16*, 177.

[92] Sushynski, E.; van Roodselaar, A.; Jordan, R. B. *Inorg. Chem.* **1972**, *11*, 1887.

[93] Buckingham, D. A.; Creaser, I. I.; Sargeson, A. M. *Inorg. Chem.* **1970**, *9*, 655.

[94] Palmer, D. A.; van Eldik, R.; Kelm, H. *Inorg. Chim. Acta* **1978**, *30*, 83.

[95] Buckingham, D. A.; Francis, D. J.; Sargeson, A. M. *Inorg. Chem.* **1974**, *13*, 2630.

[96] Dixon, N. E.; Fairlie, D. P.; Jackson, W. G.; Sargeson, A. M. To be published; see reference [90].

[97] Fairlie, D. P.; Jackson, W. G. To be published; see reference [90].

[98] Balahura, R. J.; Ferguson, G.; Ecott, L.; Siew, P. Y. *J. Chem. Soc., Dalton Trans.* **1982**, 747.

[99] Fowles, A. D.; Stranks, D. R. *Inorg. Chem.* **1977**, *16*, 1276.

[100] Fowles, A. D.; Stranks, D. R. *Inorg. Chem.* **1977**, *16*, 1282.

[101] Barcza, L.; Sillén, L. G. *Acta Chem. Scand.* **1971**, *25*, 1250.

[102] Simon, A.; Paetzold, R. *Z. Anorg. Allg. Chem.* **1960**, *303*, 46.

[103] Elder, R. C.; Ellis, P. R. *Inorg. Chem.* **1978**, *17*, 870.

[104] Pearson, R. G.; Henry, P. M.; Bergmann, J. G.; Basolo, F. *J. Am. Chem. Soc.* **1954**, *76*, 5920.

[105] Murmann, R. L.; Taube, H. *J. Am. Chem. Soc.* **1956**, *78*, 4886.

[106] Hughes, M. N.; Shrimanker, K.; Wimbledon, P. E. *J. Chem. Soc., Dalton Trans.* **1978**, 1634.

[107] Basolo, F.; Hammaker, G. S. *Inorg. Chem.* **1962**, *1*, 1.

[108] Ghazi-Bajat, H.; van Eldik, R.; Kelm, H. *Inorg. Chim. Acta* **1982**, *60*, 81.

[109] Klemenc, A.; Hayek, E. *Monatsh. Chem.* **1929**, *53/54*, 407.

[110] Stedman, G. *Adv. Inorg. Chem. Radiochem.* **1979**, *22*, 143.

[111] Benton, D. J.; Moore, P. *J. Chem. Soc. A* **1970**, 3179.

[112] Bunton, C. A.; Stedman, G. *J. Chem. Soc.* **1959**, 3466.

[113] Bunton, C. A.; Stedman, G. *J. Chem. Soc.* **1958**, 240.

[114] Turney, T. A. *J. Chem. Soc.* **1960**, 4263.

[115] Schmid, M.; Krenmayr, P. *Monatsh. Chem.* **1967**, *98*, 417.

[116] Jackson, W. G.; Lawrance, G. A.; Lay, P. A.; Sargeson, A. M. *Inorg. Chem.* **1980**, *19*, 904.

[117] Rindermann, W.; van Eldik, R.; Kelm, H. *Inorg. Chim. Acta* **1982**, *61*, 173.

[118] Grenthe, I.; Nordin, E. *Inorg. Chem.* **1979**, *18*, 1869.

[119] Jackson, W. G.; Lawrance, G. A.; Lay, P. A.; Sargeson, A. M. *Aust. J.* **1982**, *35*, 1561.

[120] Rindermann, W.; van Eldik, R. *Inorg. Chim. Acta* **1983**, *68*, 35.

[121] Mares, M.; Palmer, D. A.; Kelm, H. *Inorg. Chim. Acta* **1978,** *27,* 153.

[122] Jackson, W. G.; Lawrance, G. A.; Lay, P. A.; Sargeson, A. M. *J. Chem. Soc., Chem. Commun.* **1982,** 70.

[123] van Eldik, R.; Harris, G. M. *Inorg. Chim. Acta* **1982,** *65,* L125.

[124] Jackman, L. M.; Scott, R. M.; Portman, R. H.; Dormish, J. F. *Inorg. Chem.* **1979,** *18,* 1497, and references cited therein.

[125] Gibbs, W.; Genth, F. A. *Justus Liebigs Ann. Chem.* **1857,** *104,* 173.

[126] Werner, A. *Ber. Dtsch. Chem. Ges.* **1907,** *40,* 4098.

[127] Dixon, B. E. *J. Chem. Soc.* **1935,** 779.

[128] Buckingham, D. A.; Engelhardt, L. M. *J. Am. Chem. Soc.* **1975,** *97,* 5915.

[129] Jackman, L. M.; Dormish, J. F.; Scott, R. M.; Portman, R. H.; Minard, R. D. *Inorg. Chem.* **1979,** *18,* 1503.

[130] Buckingham, D. A.; Harrowfield, J. M.; Sargeson, A. M. *J. Am. Chem. Soc.* **1973,** *95,* 7281.

[131] Bunton, C. A.; Llewellyn, D. R. *J. Chem. Soc.* **1953,** 1692.

[132] Jordan, R. B.; Taube, H. *J. Am. Chem. Soc.* **1966,** *88,* 4406.

[133] Andrade, C.; Taube, H. *J. Am. Chem. Soc.* **1964,** *86,* 1328.

[134] Buckingham, D. A. Private communication, 1983.

[135] Miskelly, G. M.; Clark, C. R.; Simpson, J.; Buckingham, D. A. *Inorg. Chem.,* in press.

[136] Basza, G.; Nikolasev, V.; Beck, M. T. *Proc. Symp. Coord. Chem.* **1971,** *3,* 15.

[137] Dwek, R. A.; Luz, Z.; Peller, S.; Shporer, M. *J. Am. Chem. Soc.* **1971,** *93,* 77.

[138] Wharton, R. K.; Taylor, R. S.; Sykes, A. G. *Inorg. Chem.* **1975,** *14,* 33.

[139] Mercer, E. E.; Hormuth, J. A. *J. Inorg. Nucl. Chem.* **1969,** *31,* 2145.

[140] Earley, J. E.; Cannon, R. D. *Transition Met. Chem. (N.Y.)* **1965,** *1,* 33.

[141] Duffy, N. V.; Early, J. E. *J. Am. Chem. Soc.* **1967,** *89,* 272.

[142] Harrowfield, J. M.; Sargeson, A. M.; Singh, B.; Sullivan, J. C. *Inorg. Chem.* **1975,** *14,* 2864.

[143] Balahura, R. J.; Jordan, R. B. *Inorg. Chem.* **1970,** *9,* 1567.

[144] Pinnell, D.; Wright, G. B.; Jordan, R. B. *J. Am. Chem. Soc.* **1972,** *94,* 6104.

[145] Buckingham, D. A.; Keene, F. R.; Sargeson, A. M. *J. Am. Chem. Soc.* **1973,** *95,* 5649.

[146] Beech, T. A.; Lawrence, N. C.; Lincoln, S. F. *Aust. J. Chem.* **1973,** *26,* 1877.

[147] Hunt, H. R.; Taube, H. *J. Am. Chem. Soc.* **1958,** *80,* 2642.

[148] Lincoln, S. F.; Stranks, D. R. *Aust. J. Chem.* **1968,** *21,* 37, 57, 67, 1733.

[149] Ferrer, M.; Sykes, A. G. *Inorg. Chem.* **1979,** *18,* 3345.

[150] Martinez, M.; Ferrer, M. *Inorg. Chim. Acta* **1983,** *69,* 123.

[151] Coomber, R.; Griffith, W. P. *J. Chem. Soc. A* **1968,** 1128.

[152] Taylor, R. S. *Inorg. Chem.* **1977,** *16,* 116.

[153] Honig, D. S.; Kustin, K. *Inorg. Chem.* **1972,** *11,* 1.

[154] Cruywagen, J.; Rohwer, E. F. C. H. *Inorg. Chem.* **1975,** *14,* 3136.

[155] Diebler, H.; Timms, R. *J. Chem. Soc. A* **1971,** 273.

[156] Gamsjaeger, H.; Thompson, G. A. K.; Sagmueller, W.; Sykes, A. G. *Inorg. Chem.* **1980,** *19,* 997.

[157] Gilbert, K.; Kustin, K. *J. Am. Chem. Soc.* **1976,** *98,* 5502.

[158] Schwarzenbach, G.; Geier, G.; Littler, J. *Helv. Chim. Acta* **1962,** *45,* 2603.

[159] von Felten, H.; Wernli, B.; Gamsjäger, H.; Baertschi, P.; *J. Chem. Soc., Dalton Trans.* **1978,** 496.

[160] Femahashi, S.; Kato, Y.; Nakayama, M.; Tanaka, M. *Inorg. Chem.* **1981,** *20,* 1752.

[161] Saito, K.; Sasaki, Y. *Adv. Inorg. Bioinorg. Mech.* **1982,** *1,* 179.

[162] Sullivan, J. C.; French, J. E. *Inorg. Chem.* **1964,** *3,* 832.

[163] Woods, M.; Sullivan, J. C. *Inorg. Chem.* **1973,** *12,* 1459.

[164] Haight, G. P. *Inorg. Chem.* **1973**, *12*, 1461.
[165] Joshi, K. C.; Pathak, V. N. *Coord. Chem. Rev.* **1977**, *22*, 37.
[166] Kuroda, K. *Chem. Lett.* **1979**, 93.
[167] Fukuda, Y.; Ishige, M.; Ito, T.; Kuroda, K.; Sone, K.; Suzuki, Y.; Yano, S.; Yoshikawa, S. *Chem. Lett.* **1981**, 1699.
[168] Kitamura, Y.; van Eldik, R. *Transition Met. Chem.* (*N.Y.*), in press.
[169] Milstein, S.; Cohen, L. A. *J. Am. Chem. Soc.* **1972**, *94*, 9158, 9175.
[170] Hershfield, R.; Schmir, G. L. *J. Am. Chem. Soc.* **1973**, *95*, 7359, 8032.
[171] Boreham, C. J.; Buckingham, D. A.; Francis, D. J.; Sargeson, A. M.; Warner, L. G. *J. Am. Chem. Soc.* **1981**, *103*, 1975.
[172] Boreham, C. J.; Buckingham, D. A. *Inorg. Chem.* **1981**, *20*, 3112.
[173] Pocker, Y.; Sarkaren, S. *Adv. Enzymol. Relat. Areas Mol. Biol.* **1978**, *47*, 149.
[174] Pocker, Y.; Deits, T. L.; Tanaka, N. *Adv. Solution Chem.* [*Sess. Lect. Contrib. Microsymp. "Eval. Solvation Energ. Reagents Transition States,"* Int. Symp. Solute-Solute-Solvent Interact.], *5th, 1980* **1981**, 253.
[175] Huss, A.; Lim, P. K.; Eckert, C. A. *J. Phys. Chem.* **1982**, *86*, 4224, 4229, 4233.
[176] Cannon, R. D. "Electron Transfer Reactions"; Butterworth: London, 1980.
[177] Pennington, D. E. *ACS Monogr.* **1978**, *174*, 498.

ADVANCES IN INORGANIC AND BIOINORGANIC MECHANISMS, VOL. 3

Binuclear Dioxygen Complexes of Cobalt

S. Fallab and P. R. Mitchell

*Institut für Anorganische Chemie der Universität Basel
Basel, Switzerland*

I. INTRODUCTION

Although the earliest reports on the reaction of cobalt(II) salts with oxygen date from 1798[1] (Tassaert) and 1802[2] (Thenard), the first detailed study in which μ-peroxodicobalt(III) complexes were isolated was that of Frémy[3] in 1852. He isolated and analyzed the brown complex $[Co(NH_3)_{10}O_2](SO_4)_2 \cdot 3H_2O$ and found that it evolved oxygen on redissolving in water. In 1883 Maquenne[4] observed that the reaction of ozonized oxygen with ammoniacal cobalt(II) sulfate gave a green product instead of the brown complex, which Frémy[3] had obtained using only oxygen. Vortmann, in 1885, reported[5] that although the action of HNO_3 or of H_2SO_4 on the brown complex also gave the green species, the yield was better with addition of an oxidizing agent.

The problem of the structures of these two series of complexes was solved in 1898 by Werner and Mylius,[6] who were the first to suggest the presence of a bridging dioxygen. Werner[6,7] characterized the brown products as μ-peroxodicobalt(III) complexes, i.e., $[(H_3N)_5CoOOCo(NH_3)_5]^{4+}$, and the green products as salts of $[(H_3N)_5CoOOCo(NH_3)_5]^{5+}$. He also prepared[7] several other green complexes, such as $[(L)_4Co(O_2,NH_2)Co(L)_4]^{4+}$ $[(L)_4 = (NH_3)_4$ or $(en)_2]$, and assigned the oxidation states as cobalt(III)-peroxo-cobalt(IV). Gleu and Rehm[8] found that the green salt $[(H_3N)_5CoOOCo(NH_3)_5](SO_4)_2(HSO_4) \cdot 3H_2O$ was paramagnetic, with a magnetic moment $\mu = 1.67$ BM. Dunitz and Orgel[9] suggested that it was a μ-superoxodicobalt(III) complex. With a "noninnocent" ligand such as dioxygen, the assignment of oxidation states to the cobalt is not easy[10,11]; however, despite recent criticism,[11] there is now general agreement[12-17] that all of these complexes are best considered as cobalt(III) complexes, the brown species having a bridging peroxide group and the green species having a paramagnetic superoxide group as the bridging ligand.

Considerable progress in the investigation of the coordination chemistry of dioxygen, acting either as monodentate or as a bridging ligand, has been made in the past decade. In earlier years much of the interest in the μ-peroxo and μ-superoxo complexes centered on their redox reactions,[18] both from preparative[7,19] and from kinetic[20,21] points of view. However, due to their potential use as oxygen carriers[22] and to the superficial similarity between the μ-peroxodicobalt(III) complexes and some proteins that bind dioxygen,[15-17] there has been an increase in interest in such properties as the equilibrium constants of formation[17,23,24] and in the kinetics both of formation and of the decomposition back to cobalt(II) and oxygen.

It is commonly believed,[25-28] but has never been proved, that the μ-peroxodicobalt(III) complexes are intermediates in the synthesis of mononuclear cobalt(III) complexes by oxygenation of mixtures containing

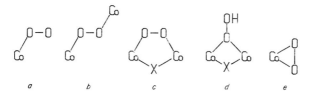

Fig. 1. Different structures of dioxygen adducts.

cobalt(II) salts and the ligand required. In some cases the rate of formation of the mononuclear cobalt(III) complex seems to depend on the rate of decomposition of the μ-peroxodicobalt(III) complex. However, although it was suggested many years ago that this aspect deserved further attention,[20] this has still been rather neglected.

Although several modes of bonding of dioxygen to cobalt complexes have been observed (Fig. 1a–e), dioxygen has a tendency to bind in an "end-on" manner, and structure e has been found only in cobalt phosphine complexes.[29] Complexes of type a will only be considered insofar as they are intermediates in the formation or decomposition of the binuclear dioxygen-bridged complexes; b and c will be discussed in detail. For d there is only one proved example, with $X = NH_2$.[30] Two other modes of bonding are known for rhodium dioxygen complexes, and these have been discussed in a recent review.[31]

Thus, to minimize overlap with the reviews already cited, this article will focus on the kinetics and mechanisms of the formation of the μ-peroxo- and μ-superoxodicobalt(III) complexes, and on the two main routes of decomposition, i.e., decomposition back to cobalt(II) and oxygen, and decomposition to form mononuclear cobalt(III) complexes. Structural as well as mechanistic features of some of the most important biological oxygen carriers have been summarized in Volume 1 of this series.[16]

II. FORMATION, STABILITY, AND PROPERTIES OF μ-PEROXO COMPLEXES

A. Formation of singly bridged complexes

The structural requirements for the formation of singly bridged binuclear dioxygen adducts are difficult to define precisely. The hydrated cobalt(II) ion obviously does not react with O_2 nor do other cobalt(II) complexes react with weak ligands such as halides, SCN^-, CO_3^{2-}, $C_2O_4^{2-}$, or other oxygen-donor chelating species. On the other hand, almost any amine complex will form a stable binuclear adduct. In a careful study on the oxygenation of aqueous ammonia solutions of cobalt(II) salts by Simplicio and Wilkins,[32] oxygen uptake was shown to be due mainly to the reaction of $[Co(NH_3)_5(H_2O)]^{2+}$.

$$2[Co(NH_3)_5(H_2O)]^{2+} + O_2 \rightleftharpoons [(NH_3)_5CoOOCo(NH_3)_5]^{4+} + 2H_2O \tag{1}$$

The hexammine ion was found to react much more slowly, and the tetrammine and lower species were unreactive. The reversibility of equilibrium (1) has been demonstrated unambiguously in dilute solution,[33] as well as with solid salts[34] of the binuclear adduct. Experiments involving $^{18}O_2$ have shown that the bridging dioxygen ligand originates entirely from molecular oxygen.[34] Replacing ammonia by polydentate amines increases the stability of the binuclear adduct, thus favoring oxygen uptake. The corresponding cobalt(II) 1,5,8,11,15-pentaazapentadecane ion, $[Co(tetren)(OH_2)]^{2+}$, forms such a stable dioxygen complex that it is difficult even to prove reversibility.[35]

$$2[Co(tetren)(H_2O)]^{2+} + O_2 \rightleftharpoons [(tetren)CoOOCo(tetren)]^{4+} + 2H_2O \tag{2}$$

In contrast to observations with the cobalt(II) ammine complexes, oxygenation of cobalt(II) chelates also takes place if the number of N-donor groups is reduced to four or three.[36] Oxygen uptake by diethylenetriamine-N_1,N_3-diacetatocobalt(II) can be measured conveniently[37] whereas with the analogous complex, in which the central imino group of the pentadentate ligand is replaced by an ether oxygen, no oxygenation has been observed.[38] $[Co(trien)(H_2O)_2]^{2+}$ and $[Co(dien)(H_2O)_3]^{2+}$ have also been proved to react with oxygen,[39,40] but no reaction of $[Co(en)(H_2O)_4]^{2+}$ has been reported and chelates with ethylenediaminediacetic acid definitely do not form singly bridged binuclear adducts,[23] although there is evidence for the formation of a μ-peroxo μ-hydroxo complex. On the other hand, $[Co(en)(NH_3)_3(H_2O)]^{2+}$ also forms a stable dioxygen complex. Clearly, the number of ammonia or amino groups coordinated to cobalt is a crucial factor for oxygenation. Although a precise structure–stability relationship cannot be formulated, the rule of thumb that a minimum of three comparatively strong donor groups is necessary[36] holds well for this class of oxygenation products. Surprisingly, few exceptions have been reported so far, and these usually involve polydentate ligands with phenolate groups.

Equilibrium constants have been determined for a number of complexes using various methods, including manometric measurements, amperometric determination of the oxygen concentration, or potentiometric pH titration.

Some results have been summarized by Martell.[24] The K_{O_2} values listed in Table 1 are defined by

$$K_{O_2} = [(Z)CoOOCo(Z)]/[O_2][Co(Z)]^2 \qquad (3)$$

where Z may represent either a pentadentate ligand or a combination of ligands occupying five coordination sites. A comparison of the value for the bis(pentaammine) complex with that for the corresponding chelate with tetraethylenepentamine (tetren) immediately reveals that, in addition to the intrinsic donor strength of the ligand atoms, some chelating factor must also be involved in the stability of the dioxygen adduct. McLendon and Martell[43] showed a rough correlation between the overall basicity of the chelating agent and K_{O_2}. For example, the considerably lower stability of the adduct of diethylenetriaminediacetatocobalt(II) compared with [(tetren)-CoOOCo(tetren)]$^{4+}$ may reasonably be explained by the comparatively low basicity of the carboxylate. Although the intrinsic donor strength of the ligands, rather than the overall basicity, is undoubtedly an important factor, it cannot account for the low stability of adducts with monodentate ligands. For example, on the basis of the linear relationship proposed,[43] K_{O_2} for [(NH$_3$)$_5$CoOOCo(NH$_3$)$_5$]$^{4+}$ would be expected to be about 10^{12} times higher than the observed value.[32] The comparison of the sum of the macro-scopic pK_a values with log K_{O_2} inevitably includes factors such as the prox-imity of basic groups and the rigidity of the chelating framework. Not even

TABLE 1

Equilibrium constants $K_{O_2} = [(Z)CoOOCo(Z)]/[O_2][Co(Z)]^2$
of singly bridged dioxygen adducts at 25°C

Complex^a	log K_{O_2}	Ref.
[(NH$_3$)$_5$CoOOCo(NH$_3$)$_5$]$^{4+}$	6.8	[32]
[(tetren)CoOOCo(tetren)]$^{4+}$	15.8	[24]
[(dtda)CoOOCo(dtda)]$^{2+}$	6.6	[37]
[(pydien)CoOOCo(pydien)]$^{4+}$	11.4	[24]
[(imdien)CoOOCo(imdien)]$^{4+}$	12.6	[24]
[(pydpt)CoOOCo(pydpt)]$^{4+}$	7.7	[24]
[(imdpt)CoOOCo(imdpt)]$^{4+}$	9.5	[24]
[(dap)$_2$CoOOCo(dap)$_2$]$^{2+}$	8.9	[47]
[(his)$_2$CoOOCo(his)$_2$]$^{2+}$	6.6	[42]

^{a} Abbreviations for ligands: tetren (tetraethylenepentamine), 1,4,7,10,13-pentaazatridecane; dtda (diethylenetriamine-N_1, N_3-diacetic acid), 1,9-bis(carboxy)-2,5,8-triazanonane; pydien, 1,9-bis(2-pyridyl)-2,5,8-triazanonane; imdien, 1,9-bis(4-imida-zolyl)-2,5,8-triazanonane; pydpt, 1,11-bis(2-pyridyl)-2,6,10-triazaundecane; imdpt, 1,11-bis(4-imidazolyl)-2,6,10-triazaun-decane; dap, DL-2,3-diaminopropanoic acid; his, DL-histidine.

within the narrow class of chelates with pentadentate amines is there a convincing correlation with ligand basicity. Although the stability of the dioxygen adduct drops when aliphatic amino groups are replaced by less basic aromatic amines[24] (compare the tetren and pydien complexes in Table 1), steric factors may also play a role. This is clearly seen in the examples where six-membered chelate rings are formed: these destabilize the adduct, probably due to increased flexibility.

The amino-acidate ligands in the complexes $[(his)_2CoOOCo(his)_2]^{2+}$ and $[(dap)_2CoOOCo(dap)_2]$ cannot be coordinated completely. Although definite structural data are not yet available for these complexes, which are at the lower end of the stability scale, it may be concluded that one of the carboxylate groups of each ligand molecule is dangling uncoordinated; the resulting steric interactions may lower the stability of the binuclear adduct.

The relationship between the equilibrium constants K_{O_2} and the structural features of the μ-peroxodicobalt(III) complexes is complicated by the possibility of isomerism, especially with ligands for which different configurations are possible. Multidentate ligands, which are capable of arranging their donor atoms around a metal ion in several different ways, have been termed facultative ligands.[44,45] Faculatative ligand disposition greatly increases the number of possible isomers, especially for a binuclear complex: thus with a linear pentadentate amine four structures are possible for each of the mononuclear fragments (Fig. 2) yielding a total of 10 isomers for the binuclear complex.

With the ligand 1,5,8,11,15-pentaazapentadecane (papd) an isomerically pure dioxygen adduct has been isolated, and an X-ray diffraction study[46] showed that both of the cobalt atoms have the last of the four configurations depicted in Fig. 2. The crystallization of this isomer from a concentrated oxygenation mixture, however, does not prove that it is also the predominant product in the dilute solutions used in equilibrium work. Lattice-packing effects can dominate solubility and thus the ease of crystallization, and isolated binuclear cations of this type may be stabilized by intramolecular hydrogen bonding in the solid state.[46] On the other hand, individual hydration effects may influence isomer equlibria in solution. Overall equilibrium constants must therefore be interpreted with caution.

Both the investigation of the temperature dependence of the equilibrium constants and calorimetric measurements have shown[47,48] that formation of a dioxygen adduct is accompanied by surprisingly large negative enthalpy changes: typical values for binuclear complexes $[(Z)CoOOCo(Z)]$ are around -120 kJ mol^{-1}. However, the apparently strong bonding of dioxygen is counteracted by a large negative entropy change which in part is easily explained by the loss of translational and rotational entropy of oxygen.

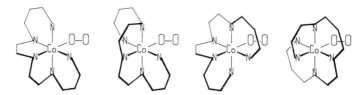

Fig. 2. Isomers of $[Co(papd)O_2]^{2+}$.

It has been found that for $[(his)_2CoOOCo(his)_2]^{2+}$ the entropy change is -283 J deg^{-1} mol^{-1} (25°C), and a similar value has been estimated[42] for $[(en)_2(H_2O)CoOOCo(H_2O)(en)_2]^{4+}$. The oxygenation of complexes with pentadentate amines such as pydien and pydpt (Table 1) has been subject of a calorimetric study.[41] Values of ΔH and ΔS appear to depend on the size of the chelate rings: when only five-membered chelate rings were present, the values obtained were $\Delta H = -140$ kJ mol^{-1} and ΔS between -200 and -240 J deg^{-1} mol^{-1}. When six-membered chelate rings were present, both enthalpy and entropy changes were lower: $\Delta H \approx -80$ kJ mol^{-1} and $\Delta S \approx -100$ J deg^{-1} mol^{-1}. These interesting results support the view that the rigidity of the coordination sphere influences the formation of an oxygen adduct significantly.

As will be discussed later, there is considerable evidence that the formation of a binuclear dioxygen-bridged complex is a two-step process in which a 1:1 adduct is formed first. In some cases, however, the mononuclear complex $[(Z)CoO_2]$ does not react further.

$$Co(Z) + O_2 \rightleftharpoons [(Z)CoO_2] \tag{4}$$

$$[(Z)CoO_2] + Co(Z) \rightleftharpoons [(Z)CoOOCo(Z)] \tag{5}$$

This is especially true for square-planar cobalt(II) complexes with tetra-dentate ligands. Among the first oxygenation studies were those with cobalt complexes of Schiff's bases formed from salicylaldehyde and aliphatic diamines.[49] The early interest in these complexes stemmed from their possible practical use for oxygen storage.[22,50] More recent investigations of the reaction with N,N'-ethylenebis(salicylideniminato)cobalt(II) and its derivatives showed that products formed depend strongly on the reaction medium and on substituents in the aromatic rings of the ligand. The majority of complexes of this type yielded the dioxygen-bridged binuclear product. However, oxygenation of the 3-methoxy derivative in basic solvents such as pyridine gave only the mononuclear adduct,[51] with pyridine coordinated trans to the dioxygen. The importance of the axial ligand was also demonstrated in studies with N,N'-ethylenediaminebis(acetylaceimineto)cobalt(II), which reacts reversibly with dioxygen at 0°C to form a 1:1 adduct if a base such as pyridine is added to a solution in toluene.[52] Similar results were obtained with cobalt(II) complexes of macrocyclic ligands of the porphyrin type[53]: the oxygenation equilibrium (4) was very dependent on the solvent and on the base added. In some cases significant amounts of the oxygen adduct were only formed below 0°C.

Although the existence of mononuclear adducts has also been demonstrated in several complexes that do not contain macrocyclic ligands, only a few have been isolated, e.g., $[Co(CN)_5O_2]^{3-}$. Whereas the normal binuclear dioxygen-bridged adduct can be prepared by oxygenation of $[Co(CN)_5]^{3-}$ in aqueous solution, the mononuclear product can be crystallized from solutions in DMF.[54] The structure of $[Co(bipy)_2(H_2O)O_2]^{2+}$, which has recently been determined by X-ray diffraction,[55] is unusual because the metal is coordinated by only four strong donor groups. The complex—like $[Co(CN)_5O_2]^{3-}$—has structure type *a* (Fig. 1) and may be stabilized by forming a hydrogen bond between O_2 and H_2O coordinated cis to each other (Fig. 3). A surprising claim for a kinetically stabilized 1:1 complex has been reported[56]. The ligand, a dioxocyclam-14 derivative, has a bulky naphthyl group that is supposed to prevent formation of the 2:1 complex. Lever *et al.* found that the complex $[Co(dmen)_2Cl_2]$ (dmen, *N,N'*-dimethyl-ethylenediamine) forms a 1:1 adduct that is at least stable enough for spectroscopic characterization.[57]

Various explanations have been suggested for the apparent stabilization of the 1:1 adduct: (1) inhibition of the second step [equilibrium (5)] by steric crowding in the bridging region[13] and (2) thermodynamic factors[58] or solvent effects.[59] Whichever the more important factor may be, the relative stability of the two types of adduct may be affected by small changes in the ligand. Subtle modifications, such as the introduction of a methoxy group in a remote position of the chelating ligand, may shift equilibria (4) and (5) enough for the formation of either the 1:1 or the 2:1 type of adduct to be favored.

Steric inhibition of the formation of the binuclear complex was clearly demonstrated in a study of complexes of alkyl-substituted triethylenetetra-

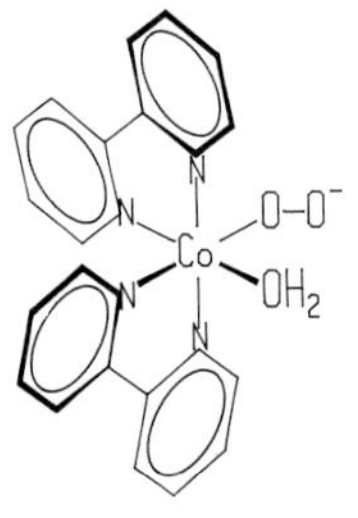

Fig. 3. Schematic representation of $[Co(bipy)_2(H_2O)O_2]^{2+}$.

mine ligands.[60] Oxygenation of $[Co(trien)(H_2O)_2]^{2+}$ in aqueous solution
leads to stable binuclear dioxygen-bridged complexes, but no 1:1 adduct has
ever been isolated. Substitution of a single methyl or ethyl group in either
the primary or the secondary amino groups does not prevent oxygenation.
But cobalt(II) complexes of the corresponding ligands with an isopropyl
substituent or with dimethyl or diethyl substituents at the terminal amino
groups do not form dioxygen adducts.[60]

Simple bonding schemes have been proposed,[61] based mainly on the
results of X-ray crystallography and on the information gained from the
ESR spectra of 1:1 adducts. Early determination of the magnetic suscep-
tibilities of the 2:1 adducts showed them to be diamagnetic, and suggested
a more or less complete charge transfer from cobalt(II) to the bridging
dioxygen, leading to a μ-peroxodicobalt(III) complex. This picture is consistent with the O—O distances determined by X-ray crystallography, which
are always found to be close to 148 pm; this is intermediate between the
values obtained for H_2O_2 and for Na_2O_2 (Table 2). Likewise, the O—O
stretching frequencies observed by Raman spectroscopy resemble those in
H_2O_2 and Na_2O_2. There is a similar correspondence of the same properties
for potassium superoxide and the μ-superoxodicobalt(III) complexes that
are discussed on p. 362.

The bridging ligand apparently has the electronic structure of a peroxo
group with completely filled π^* orbitals, and, defining the Co–O direction
as the z-axis, the bonding of O_2^{2-} to cobalt can be described as a σ-type

TABLE 2

The O—O distance and stretching frequency in μ-peroxo-
and μ-superoxodicobalt(III) complexes

Complex	O—O distance (pm)	Ref.	v_{O-O} (Raman)[a] (cm^{-1})	Ref.
$[(H_3N)_5CoOOCo(NH_3)_5]^{4+}(SO_4^{2-})_2 \cdot 4H_2O$	147	[62]	788	[63]
$[(H_3N)_5CoOOCo(NH_3)_5]^{5+}(NO_3^-)_5$	132	[64]	1123	[63]
$(K^+)_8[(NC)_5CoOOCo(CN)_5]^{6-}(NO_3^-)_2 \cdot 4H_2O$	145	[65]	808	[63][b]
$(K^+)_5[(NC)_5CoOOCo(CN)_5]^{5-} \cdot H_2O$	126	[66]	1099	[63]
O_2	121	[67]	1555	[68]
$K^+O_2^-$	128	[67]	1145	[69]
H_2O_2	145	[70]	880	[71]
$(Na^+)_2O_2^{2-}$	149	[67]	738, 794	[72]

[a] As these complexes have either been shown to be centrosymmetric in the solid state, or may
be assumed to be centrosymmetric in solution, the O—O stretching is infrared inactive.
[b] In $(K^+)_6[(NC)_5CoOOCo(CN)_5]^{6-} \cdot H_2O$.

interaction between π^* and d_{z^2} (see Fig. 6a). Several investigators have pointed out that the primary effect of a strong σ-donor ligand in the trans position to dioxygen will be to increase the energy of the d_{z^2} level and thus influence the addition reaction considerably.[73–76]

Although the formulation of the 1:1 and 2:1 adducts as superoxo- or μ-peroxocomplexes, respectively, has gained wide acceptance, the estimation of the polarity of the Co–O$_2$ bond from ESR measurements is not unambiguous.[77] Various estimates of the charge on the O–O group range from 0.1 to 0.8 e^-. An alternative view has been taken by Drago *et al.* that stresses the spin pairing of d_{z^2} and π^* electrons.[78,79] The higher the energy of the d_{z^2} orbital, the more energy there is to be gained when an electron of Co(II) drops into the σ MO formed from d_{z^2} and π^*. The charge on the O–O group is then defined by the LCAO coefficients, and an increase in d_{z^2} energy would also increase the transfer of charge onto O$_2$. This bonding scheme explains all the known properties of dioxygen adducts without necessarily implying a large charge transfer. The designation of the adducts as superoxocobalt(III) or μ-peroxodicobalt(III) complexes must therefore be regarded as a purely formal one.

B. Formation of doubly bridged complexes

The binuclear μ-peroxo complexes discussed so far are of structure type *b*. They are formed when the coordination sites cis to the bridging dioxygen are occupied by strong ligands. If, however, one of the cis positions on each cobalt is occupied by a labile ligand, in aqueous solution a more stable oxygenation product is formed in which the two metal centers are also linked by a hydroxide bridging ligand. The formation of μ-peroxo μ-hydroxo complexes has long been suspected on the basis of pH titration results and kinetic observations,[39] and the existence of equilibrium (6) has been established.[80]

$$[(en)_2(NH_3)CoOOCo(NH_3)(en)_2]^{4+} + OH^- \rightleftharpoons [(en)_2Co(O_2,OH)Co(en)_2]^{3+} + 2NH_3 \quad (6)$$

However, Saito and co-workers[81] were the first to report the isolation of a hydroxo-bridged complex, $[(en)_2Co(O_2,OH)Co(en)_2](ClO_4)_3$, and the structure was confirmed by Thewalt.[82]

Formation of doubly bridged μ-peroxo μ-hydroxo complexes in aqueous solution seems to be favored thermodynamically: calorimetric measurements[42] by Powell and Nancollas showed that for reaction (7) $\Delta H = -123$ kJ mol^{-1} and $\Delta S = -272$ J deg^{-1} mol^{-1}. Based on a comparison with thermodynamic data for the diolation of Fe^{3+}, the values $\Delta H \approx +40$ kJ mol^{-1} and $\Delta S \approx +60$ J deg^{-1} mol^{-1} were estimated for the olation step [reaction (8)] separately.

$$2[Co(en)_2(H_2O)_2]^{2+} + O_2 \rightarrow [(en)_2Co(O_2,OH)Co(en)_2]^{3+} + 2H_2O + H_3O^+ \quad (7)$$

$$[(en)_2(H_2O)CoOOCo(H_2O)(en)_2]^{4+} \rightarrow [(en)_2Co(O_2,OH)Co(en)_2]^{3+} + H_3O^+ \quad (8)$$

Therefore, it was suggested[42] that stabilization by an additional hydroxide bridge was due mainly to a favorable entropy term. However, a recent and more detailed reexamination[82a] of the Co^{2+}/en/O_2 system, using a combination of calorimetric, spectrophotometric, oxygraph, and pH measurements, confirmed the basic calorimetric results of Powell and Nancollas, but suggests that the reasons for the stabilization are more complicated and are not fully explicable.

The structural features of the bridging ring system formed in the olation reaction are very similar to those found for μ-peroxo μ-amido complexes[83]: the O–O bond distance is always about 148 pm, and the Co–O–O–Co group is nonplanar, with a dihedral angle of $\sim 60°$. The μ-peroxo μ-amido complexes are formed by condensation from bis(ammino) species [reaction (9)] in strongly alkaline solution.[84]

$$[(en)_2(NH_3)CoOOCo(NH_3)(en)_2]^{4+} + OH^- \rightarrow$$

$$[(en)_2Co(O_2,NH_2)Co(en)_2]^{3+} + H_2O + NH_3 \quad (9)$$

Their reactions have been studied extensively by Sykes and co-workers.[20] An unusual feature of this binuclear system is the apparent isomerization to a protonated form in which the O–O group sticks out of the bridging ring[85] (structure type *d*, with X = NH_2); however, no further example of this unusual type of μ-hydroperoxo complex has been proved to exist.

Azide can also function as a second bridging ligand, and the structure of $[(tren)Co(O_2,N_3)Co(tren)]^{3+}$ has been determined recently.[55] Azide forms a one-atom bridge and the conformation of the bridging ring is again identical with that of the μ-peroxo μ-amido complexes, within the errors of the crystallographic data. A doubly bridged μ-peroxo complex with an 11-membered bridging ring has been prepared [reaction (10)] by the reaction of tren with a bis(ammino) complex[86]:

$$[(tren)(NH_3)CoOOCo(NH_3)(tren)]^{4+} + tren \rightarrow$$

$$[(tren)Co(O_2,tren)Co(tren)]^{4+} + 2NH_3 \quad (10)$$

Although oxygenation of dilute solutions of Co^{2+}/tren yields[87] only $[(tren)Co(O_2,OH)Co(tren)]^{3+}$, a ^{13}C NMR study[88] has shown that the tren-bridged complex is also formed by the oxygenation of concentrated solutions containing Co^{2+}/tren with a 1:1.5 molar ratio. The crystal structure (Fig. 4) shows[86] that the bridging tren is coordinated by two of its

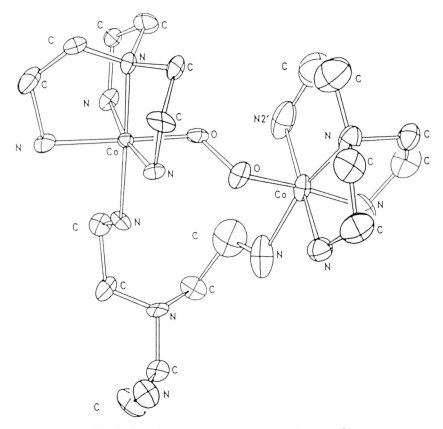

Fig. 4. ORTEP drawing of $[(tren)Co(O_2,tren)Co(tren)]^{4+}$.

primary amino groups, with a transoid Co–O–O–Co group, and a dihedral angle of 160°. The complex is quite stable in solution, although in the absence of an excess of free ligand the thermodynamically more favorable μ-peroxo μ-hydroxo species is slowly formed.

C.   Electronic spectra

Dioxygen adducts generally exhibit intense absorption bands in the near UV. Apart from charge-transfer bands between 220 and 240 nm, common to all cobalt(III) complexes, their absorption spectrum is characterized by broad bands reaching into the visible and thus giving rise to yellow or brown

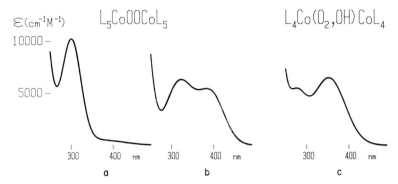

Fig. 5. Types of UV/visible spectra of μ-peroxodicobalt(III) complexes. (a and b) Singly bridged complexes; (c) doubly-bridged complexes.

colors in solution. Somewhat conflicting views concerning the rules relating the number of bands observed between 250 and 450 nm with the structure have been published.[17,80,89,90] Three main types of spectra can be distinguished empirically (Fig. 5):

1. Most singly bridged μ-peroxodicobalt(III) complexes have a strong band at about 300 nm, with a molar extinction $\varepsilon \approx 11,000$. In addition, a broad shoulder of lower intensity around 400 nm is frequently observed. Numerous X-ray structure determinations have shown that singly bridged complexes (type *b* in Fig. 1) normally have a planar arrangement of the Co–O–O–Co group. Of the two orthogonal π^* orbitals, only one can be expected to contribute to strong σ bonding in this case; the other one is perpendicular to the CoOOCo plane and can give rise only to a weak π

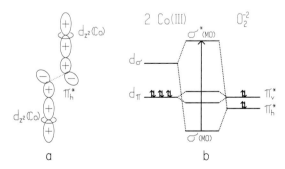

Fig. 6. (a) Schematic illustration of the orbitals involved in the bonding in singly bridged μ-peroxodicobalt(III) complexes with a transoid Co–O–O–Co group. (b) Qualitative representation of the energy levels of the interacting orbitals.

interaction (Fig. 6). The band observed at 300 nm has been assigned to a $\pi^* \rightarrow d_{\sigma^*}$ transition.[90]

2. In some cases absorption bands of lower and approximately equal intensity ($\varepsilon \approx 5000$) are observed, with maxima between 300 and 400 nm. This type of spectrum is seen whenever the ligand field is very unsymmetrical. Amine complexes with five similar N-donor groups have a type (a) spectrum (Fig. 5), but partial replacement by ligands of different donor strength leads to a spectrum of type (b), e.g., [(tren)(CN)CoOOCo(CN)(tren)]$^{2+}$, [(his)$_2$-CoOOCo(his)$_2$], and [(dien)(ox)CoOOCo(ox)(dien)]. It has been suggested that this marked difference in absorption behavior involves differing geometries of the Co–O–O–Co group[91]: a substantial deviation from planarity could give rise to two $\pi^* \rightarrow d_{\sigma^*}$ transitions because the distinction between an in-plane and an out-of-plane dioxygen π^* orbital would then be lost and both would become strongly σ-bonding. However, there is no experimental evidence in favor of this theory, and as for one of the complexes, [(tren)(CN)-CoOOCo(CN)(tren)]$^{2+}$, a planar geometry was found in a recent X-ray structure determination[92]; the splitting may well be due to the unsymmetrical ligand field. On the other hand, quite a few complexes reported in a recent review[311] were found to have a nonplanar Co–O–O–Co group; however, they are mostly of the salen type and no spectra have been reported.

3. A two-band spectrum is also observed with complexes of type c (Fig. 1), in which X is a μ-hydroxo or μ-azido group. The additional bridging ligand forces the Co–O–O–Co group into a cisoid nonplanar arrangement with a dihedral angle close to 60°. One of the absorption bands is usually well below 300 nm, and could, however, be due to the second bridging ligand since bis(μ-hydroxy)dicobalt(III) complexes also have a strong charge-transfer band below 300 nm.[93]

The charge-transfer absorption bands of μ-peroxo μ-amido complexes are less pronounced: the complex [(en)$_2$Co(O$_2$,NH$_2$)Co(en)$_2$]$^{3+}$ has a band at 337 nm with $\varepsilon = 3300$ and a barely visible shoulder around 280 nm. Oxygenation of the cobalt(II) complexes of 1,1,8,8-tetraacetyl-3,6-diazaoctane and 1,1,9,9-tetraacetyl-3,7-diazanonane in solutions containing pyridine[94] leads to very unusual absorption spectra. Formation of 1:1 as well as singly bridged 2:1 complexes is observed, and the tetradentate ligands are assumed to occupy four equatorial positions. Pyridine occupies an apical position and thus there are three N- and two O$^-$-donor groups that give rise to a spectrum that could be described as an extreme example of type (b) (Fig. 5), with the absorption bands much farther into the visible. The main absorption band at 525 nm is assigned by the authors to a $\pi_v^* \rightarrow d_{x^2-y^2}$ transition.

III. KINETICS OF FORMATION OF μ-PEROXO COMPLEXES

More detailed information about adduct formation can be obtained from kinetic measurements. The uptake of oxygen can be monitored either using an O_2-sensitive electrode or spectrophotometrically by following the intense charge-transfer bands of the products of oxygenation. Wilkins and co-workers[95] have shown in comprehensive kinetic investigations that the rate-determining step in the formation of the binuclear adduct is a second-order reaction between the cobalt(II) complex and dissolved oxygen [Eq. (11)]; if a large excess of the cobalt(II) complex is used, pseudo first-order kinetics are observed and a macroscopic rate constant m can be evaluated. However, if [Co(II)(Z)] is varied over a wide range, m is no longer constant and at least two parameters are needed to fit the observed concentration dependence of m, which then involves both linear and quadratic terms [Eq. (12)].

$$d(product)/dt = m[Co(II)(Z)][O_2] \qquad (11)$$

$$m = p_1[Co(Z)]^2/(1 + p_2[Co(Z)]) \qquad (12)$$

Assigning rate constants k_1, k_{-1}, k_2, and k_{-2} to the forward and reverse steps of the reaction scheme of equilibria (4) and (5) (p. 317), and assuming steady-state conditions for the 1:1 intermediate and that equilibrium (5) is practically irreversible, approximate values for k_1 and k_2/k_{-1} can be computed from the empirical parameters p_1 and p_2 [Eq. (12)].

Values of k_1 may be used as a measure of the ability of [Co(II)(Z)] to add dioxygen. For the majority of systems studied so far the values lie in a surprisingly small range between 10^4 and 5×10^5 M^{-1} s^{-1} (see Table 3 for some typical values for k_1). Species such as $[Co(NH_3)_5(H_2O)]^{2+}$ and $[Co(en)_2(H_2O)_2]^{2+}$ with fewer than six nitrogen donors are generally found at the upper end of this range. In these cases oxygenation involves replacement of coordinated water. When all six coordination sites are occupied by amines or amino acids, somewhat lower values are found for k_1, indicating that dissociation of a more strongly bound ligand must precede formation of an oxygen adduct. It has therefore been suggested that oxygenation of cobalt(II) complexes is dominated by their ligand exchange rates.[99]

Values of the rate of exchange of coordinated water fit in well with this idea for many complexes. Merbach found the rate constant $k(H_2O) = 3 \times 10^6$ s^{-1} for the substitution of a particular water molecule in $[Co(H_2O)_6]^{2+}$ using high-pressure NMR measurements.[103] The data indicated a dissociative interchange mechanism for octahedral high-spin Co(II). Assuming that small neutral molecules have roughly equal rates of entry, the probability of oxygenation should be $[O_2]/[H_2O]$ times less than the probability of reentry of water. Second-order rate constants of the order of $5 \times 10^4\,M^{-1}\,s^{-1}$ can therefore be expected for the substitution of coordinated water by oxygen.

In many of the complexes cited in Table 3, however, there is a considerable deviation from octahedral symmetry that may influence substitution rates to some extent. The unusually high reactivity exhibited by the first few species in the table needs an additional explanation: $[Co(cyclam-14)-(H_2O)_2]^{2+}$ has been shown to be low spin,[104] and the elongation of the $Co-(OH_2)$ bonds along the z-axis suggests that the unpaired electron is in the d_{z^2} orbital, which should increase reactivity in the apical positions. However, this cannot be the explanation for the reactivity of $[Co(papd)-(H_2O)]^{2+}$, as this is high spin.[96]

The surprisingly low value of k_1 for the oxygenation of $[Co(tren)]^{2+}$ has been explained[105] by the particular geometry of the ligand that favors the formation of penta-coordinated complexes.[106] A fast preequilibrium, $[Co(tren)(H_2O)]^{2+} \rightleftharpoons [Co(tren)(H_2O)_2]^{2+}$, is postulated in which only the hexa-coordinated diaquo species is active toward oxygenation. Of the three possible geometrical isomers of $[(tren)Co(O_2,OH)Co(tren)]^{3+}$, the one in which both tertiary amino groups are cis to the bridging dioxygen is formed exclusively.[87]

TABLE 3

Observed second-order rate constants (25°C) for the formation of μ-peroxodicobalt(III) complexes from various cobalt(II) species

Complex	$k_1\ (M^{-1}\,s^{-1})$	Ref.
$[Co(papd)]^{2+}$	1.4×10^7	[96]
$[Co(cyclam-14)(H_2O)_2]^{2+}$	5×10^5	[97]
$[Co(en)_2(H_2O)_2]^{2+}$	4.7×10^5	[95]
$[Co(tetren)(H_2O)]^{2+}$	10^5	[95]
$[Co(trien)(H_2O)_2]^{2+}$	2.5×10^4	[98]
$[Co(NH_3)_5(H_2O)]^{2+}$	2.5×10^4	[32]
$[Co(en)_3]^{2+}$	10^4	[95]
$[Co(gly-gly)_2]^{2-}$	7×10^3	[99]
$[Co(L-his)_2]$	3.5×10^3	[100]
$[Co(dien)_2]^{2+}$	1.2×10^3	[95]
$[Co(NH_3)_6]^{2+}$	10^3	[32]
$[Co(dmtr)(H_2O)_2]^{2+}$	5×10^2	[101]
$[Co(tren)(H_2O)]^{2+}$	2×10^2	[102]

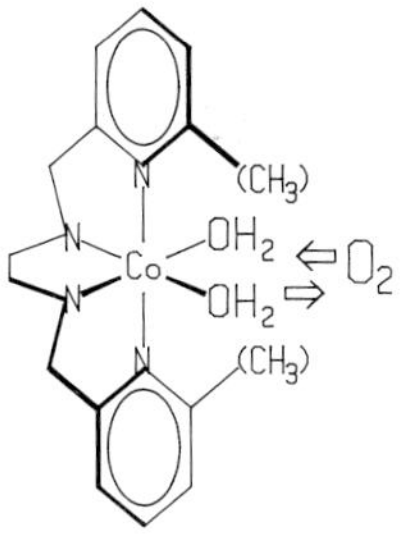

Fig. 7. Possible steric effects on the substitution of H_2O by O_2.

As an interchange mechanism implies that ligands leave and enter simultaneously, the size of neighboring groups can be expected to influence the rates. Chelates with bulky groups cis to the entering dioxygen react more slowly (Fig. 7); e.g., replacing the terminal aminomethyl groups in trien by pyridine lowers k_1 by a factor of 7, and introducing methyl groups into the heteroaromatic ring leads to a further drop in k_1 (Table 4). No such effect was found, however, with corresponding chelates of pentadenate amines.[107] A surprisingly low value of k_1 has also been found[101] for $[Co(dmtr)(H_2O)_2]^{2+}$. Although the methyl substituents may hinder ligand exchange to some extent, the large drop in reactivity compared with $[Co(trien)(H_2O)_2]^{2+}$ must also be due partially to electronic effects: oxygen is forced to enter trans to a tertiary amino group.

As formation of a cobalt(II) complex that is capable of reacting with oxygen is usually pH dependent, the observed oxygen uptake likewise

TABLE 4

Steric factors in the kinetics of adduct formation:
Second-order rate constants (25°C)

Complex[a]	k_1 (M^{-1} s^{-1})	Ref.
$[Co(trien)(H_2O)_2]^{2+}$	2.5×10^4	[98]
$[Co(dpdh)(H_2O)_2]^{2+}$	3×10^3	[107]
$[Co(mpdh)(H_2O)_2]^{2+}$	4×10^2	[107]
$[Co(dmtr)(H_2O)_2]^{2+}$	5×10^2	[101]
$[Co(tetren)(H_2O)]^{2+}$	10^5	[95]
$[Co(dptn)(H_2O)]^{2+}$	1.6×10^5	[107]

[a] Abbreviations for ligands: dpdh, 1,6-di(2-pyridyl)-2,5-diazahexane; mpdh, [1-(6-methyl)-2-pyridyl]-6-(2-pyridyl)-2,5-diazahexane; dmtr, 4,7-dimethyl-1,4,7,10-tetraazadecane; dptn, 1,9-di(2-pyridyl)-2,5,8-triazanonane.

depends on the acidity of the solution. In the upper pH range, in which formation of [Co(II)(Z)] is practically complete, pH-independent values for m [Eq. (11)] could be expected. However, if there is also coordinated water a further pH dependence may be observed. An increase in k_1 with pH is found with complexes of the type $[Co(II)(Z)(H_2O)_2]^{2+}$ in which Z is a tetradentate amine; it has been suggested that the hydroxo aquo complex is more reactive (Table 5). The opposite is observed with $[Co(Z)(H_2O)]^{2+}$ in which Z is a pentadentate amine. This is to be expected as replacement of a coordinated hydroxo group, e.g., in $[Co(Z)(OH)]^+$, is generally much slower than replacement of an aquo ligand, e.g., in $[Co(Z)(OH_2)]^{2+}$.

The second forward step in the reaction scheme of equilibria (4) and (5) should also be dominated by the substitution of H_2O in a cobalt(II) species. The entering ligand in this case is the mononuclear complex $[Co(Z)O_2]^{2+}$. Although this has a polarized O–O bond that may increase its nucleophilicity compared with that of dioxygen, the exchange process is dissociative in nature and k_2 cannot be significantly larger than k_1. On the other hand, steric crowding in the bridging region and stricter orientation requirements of the nucleophile should decrease k_2. Furthermore, when Z is an amine, both reactants have a charge of $+2$, which should lead to an unfavorable activation entropy. Thus k_2 can be expected to be either close to k_1 or somewhat less than k_1.

Unfortunately, experimental information about k_2 is scarce because more refined treatment of the kinetic data is needed. Model calculations show that the steady-state approach outlined above is questionable and in any case is not suitable for the determination of k_2. Endicott and co-workers[97] have estimated both second-order rate constants for the formation of [(cyclam-14)-CoOOCo(cyclam-14)]$^{4+}$: $k_1 \approx 5 \times 10^5$, $k_2 \approx 5 \times 10^5$ M^{-1} s^{-1}.

ESR measurements provide important experimental evidence in support of the two-step mechanism [equilibria (4) and (5)]. The formation of an

TABLE 5

Comparison of the rates of oxygenation of aquo and hydroxo complexes: Second-order rate constants (25°C)

Complex	k_1 $(M^{-1}$ $s^{-1})$	Ref.
$[Co(trien)(H_2O)_2]^{2+}$	2.5×10^4	[98]
$[Co(trien)(H_2O)(OH)]^+$	2.8×10^5	[98]
$[Co(dmtr)(H_2O)_2]^{2+}$	5×10^2	[101]
$[Co(dmtr)(H_2O)(OH)]^{2+}$	4×10^4	[101]
$[Co(dptn)(H_2O)]^{2+}$	1.6×10^5	[107]
$[Co(dptn)(OH)]^+$	2×10^3	[107]

intermediate with an eight-line signal corresponding to the interaction of an unpaired electron with a single ^{59}Co nucleus of spin $\frac{7}{2}$ has been found in studies with various cobalt(II) amine complexes.[108] However, depending on the reaction conditions it may be difficult to observe the signal at all,[109] as was the case in the formation of $[(CN)_5CoOOCo(CN)_5]^{6-}$.

In certain systems, measurements of the oxygen uptake using an O_2-sensitive electrode provide further evidence for the two-step mechanism.[96] Oxygenation of aqueous solutions of $Co(Z)^{2+}$, where Z represents the pentadentate amine papd (Table 3) or the combination of tridentate tris-(aminoethyl)ethane and bidentate ethylenediamine, clearly shows the formation of an intermediate that contains more than $\frac{1}{2}$ equivalent of dioxygen/cobalt(II) (Fig. 8a). An eight-line ESR spectrum indicates that this is a mononuclear adduct $[(Z)CoO_2]^{2+}$. Parallel measurements of the absorption spectra, using a diode array spectrophotometer, show that in both systems within the same time range the charge-transfer band is shifted 10–15 nm toward shorter wavelengths (Fig. 8b). Using a more elaborate procedure for the analysis of rate data, Maeder[96] has determined the consecutive second-order rate constants for the formation of $[(papd)CoOOCo-(papd)]^{4+}$: $k_1 = 1.4 \times 10^7$, $k_2 = 6 \times 10^6 \ M^{-1} s^{-1}$ (25°C).

With some ligands, after the oxygen uptake and formation of the binuclear adduct is practically complete, a slower third phase has been observed.[98] In spectrophotometric investigations this is generally seen as a decrease in absorption, from which a first-order rate constant k_3 can be calculated. Thus oxygenation of dilute solutions of $[Co(trien)(H_2O)_2]^{2+}$, when monitored by fast-scan spectrophotometry, first shows formation of a band in the 300-nm region, followed by a shift of the band to 355 nm. Comparison of the spectrum with that of analogous products, which have had their structure determined by X-ray crystallography, indicate[98] that this last

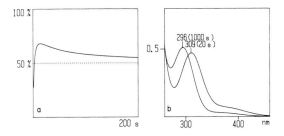

Fig. 8. Experimental evidence for the stepwise formation of binuclear adducts. (a) Amperometric recording of O_2 consumption vs time; $[Co(tame)(en)]^{2+} \approx [O_2] \approx 2.5 \times 10^{-4} \ M$. (b) Extinction; spectra recorded after 20 and 1000 s in an oxygenation experiment, with $[Co(tame)-(en)]^{2+} = 2 \times 10^{-4} \ M$ and $[O_2]$ saturation; 25°C.

step with $k_3 = 2 \ M^{-1} \ s^{-1}$ leads to [(trien)Co(O$_2$,OH)Co(trien)]$^{3+}$. Formation of a doubly bridged μ-peroxo μ-hydroxo species is always observed if the singly bridged μ-peroxo complex has a labile ligand cis to the dioxygen.

This bridging reaction apparently involves rearrangement of the singly bridged bis(aquo) intermediate and formation of a second bridge by dissociation of coordinated H$_2$O on one cobalt and deprotonation of that on the adjacent center [Eq. (13)].

$$(13)$$

$$(14)$$

In the oxygenation of diethylenetriaminediacetatocobalt(II), where the singly bridged complex [(dtda)CoOOCo(dtda)] can be isolated,[37] the subsequent OH bridging is much slower ($k_3 \approx 10^{-3} \ s^{-1}$). In this case dissociation of coordinated carboxylate group has to precede formation of the doubly bridged species [Eq. (14)], which may explain the much lower k_3 value (Table 6).

Only a few rate constants for the formation of a hydroxo bridge have been reported so far because, owing to the facultative ligand disposition, the

TABLE 6

Apparent first-order rate constants for the formation
of a hydroxo bridge (25°C)

Complex	$k_3 \ (s^{-1})$	Ref.
[(tren)(H$_2$O)CoOOCo(OH)(tren)]$^{3+}$	3.3	[102]
[(tren)(OH)CoOOCo(OH)(tren)]$^{2+}$	10^{-2}	[102]
[(trien)(H$_2$O)CoOOCo(OH)(trien)]$^{3+}$	2.0	[98]
[(trien)(OH)CoOOCo(OH)(trien)]$^{2+}$	1.2×10^{-2}	[98]
[(dmtr)(H$_2$O)CoOOCo(OH)(dmtr)]$^{3+}$	1.8×10^{-3}	[101]
[(dtda)CoOOCo(dtda)]	10^{-3}	[110]

common polydentate amines usually lead to a mixture of isomers that may differ significantly in their reactivity. In addition, the linking of the two metal centers by a second bridging ligand increases the number of isomers.

Kinetic measurements[111] on the oxygenation product of $[Co(trien)]^{2+}$ have shown the presence of at least three isomers of $[(trien)Co(O_2,OH)Co(trien)]^{3+}$, and this has been confirmed by ^{13}C NMR spectroscopy.[88] It must be assumed that the value reported[98] for k_3 ($2\ s^{-1}$) refers to the most abundant isomer. It is interesting to note that introduction of N-methyl groups in the tetradentate amine slows down oxygen uptake and OH bridging by the same order of magnitude. An X-ray structure determination of crystals grown in an oxygenation mixture of $[(dmtr)Co(O_2,OH)Co(dmtr)]^{3+}$ yielded a cation in which the ligands had (cis)-$\Delta\Delta/\Lambda\Lambda$ configuration.[112] For the olation step of Eq. (13), which was followed spectrophotometrically, two reaction phases were observed: $k_3 \approx 0.2$ and $1.8 \times 10^{-3}\ s^{-1}$ ($25°C$), respectively, the latter constant belonging to the larger amplitude.[101]

Formation of the μ-peroxo μ-hydroxo complex requires deprotonation of a coordinated H_2O molecule [Eq. (13)]. Whereas a direct and precise determination of the equilibrium steps [equilibrium (15)] is not feasible, values of the pK_a have been estimated for several systems, based mainly on kinetic data.

$$
\begin{array}{ccccc}
\overset{\displaystyle OH_2}{\underset{\displaystyle OH_2}{[(Z)Co-O-O-Co(Z)]}} & \overset{K_1}{\rightleftharpoons} & \overset{\displaystyle OH_2}{\underset{\displaystyle OH}{[(Z)Co-O-O-Co(Z)]}} & \overset{K_2}{\rightleftharpoons} & \overset{\displaystyle OH}{\underset{\displaystyle OH}{[(Z)Co-O-O-Co(Z)]}}
\end{array}
\tag{15}
$$

pK_1 is generally believed to be ~ 6–7 and various kinetic observations indicate a pK_2 of 10–11.[98] The rate of OH bridging should therefore be independent of pH in the range 8–9, and as the pH approaches pK_2, k_3 should decrease because the fully deprotonated bis(hydroxy) species of equilibrium (15) is expected to be less reactive as displacement of OH^- is required (Table 6).

It has usually been assumed that the singly bridged binuclear μ-peroxo complex is an intermediate and that it reacts further by intramolecular reaction steps to yield the μ-peroxo μ-hydroxo species of Eq. (13), which is considerably more stable. Doubts remain about this mechanism, however, because it involves a comparatively fast substitution at a cobalt(III) center. No such reactions have been reported yet with rate constants $> 1\ s^{-1}$.

An alternative mechanism is based on the observation that peroxo-bridged bis(aquo), aquohydroxy, and bis(hydroxy) complexes decompose to cobalt(II) and oxygen with rates equal to or even greater than those of the apparent intramolecular formation of an OH bridge (Section IV). Under conditions where the μ-peroxo μ-hydroxo complex is the main product,

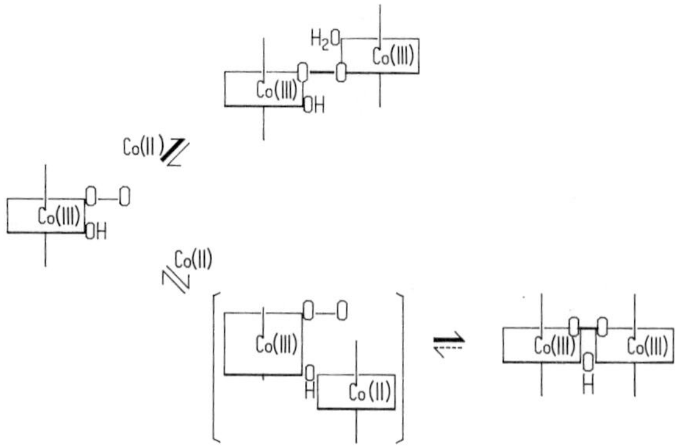

Fig. 9. An alternative mechanism for the formation of μ-peroxo μ-hydroxo complexes.

the singly bridged species may well be kinetically favored. However, it remains in a rapid equilibrium with the mononuclear superoxo species (Fig. 9) and can be considered to undergo further substitution processes simultaneously. The final product could be formed either in one step or via a labile OH-bridged intermediate. In the latter case both steps can be fast because they involve substitution at cobalt(II). On purely kinetic grounds it is virtually impossible to decide between this alternative reaction scheme and the mechanism proposed by Miller and Wilkins.[98] For neither mechanistic model is there positive kinetic proof. However, fast substitution at cobalt(III) cannot be accepted as part of an established mechanism before other models have been properly tested.

IV. DECOMPOSITION OF μ-PEROXO COMPLEXES TO COBALT(II) AND DIOXYGEN

A. Singly bridged complexes

The reactivity of μ-peroxo complexes in solution is dominated by their tendency to decompose to cobalt(II) and dioxygen:

$$[(Z)Co(III)OOCo(III)(Z)] \rightarrow [(Z)Co(III)O_2] + Co(II)(Z)$$

$$\rightarrow 2[Co(II)(Z)] + O_2 \qquad (16)$$

Between pH 2 and 12, this is usually by far the fastest reaction to be observed with singly bridged μ-peroxo complexes. Although the equilibrium constants for the formation of the μ-peroxo complexes vary by a factor of $\sim 10^{10}$ (see Table 1), the rates of formation vary by a factor of only $\sim 10^3$ (see Table 3). As these are related by the equation $K = k_1 k_2 / k_{-1} k_{-2}$, the large difference in stability must be due mainly to differences in the decomposition rates. Thus a wide range of rate constants (k_{-2}) can be expected for reaction (16) which can be brought about by changing the conditions, e.g., the oxygen pressure, the pH, or $[Co^{2+}]$. Accurate rate measurements should be carried out under conditions where reoxygenation cannot occur, and this can be achieved by capturing one of the decomposition products. Three main methods have been useful:

1. The cobalt(II) ion can be captured by complexing with a ligand stronger than Z; in most cases edta is efficient enough. However, in neutral or weakly basic solution edta may form an ion pair with the binuclear adduct. In all cases investigated so far, ion pairing stabilizes the μ-peroxo complex[113]; e.g., this stabilization is greater for $[(en)_2(NH_3)CoOOCo(en)_2(NH_3)]^{4+}$ than for $[(en)_2(SCN)CoOOCo(en)_2(SCN)]^{2+}$ owing to the higher positive charge, which should lead to a more stable ion pair.[113]

2. As Z is usually strongly basic, oxygenation can also be stopped by mixing a solution of the adduct with acid. On the other hand, in strongly acid solution the μ-peroxo complex may be protonated, which could promote decomposition and/or lead to completely different decomposition products, such as the mononuclear cobalt(III) complexes or isomeric binuclear species.[114]

3. Addition of a reagent that consumes oxygen rapidly (e.g., dithionite) can also be used to prevent reoxygenation[100]; however, if meaningful kinetic results are to be obtained, the removal of the oxygen must be substantially faster than either reaction (16) or reoxygenation.

The stoichiometry of reaction (16) can be checked using the oxygen-sensitive electrode. Some oxygen adducts of cobalt(II) amines such as $[(tren)(NH_3)CoOOCo(NH_3)(tren)]^{4+}$ will release their oxygen[115] almost quantitatively in dilute solution $(10^{-4}\ M)$. In other cases the reversibility of oxygen carrying is lower, and decomposition also yields mononuclear cobalt(III) species. In general, the proportion of this side reaction increases with increasing acidity of the solution and with increasing total concentration. Some anions (e.g., sulfate) also increase the amount of Co(III) product formed on acidification.[116]

Spectrophotometry is by far the best method for the determination of rate constants. As the decomposition products have a very much lower absorption in the 300- to 400-nm region, concentration changes down to 10^{-5} M can be measured accurately. Rigorous analysis of the variation of absorbance with time using a least-squares treatment[117] allows the determination of up to four reaction phases corresponding to simultaneously decomposing species or of consecutive reaction steps. From the calculated amplitudes of these phases the ratio of the absorbances of the reacting complexes may be computed. Multiwavelength kinetics has proved to be very useful in cases where μ-peroxo complexes of different structural types are present. Matrix operations may then be used for the computation of the spectra of the reacting species.[118]

Studies of the decomposition of isomerically pure complexes allow a more detailed examination of the mechanism. The kinetics of the decomposition of $[(en)_2(NH_3)CoOOCo(NH_3)(en)_2]^{4+}$ and of related complexes have been reported by Saito and co-workers[119]: spectrophotometric results indicate one main reaction phase with a macroscopic rate constant of $\approx 5 \times 10^{-3}\,s^{-1}$ (25°C) in which the charge-transfer band at 300 nm disappears completely. However, during the first few minutes the fall in absorption is less than expected, and as this cannot be explained by postulating the simultaneous reaction of an isomeric species, it has been suggested[119] that it is a dissolution step. This seems unlikely, as different solids, a perchlorate, a thiocyanate, and a hexafluorophosphate, gave identical results.[113] A more detailed analysis of the spectrophotometric data shows that this initial phase, with a small, negative amplitude, does not fit a mechanistic model in which a strongly absorbing species is produced from a suspended solid. The obvious explanation of this observation of two reaction phases is the consecutive breaking of the two Co(III)–OO bonds [reactions (17) and (18)].

Based on this model, macroscopic rate constants $m_1 = 4.3 \times 10^{-3}$ and $m_2 = 1.2 \times 10^{-2}$ have been computed for the decomposition in acid solution.[113] There is no significant dependence either on pH in the range 1–11 or on the ionic strength. There are in principle two possible mechanistic interpretations: either m_1 or m_2 can represent the first decomposition step, reaction (17). In the presence of edta, m_1 varies from 1.3×10^{-3} (0.3 M edta) to $2.9 \times 10^{-3}\,s^{-1}$ (0.005 M edta), whereas m_2 does not depend significantly on [edta]. Because of the higher positive charge of the binuclear complex this observation supports the view that $m_1 = k_{-2}$ and that the first decomposition step, reaction (17), is rate determining, as has always been assumed.[119]

$$[(en)_2(NH_3)CoOOCo(NH_3)(en)_2]^{4+} \xrightarrow{m_1 = k_{-2}}$$

$$[(en)_2(NH_3)CoOO]^{2+} + [Co(en)_2(NH_3)]^{2+} \quad (17)$$

$$[(en)_2(NH_3)CoOO]^{2+} \xrightarrow{m_2 = k_{-1}} [Co(en)_2(NH_3)]^{2+} + O_2 \quad (18)$$

On the basis of this assignment a molar extinction $\varepsilon_{300} \approx 4 \times 10^3 \; M^{-1} \, cm^{-1}$ can be calculated for the intermediate $[(en)_2(NH_3)CoOO]^{2+}$. Similar results for the consecutive decomposition steps have been obtained[113] both for $[(tren)(NH_3)CoOOCo(NH_3)(tren)]^{4+}$ and for *trans*-$[(SCN)(en)_2CoOOCo(en)_2(SCN)]^{2+}$. A much larger difference between the rate constants of the successive steps has been observed[110] with $[(tame)(en)CoOOCo(en)(tame)]^{4+}$: $m_1 \approx 10^{-5}$, $m_2 = 1.5 \times 10^{-2} \; s^{-1}$. Similarly, Maeder[96] has measured the rate constants $k_{-2} \approx 2 \times 10^{-6}$ and $k_{-1} = 0.65 \; s^{-1}$ (25°C) for the successive steps of the decomposition of $[(papd)CoOOCo(papd)]^{4+}$. Analysis of spectrophotometric rate data reveals similar spectra for the 2:1 and the 1:1 adduct. Absorption maxima are found at 303 ($\varepsilon \approx 13{,}000$) and 313 nm ($\varepsilon \approx 6600$), respectively.

In the majority of cases only values for the rate of the first step of the decomposition (k_{-2}) have been reported. Rate constants of μ-peroxo complexes having five N-donor groups at each metal center are listed in Table 7. Chelate structure is obviously the main factor dominating the kinetic stability; the intrinsic donor strength of the ligands seems to be less important. The remarkably large difference between the constants for $[(NH_3)_5CoOOCo(NH_3)_5]^{4+}$ and $[(papd)CoOOCo(papd)]^{4+}$ can be explained by the fact that decomposition is coupled with a change in oxidation state. Although the reductive elimination of O_2 from the binuclear adduct is not a full-fledged redox process, breaking of the Co(III)–OO bonds is to a

TABLE 7

Observed first-order rate constants for the decomposition of μ-peroxo complexes with N-donor groups (25°C)

Complex	$k_{-2} \, (s^{-1})$	Ref.
$[(NH_3)_5CoOOCo(NH_3)_5]^{4+}$	56	[32]
$[(en)(NH_3)_3CoOOCo(NH_3)_3(en)]^{4+}$	0.5	[113]^a
$[(dien)(NH_3)_2CoOOCo(NH_3)_2(dien)]^{4+}$	2.9×10^{-3}	[110]
$[(tren)(NH_3)CoOOCo(NH_3)(tren)]^{4+}$	2.0×10^{-3}	[115]
$[(en)_2(NH_3)CoOOCo(NH_3)(en)_2]^{4+}$	4.3×10^{-3}	[110]
$[(tn)_2(NH_3)CoOOCo(NH_3)(tn)_2]^{4+}$	1.5×10^{-2}	[110]^a
trans-$[(SCN)(en)_2CoOOCo(en)_2(SCN)]^{2+}$	4.9×10^{-3}	[113]
$[(dien)_2CoOOCo(dien)_2]^{4+}$	1.6×10^{-2}	[95]
$[(dien)(en)CoOOCo(en)(dien)]^{4+}$	2.9×10^{-5}	[110]^b
$[(papd)CoOOCo(papd)]^{4+}$	2×10^{-6}	[96]^c

^{a} Configuration unknown.

^{b} Configuration as determined in Ref. [120].

^{c} No appreciable decomposition within 24 h at 25°C.

certain extent an electron transfer reaction. Its readiness will depend strongly on the flexibility of the coordination spheres, and the transfer of electron charge onto the metal center will take place only if the complex is vibrationally conditioned. The pentadentate amine papd forms two five-membered and two six-membered chelate rings (Fig. 2). The X-ray structure[46] shows that both the Co–N and the Co–O bond distances are equal within experimental error to those found in the decammine μ-peroxo complex and gives no clues as to the origin of the kinetic stability of the μ-peroxo complex. If only some of the monodentate ligands are replaced by chelating amines, as in $[(en)(NH_3)_3CoOOCo(NH_3)_3(en)]^{4+}$ and $[(dien)(NH_3)_2CoOOCo(NH_3)_2-(dien)]^{4+}$, then intermediate rate constants are observed. The comparatively low kinetic stability of $[(dien)_2CoOOCo(dien)_2]^{4+}$ may be explained by incomplete coordination of the amines,[95] which increases the mobility of the coordination sphere.

Decomposition rates also depend on the nature of coordinated groups. This is shown in a comparison of complexes with the pentadentate chelating agents tetraethylenepentamine (tetren) and diethylenetriamine-N_1,N_3-diacetate (dtda). Replacing the aminoethyl ends of the linear pentadentate amine by acetato groups leads to a kinetic destabilization of the binuclear framework CoOOCo by several orders of magnitude (Table 8). The precise extent of these electronic effects, however, can be estimated only when the location of all the donor atoms is known and complexes are structurally comparable. Figure 10 depicts schematically two similar μ-peroxo complexes, the precise geometry of which has been determined by X-ray crystal-

TABLE 8

Observed first-order rate constants for the decomposition of singly bridged complexes: Influence of the donor atoms (25°C)

Complex	k_{-2} (s^{-1})	Ref.
$[(tetren)CoOOCo(tetren)]^{4+}$	10^{-5}	[35][a]
$[(dtda)CoOOCo(dtda)]^{2+}$	0.18	[110][b]
	0.9	[110][b]
$[(tren)(NH_3)CoOOCo(NH_3)(tren)]^{4+}$	2.0×10^{-3}	[115]
$[(dtma)(NH_3)CoOOCo(NH_3)(dtma)]^{3+}$	0.5	[110]
$[(tame)(en)CoOOCo(en)(tame)]^{4+}$	10^{-5}	[110]
$[(tame)(gly)CoOOCo(gly)(tame)]^{2+}$	8×10^{-5}	[110]
$[(tame)(ox)CoOOCo(ox)(tame)]$	1.5×10^{-3}	[110]
$[(tame)(H_2O)_2CoOOCo(H_2O)_2(tame)]^{4+}$	2	[110]

[a] Ligand configuration unknown.

[b] Oxygenation of $[Co(dtda)H_2O]$ in dilute solution yields two predominant isomers, the detailed structure of which is unknown.[110]

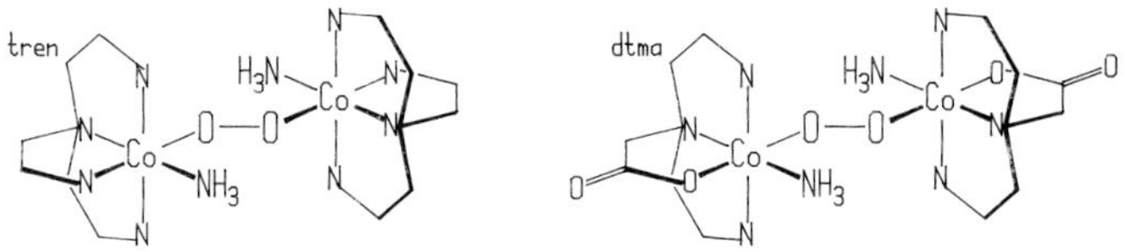

Fig. 10. Schematic drawing of $[(\text{tren})(\text{NH}_3)\text{CoOOCo}(\text{NH}_3)(\text{tren})]^{4+}$ ($k_{-2} = 2.0 \times 10^{-3}\,\text{s}^{-1}$) and $[(\text{dtma})(\text{NH}_3)\text{CoOOCo}(\text{NH}_3)(\text{dtma})]^{2+}$ ($k_{-2} = 0.5\,\text{s}^{-1}$).

lography.[55,115] They differ only in the position trans to the peroxo group; replacing the amino group by a carboxylate group destabilizes the complex by a factor of 250.

A suitable model for further studies of this kind has been found in complexes with terdentate amines of the tripod type. Tris(aminoethyl)ethane (tame) guarantees facial coordination and therefore only one singly bridged isomer is possible for complexes of the type $[(\text{tame})(\text{LL})\text{CoOOCo}(\text{tame})\text{-}(\text{LL})]^{4+}$, where LL represents a symmetrical bidentate ligand. Oxygenation of $\text{Co}(\text{tame})^{2+}$ in the presence of additional bidentate ligands[1] such as ethylenediamine or oxalate leads to a solution containing only one singly bridged mixed ligand μ-peroxo complex. Differences in kinetic stability therefore must in this case be fully ascribed to electronic effects. Each O-donor group labilizes the Co–OO bond by a factor of about 10 (Table 8). When oxalate is replaced by solvent molecules (H_2O), the dissociation rate increases markedly, showing again that the chelate effect predominates. The tetraaquo species can be produced in solution by acidifying the doubly bridged complex $[(\text{tame})(\text{H}_2\text{O})\text{Co}(\text{O}_2,\text{OH})\text{Co}(\text{OH}_2)(\text{tame})]^{3+}$ (see p. 339).

Little is also known about the effect of coordinated water in complexes of the type $[(\text{N})_4(\text{H}_2\text{O})\text{CoOOCo}(\text{H}_2\text{O})(\text{N})_4]^{4+}$. Endicott[97] found $k_{-2} = 0.6$ s^{-1} for a complex with the tetradentate ligand cyclam-14, in which H_2O is trans to the peroxo group. The bis(*cis*-aquo) complexes are difficult to study; although usually formed by direct oxygenation, they rapidly change into the more stable doubly bridged μ-hydroxo species. However, formation of a bis(*cis*-hydroxo) complex with 4,7-dimethyl-1,4,7,10-tetraazadecane as a tetradentate ligand has been observed in strongly alkaline solution.[121] The pH dependence of the decomposition rate of this complex indicates the occurrence of two protonation equilibria with $pK_a = 10.3$ and 5.2, leading to the aquohydroxo and the bis(aquo) complexes, respectively.

The bis(*cis*-hydroxo) complex ($k_{-2} = 6 \times 10^{-2}\,\text{s}^{-1}$) is more stable than either the aquohydroxo complex or the bis(aquo) complex, which have

---

[1] With an unsymmetrical bidentate ligand such as glycine, although racemic and meso isomers are possible, their reaction rates would be expected to be very similar, and indeed only one reaction phase is observed.

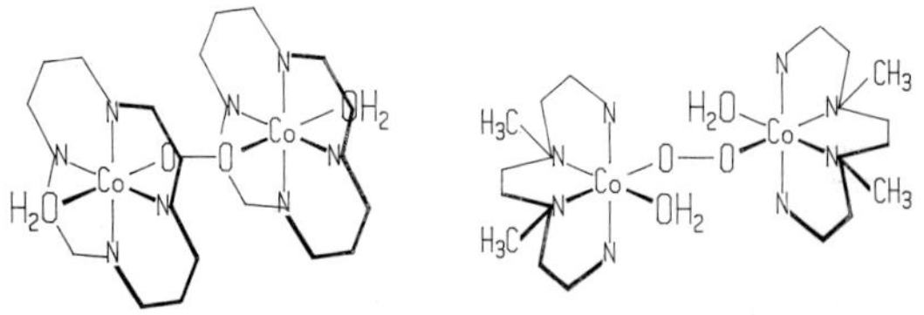

Fig. 11. Schematic drawing of $[(H_2O)(cyclam-14)CoOOCo(cyclam-14)(H_2O)]^{4+}$ ($k_{-2} = 0.6\ s^{-1}$) and $[(dmtr)(H_2O)CoOOCo(H_2O)(dmtr)]^{4+}$ ($k_{-2} = 1.8\ s^{-1}$).

$k_{-2} = 0.4$ and $1.8\ s^{-1}$, respectively. Although experimental data are insufficient yet, the labilizing effect of an aquo ligand either in a trans or a cis position (Fig. 11) is beyond doubt.

B. Doubly bridged complexes

Doubly bridged μ-peroxo species, such as the μ-peroxo μ-hydroxo and especially the μ-peroxo μ-amido complexes, are much more stable kinetically than the singly bridged species. The kinetics of decomposition of the μ-peroxo μ-hydroxo complexes in acid solution [reaction (19)] has been studied extensively.[111]

$$[(Z)Co(III)(O_2,OH)Co(III)(Z)] + H^+ \rightarrow 2[Co(II)(Z)] + O_2 + H_2O \qquad (19)$$

In contrast to the decomposition of singly bridged species, the rate of decomposition of the μ-hydroxo μ-peroxo complexes depends on the pH of the solution (see Fig. 12). Decomposition in neutral or weakly alkaline solution in the presence of edta is generally extremely slow ($t_{1/2} > 10^4\ s$), and in acidic solution there is a marked proton catalysis.[39]
Experiments on the decomposition of the doubly bridged cation [(dmtr)-Co(O_2,OH)Co(dmtr)]$^{3+}$, the structure of which is known by X-ray crystallography,[112] furnished further clues as to the mechanism of reaction (19). When a solution of this complex is mixed with a buffered edta solution of pH ≈ 10 a slow first-order decrease of the charge-transfer absorption is observed.[121] If this same experiment is carried out with an aged solution of the complex an additional reaction phase is observed; its rate constant was found to be identical with k_{-2} for the decomposition of the singly bridged species $[(dmtr)(H_2O)CoOOCo(H_2O)(dmtr)]^{4+}$ or its deprotonated analogs (Fig. 11).
These observations can be interpreted in terms of a two-step mechanism in which opening of the bridging ring at Co–OH–Co [reaction (20)] is the

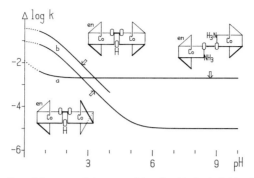

Fig. 12. pH profiles of the rates of decomposition for (a) singly bridged complexes and (b) doubly bridged μ-peroxo μ-hydroxo complexes.

rate-determining step. The intermediate will in most cases not be observable because reaction (21) is much faster than reaction (20).

$$[(Z)Co(O_2,OH)Co(Z)]^{3+} + H_3O^+ \rightarrow [(Z)(H_2O)CoOOCo(H_2O)(Z)]^{4+} \qquad (20)$$

$$[(Z)(H_2O)CoOOCo(H_2O)(Z)]^{4+} + 2edta^{4-} \rightarrow 2[Co(II)(edta)]^{2-} + O_2 + 2Z + 2H_2O \qquad (21)$$

However, the reaction can become biphasic if the rates of the two steps are closer. This has been observed[110] in the decomposition of the unusually labile μ-peroxo μ-hydroxo species [(tame)(H$_2$O)Co(O$_2$,OH)Co(H$_2$O)-(tame)]$^{3+}$, which can be formed by oxygenation of [Co(tame)(H$_2$O)$_3$]$^{2+}$ in the absence of excess ligand. A careful study of the decomposition of this species by the edta method, using multiwavelength kinetics and a least-squares treatment of absorbance/time data,[117] yields two reaction phases that must be interpreted as consecutive reaction steps.[118] The calculated spectrum of the intermediate has only one intense absorption band near 310 nm that is attributed to the singly bridged complex [(tame)(H$_2$O)$_2$-CoOOCo(H$_2$O)$_2$(tame)]$^{4+}$.

Decomposition of μ-peroxo μ-hydroxo complexes in acidic solution has been found[39] to be proton catalyzed [Eq. (22)], and the observed rate constants usually become proportional to [H$^+$] below pH 4 or 5.

$$k_{obs} = p_0 + p_1[H^+]$$

The variation of k_{obs} with pH for [(en)$_2$(NH$_3$)CoOOCo(NH$_3$)(en)$_2$]$^{4+}$ and the isomers of [(en)$_2$Co(O$_2$,OH)Co(en)$_2$]$^{3+}$ as shown in Fig. 12 is typical. In some cases the pH profile flattens again below pH 1, indicating a protonation equilibrium.[122] Kinetic data for *rac*-[(en)$_2$Co(O$_2$,OH)-Co(en)$_2$]$^{3+}$ fit a function of the type $k_{obs} = p_2p_3[H^+]/(p_3 + [H^+])$, with $p_3 = 1.5 \times 10^{-1}$ (25°C and 0.1 M KCl). The interpretation of p_3 as the acid dissociation constant of the protonated μ-peroxo complex seemed reasonable in view of similar values reported earlier: Mori and Weil, in their pioneering work on the structurally related μ-peroxo μ-amido complexes,[85] calculated almost identical K_a values from protonation and isomerization equilibria. In other cases no deviation from linearity can be detected even at pH 0, and the pK_a of the μ-peroxo complex is presumably lower. In terms of the proposed mechanism of decomposition [reactions (20) and (21)], obviously the slower first step, in which the bridging ring is opened up [reaction (20)], must be responsible for the observed [H$^+$] dependence. These results are quite in contrast to the behavior of singly bridged peroxo complexes, which usually show no pH dependence of the decomposition rate even over a wide pH range (curve a in Fig. 12).

In μ-peroxo μ-hydroxo complexes two different protonation sites are possible: (1) the peroxo group, which is more exposed to the medium than in the transoid-planar singly bridged complexes, due to the conformation of the Co—O—O—Co group, and (2) the OH bridge. Although significant basicity of a bridging hydroxo group is perhaps unexpected, protonation of the peroxo group is more likely to promote hydrolytic cleavage of the Co–OO bond, leading to mononuclear cobalt(III) complexes; this has indeed been proposed[123] as the first step in their formation from the μ-peroxo complexes (see also p. 346).

Moreover, the decomposition of the μ-azido μ-peroxo complex is not proton catalyzed[125]; this provides further evidence that protonation of the μ-hydroxy group rather than of the peroxo group promotes ring opening. The structural parameters of the bridging ring system Co(O$_2$,N$_3$)Co in which an azide ion acts as a one-atom bridge are quite similar to those of μ-hydroxo μ-peroxo complexes.[55] Sykes[126] also proposes protonation of the μ-hydroxy group in the case of a proton-catalyzed ring opening of a μ-amido μ-hydroxo dicobalt(III) complex, leading to the corresponding bis(aquo) species.

The decomposition rates of μ-peroxo μ-hydroxo complexes change considerably if salts are added to the reaction medium; this is also in contrast to the behavior of singly bridged complexes. Especially in the presence of chlorides k_{obs} is markedly increased in acidic solutions. Since the rate-determining step is preceded by protonation, it has been suggested that this salt effect is due to a shift in the protonation equilibrium.[122] Ion pairing can be expected to favor the protonated species, as this has a higher charge.

The structures of both diastereoisomers of [(en)$_2$Co(O$_2$,OH)Co(en)$_2$]$^{3+}$ have been determined by X-ray crystallography and the rates of decomposition have been measured for both isomers.[127] The differences observed in the rates of decomposition (Fig. 12) are due partially to different reactivities of the protonated species and partially to differences in k_a.[122] In experiments on the decomposition of freshly oxygenated solutions, biphasic kinetics are found with rate constants identical to those determined using pure samples. The reaction amplitudes observed indicate that approximately equal amounts of the two diastereoisomers are formed in dilute solution.

Detailed analysis of the kinetics of decomposition of $[(trien)Co(O_2,OH)Co(trien)]^{3+}$, produced by oxygenation of solutions of $[Co(trien)(H_2O)_2]^{2+}$, indicates that at least three of the eight isomers theoretically possible for complexes of linear tetradentate amines are formed.[111] Their rates of decomposition vary by a factor of 50, showing that even with identical donor groups the symmetry of the binuclear doubly bridged structure influences rates significantly.

V. SUBSTITUTION IN μ-PEROXO COMPLEXES

In view of the proverbial kinetic inertness of cobalt(III) species with a low-spin d^6 electronic configuration, ligand substitution in μ-peroxo dicobalt(III) complexes would also be expected to be slow. However, some of the observed reactions of binuclear adducts are surprisingly fast: e.g., singly bridged μ-peroxo complexes with an ammonia or another monodentate ligand bound to each cobalt undergo an apparent substitution with rates ranging from 10^{-3} to 10^{-1} s^{-1}. In basic medium they lose the monodentate ligands and form the corresponding μ-peroxo μ-hydroxo species.

A good example of such a reaction is equilibrium (22).

$$[(tren)(NH_3)CoOOCo(NH_3)(tren)]^{4+} + OH^- \rightleftharpoons$$

$$[(tren)(Co)(O_2,OH)Co(tren)]^{3+} + 2NH_3 \quad (22)$$

In this case, the structures of both the starting material and the product have been determined by X-ray diffraction.[87,115] Two coordinated NH_3 molecules are apparently substituted by a bridging hydroxide. Although the reactivity of peroxo-bridged binuclear complexes such as $[(N)_5CoOOCo(N)_5]$ is not exactly comparable with that of mononuclear complexes such as $[Co(N)_5X]$, it is unlikely that equilibrium (22) starts with the breakage of the Co(III)–N bond. Closer examination has revealed quite a different mechanism.[128] The spectrophotometrically determined kinetics of the reaction are practically monophasic and the first-order rate constants are pH independent; this suggests that a dissociative process is the rate-determining step. The reaction is slowed down by the addition of NH_3 and the equilibrium is shifted to the left (Fig. 13). At a low $[NH_3]$ the rate constant approaches a limiting value of $\sim 2 \times 10^{-3}$ s^{-1}, equal to the rate of decomposition of the bis(ammino) complex to cobalt(II) and oxygen (Table 7).

Experiments carried out with the analogous bis(methylamine) complex labeled with $^{18}O_2$ showed that coordinated $^{18}O_2$ exchanged with $^{16}O_2$ present in the reaction medium at the same rate as the formation of a hydroxo bridge. Even in the presence of a high concentration of CH_3NH_2, when no reaction is observed spectrophotometrically, $^{18}O_2$ exchanges at the same rate. The hydroxo-bridged complex, however, exchanges about 100 times more slowly. These observations led to the simple reaction scheme (Fig. 13) in which labile Co(II) is an intermediate. As formation of the μ-peroxo μ-hydroxo complex under the same conditions is comparatively fast (see p. 329), the whole reaction becomes practically monophasic and the apparent substitution is in fact initiated by the breaking of a Co(III)–OO bond, i.e., by the reductive elimination step.

Similar reactions of binuclear μ-peroxo complexes in which NH_3 is apparently substituted by a μ-hydroxo group have been reported by Saito and co-workers.[119] The geometrical isomerization of singly bridged [(dien)(pn)CoOOCo(pn)(dien)]$^{4+}$ has also been shown[129] to proceed via labile Co(II). A study on the reactivity of trans-[(SCN)(en)$_2$CoOOCo(en)$_2$-(SCN)]$^{2+}$ [reaction (23)]

trans-[(SCN)(en)$_2$CoOOCo(en)$_2$(SCN)]$^{2+}$ + OH$^-$ →

$$[(en)_2Co(O_2,OH)Co(en)_2]^{3+} + 2SCN^- \quad (23)$$

furnishes further evidence for the mechanism depicted in Fig. 13. Since the product must have the cis configuration, isomerization must also occur. The rate of reaction (23) again equals the rate of decomposition of the trans complex (Table 7). If the reaction is carried out in the presence of added SCN$^-$ the kinetic results suggest the intermediate formation of the cis isomer.[113]

Although dissociation of NH_3, SCN$^-$, or NO_2^- from a cobalt(III) center is very much slower than the rates observed in reactions of μ-peroxo complexes, the replacement of O-donor groups by solvent molecules is faster. Typical rate constants[130] for the aquation of complexes of the type [Co(III)-(N$_5$)X] range from 10^{-6} to 10^{-5} s^{-1} (25°C). Such substitution reactions are generally believed to have a D or an I$_d$ mechanism.[25] However,

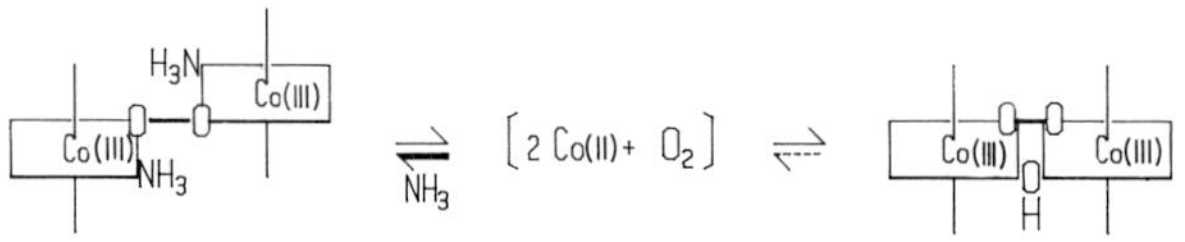

Fig. 13. Mechanism of formation of a μ-peroxo μ-hydroxo complex from a singly bridged species.

certain unusual nonleaving ligands such as S-bonded sulfite may promote substitution,[131] and the possibility that a μ-peroxo group exerts a similar effect cannot be ignored.

As was discussed in Section IV, decomposition of μ-peroxo μ-hydroxo complexes leading to cobalt(II) and dioxygen proceeds in two steps. The first step, in which the bridging hydroxo group is replaced by water, must be considered as a true substitution process in which the oxidation state of the cobalt centers is unchanged. First-order rate constants for this type of substitution range from 10^{-5} to 10^{-2} s^{-1} (Table 9), but the subsequent decomposition of the singly bridged intermediate is coupled to a redox process. Although this second step is usually faster than the aquation of the Co–OH–Co bridge, it is the aquation that initiates decomposition. The additional OH bridge can therefore be said to block the electron transfer reaction. The comparative ease with which the hydroxo bridge hydrolyzes in some cases may, on the other hand, be due either to the neighboring peroxo group or to the fact that the two bridging ligands are connected by a strained bimetallic ring.

From the limited kinetic data known so far (Table 9) two general conclusions may be drawn.

1. The range of rates for the aquation of the Co–OH–Co bridge is smaller than that for the decomposition of singly bridged μ-peroxo complexes (Tables 7 and 8), as is to be expected for a heterolytic cleavage of the coordinate bond.

2. The rate of aquation of the Co–OH–Co bridge varies with the type of ligand in a manner similar to that of the decomposition of singly bridged μ-peroxo complexes (Tables 5 and 6), with chelating ligands stabilizing the

TABLE 9

First-order rate constants for the opening of the
bridging ring (25°C, 0.1 M KCl)

Complex	k (s^{-1})	Ref.
$\Delta\Delta/\Lambda\Lambda$-[(en)$_2$Co(O$_2$,OH)Co(en)$_2$]$^{3+}$	4×10^{-5}	[127]
[(tn)$_2$Co(O$_2$,OH)Co(tn)$_2$]$^{3+}$	9×10^{-4}	[110]^a
[(tren)Co(O$_2$,OH)Co(tren)]$^{3+}$	10^{-5}	[110]
[(tren)Co(O$_2$,N$_3$)Co(tren)]$^{3+}$	10^{-3}	[125]
[(dmtr)Co(O$_2$,OH)Co(dmtr)]$^{3+}$	1.5×10^{-3}	[121]
[(NH$_3$)$_4$Co(O$_2$,OH)Co(NH$_3$)$_4$]$^{3+}$	8×10^{-3}	[125]

^{a} Configuration unknown.

doubly bridged framework and six-membered chelate rings labilizing it. The labilizing effect of the six-membered chelate ring with trimethylenediamine (tn) (see Table 7) has also been observed[132] in the aquation of $[Co(tn)_2Cl_2]^+$, compared with that of $[Co(en)_2Cl_2]^+$.

With μ-peroxo μ-amido complexes the situation seems to be reversed. Cleavage of the $Co-NH_2-Co$ bridge is slower than aquation of the $Co-OO$ bond; this may be the reason why μ-peroxo μ-amido complexes decompose to cobalt(III) species rather than to cobalt(II) and dioxygen. Reductive elimination of O_2 has in fact never been reported.

VI. DECOMPOSITION OF μ-PEROXO COMPLEXES TO MONONUCLEAR COBALT(III) SPECIES

Although all μ-peroxo dicobalt(III) complexes are unstable with respect to decomposition to mononuclear cobalt(III) complexes (but see p. 354), the rate of this reaction varies widely. Half-lives in solution ranging from 3 min for a glycylglycinamide complex to 12 days for a histamine complex have been reported,[133] but the range of rates is now known to be even greater. Complexes with a bridging hydroxo group decompose in alkaline solution to mononuclear cobalt(III) complexes more slowly than do singly bridged complexes.[134] For example, the half-life of decomposition of $[(dtda)CoOOCo(dtda)]$ is 2.5×10^3 s (pH 7, 25°C), whereas that of $[(dtda)Co(O_2,OH)Co(dtda)]^-$ is 10^6 s (pH 10.5, 25°C).[37]

The majority of μ-peroxo complexes are most stable in weakly alkaline solution; the rate of decomposition to mononuclear cobalt(III) increases slowly but significantly with increasing pH.[88,135] Thus for DL-$[(en)_2Co(O_2,OH)Co(en)_2]^{3+}$, the rate of decomposition[88] increases from 5.9×10^{-5} at pH 8.4 to 8.9×10^{-4} at pH 13.6 (50°C). It has been suggested that the acceleration of rate by hydroxide ions occurs only with doubly bridged μ-hydroxo μ-peroxo complexes, and that the mechanism involves the opening of the central bimetallic ring[134]:

$$[L_4Co(O_2,OH)CoL_4]^{3+} + OH^- \rightleftharpoons [(HO)L_4CoOOCoL_4(OH)]^{2+} \qquad (24)$$

In acidic solution dissociation to Co_{aq}^{2+}, protonated ligand, and O_2 predominates (see p. 333), although in strongly acidic solution (pH 1) the proportion that decomposes to mononuclear cobalt(III) complexes again increases.[116] The rate of this acid-catalyzed decomposition also depends on the anion present,[97,116] and especially on the charge of the anion; this reaction probably involves protonation of the μ-peroxo group as the first step, and thus the mechanism is not necessarily the same as that occurring in alkaline solution. Complexes with a protonated μ-peroxo group have been isolated; if Frémy's brown μ-peroxo complex $[(H_3N)_5CoOOCo(NH_3)_5]^{4+}$ is added to 9 M sulfuric acid, a red complex is obtained[5,85,136] that is more stable in acidic solution than the μ-peroxo complex. It has been claimed[136] that this complex is $[(H_3N)_5Co(O_2H)Co(NH_3)_5]^{5+}(HSO_4^-)_3(SO_4^{2-})$. The crystal structure of the analogous complex $[(en)_2Co(O_2H,NH_2)Co(en)_2](NO_3)_4$ has been determined.[30]

With some ligands, such as edta, the peroxo complex, which has only two nitrogen donor atoms per cobalt, is so unstable that there is only spectroscopic evidence[137] for a μ-peroxo complex as a transitory intermediate that rapidly decomposes to $[Co(III)(edta)(OH)]^-$.

μ-Peroxo complexes formed by dipeptides are usually rather unstable, and only in special cases can they be isolated.[123] However, an interesting trend in the stability has been noted[133,138]: for peptides containing a C-terminal substituent, the larger the substituent, the slower the decomposition of the μ-peroxo complexes. As for dipeptides with an N-terminal substituent, no such differences are seen[138]; it is interesting to speculate that ring closure of the C-terminal amino acid residue might be involved in the rate-determining step.

μ-Peroxo complexes of di- and polyamine ligands are much more stable; e.g., the decomposition of $[(en)_2Co(O_2,OH)Co(en)_2]^{3+}$ takes weeks at room temperature and it must be heated to achieve a reasonable rate of decomposition.[88] Even the singly bridged complexes $[(NNNNN)CoOOCo(NNNNN)]^{4+}$ (NNNNN is tetren or papd) decompose only slowly in alkaline solution.[46,139]

In no case yet studied has the rate of formation of mononuclear cobalt(III) complexes in alkaline solution been found to be *faster* than the decomposition to cobalt(II) and oxygen, and in general those complexes that are more stable in acidic solution also decompose more slowly to mononuclear cobalt(III) complexes in alkaline solution. The stabilization by a bridging hydroxo group mentioned earlier also parallels the stabilization observed in acidic solution (see p. 338).

It is thus not to be expected that the decomposition of a pure, isolated isomer of a μ-peroxo complex will necessarily yield a pure isomer of the mononuclear cobalt(III) complex, as isomerization takes place simultaneously and is faster than the formation of mononuclear cobalt(III).

Likewise, it is not surprising that the same by-products are observed in the formation of mononuclear cobalt(III) complexes by oxygenation of cobalt(II) and by decomposition of a μ-peroxo dicobalt(III) complex. Thus "hexol," $[Co\{(OH)_2Co(en)_2\}_3]^{6+}$, which is normally made by the direct oxygenation[140] of a mixture of Co^{2+} and ethylenediamine, is also formed as a by-product in the decomposition[88] of $[(en)_2Co(O_2,OH)Co(en)_2]^{3+}$; at higher temperatures ($\gtrsim 50°C$) and concentrations (0.1 M) the amount of hexol formed is considerable.[88] Similarly, a black insoluble precipitate of $Co(O)(OH)$ is sometimes formed in the decomposition[88] of $[(en)_2Co-(O_2,OH)Co(en)_2]^{3+}$ and in the synthesis of cobalt(III) complexes by oxygenation of cobalt(II) if there is insufficient ligand and the solution is too alkaline. The presence of such by-products could be taken as evidence for the intermediacy of Co^{2+} in the decomposition of μ-peroxo complexes to mononuclear cobalt(III) complexes, which will be discussed later (Section VI,C).

However, despite the presence of such by-products, there is no doubt that the main reaction products from $[L_5CoOOCoL_5]^{4+}$ are $[L_5CoOH]^{2+}$ and $[L_5CoOH_2]^{3+}$, depending on the pH of isolation. These products can sometimes be isolated in almost quantitative yields.

A. Stoichiometry of the decomposition

The traditional view[141] of the reaction is that this so-called irreversible[2] oxidation is a hydrolytic process that can be envisaged as occurring in two steps; e.g., for a singly bridged complex:

$$[L_5CoOOCoL_5]^{4+} + H_2O \rightarrow [L_5Co(OH_2)]^{3+} + [L_5CoOOH]^{2+} \quad (25)$$

and

$$[L_5CoOOH]^{2+} + H_2O \rightarrow [L_5Co(OH_2)]^{3+} + H_2O_2 \quad (26)$$

However, seldom has the formation of hydrogen peroxide been conclusively proved, and more often oxygen gas is released,[135] although not in the amount required by the overall stoichiometry

$$2[L_5CoOOCoL_5]^{4+} + 2H_2O \rightarrow 4[L_5CoOH]^{2+} + O_2 \quad (27)$$

Dehydrogenation of a free or a coordinated ligand has also been suggested[144,145]:

$$[L_5CoOOCoL_5]^{4+} + H_2O \rightarrow [L_4Co(L-2H)(OH)]^{2+} + [L_5CoOH]^{2+} + H_2O \quad (28)$$

where $L-2H$ represents a molecule of ligand from which two hydrogen

---

[2] This reaction has often been referred to as an irreversible oxidation[142] to distinguish it from "reversible oxidation," i.e., the formation of the μ-peroxo complex from cobalt(II) and dioxygen. However, the reverse process has been reported[143]: certain μ-peroxo complexes, such as $[(en)_2Co(O_2,OH)Co(en)_2]^{3+}$, can be prepared by the reaction of hydrogen peroxide with cobalt(III) aquo complexes, e.g., $[(en)_2Co(OH)(OH_2)]^{2+}$.

atoms have been removed, for example, by formation of a double bond. Alternatively, complete oxidation of some of the ligands could occur to give carbon dioxide and ammonia, although attempts to detect these have usually failed.[123]

B. Evidence for the formation of hydrogen peroxide

Although there have been several claims to have identified hydrogen peroxide as a product of the decomposition of various μ-peroxo complexes, many of these findings have been refuted by subsequent investigators. Indeed, only with $[(NC)_5CoOOCo(CN)_5]^{6-}$ and with $[(NO_2)(en)_2CoOOCo-(en)_2(NO_2)]^{2+}$ has the formation of hydrogen peroxide been unambiguously demonstrated,[141,146] and in both cases the decomposition was studied in acidic solution and may not be typical of the more usual decomposition in alkaline solution.

For example, it has been claimed[147] that 70% of the theoretical quantity of hydrogen peroxide can be detected by permanganate titration of the solution obtained by the acid decomposition of *trans*-[X(cyclam-14)-CoOOCo(cyclam-14)X]$^{2+}$ (X = Cl, NO_2, or NCS). However, more recent results[148] show that the oxidizing species is more likely to be a mononuclear hydroperoxo complex such as $[ClCo(cyclam-14)(OOH)]^+$, as it is absorbed by cation-exchange resins.

On the basis of polarographic measurements, Caglioti and co-workers[144] suggested that H_2O_2 is formed in the decomposition of μ-peroxo complexes of histidine and of glycylglycine. Although Munakata[149] observed a similar polarographic wave after the decomposition of μ-peroxo complexes containing amino acids, Martell and co-workers[145] found no satisfactory evidence for H_2O_2 and suggested that the polarographic wave was an artifact.

Using a manometric method and the addition of catalase, Gillard and Spencer[123] were unable to detect H_2O_2 after the decomposition of μ-peroxo complexes containing various dipeptides such as glycylglycine and glycyl-L-tyrosine; they also found[150] that the chemiluminescent test for hydrogen peroxide, using luminol, was also positive with several peroxo complexes. It is, however, unclear to what extent this luminescence is due to the peroxo complexes and to what extent it is due to their decomposition products.

The problem with most of the claims for detection of hydrogen peroxide lies in the inability of most methods to distinguish between free H_2O_2, μ-peroxodicobalt(III) complexes, and mononuclear peroxo complexes (which may well be intermediates[148] in this decomposition).

Moreover, not only is hydrogen peroxide catalytically decomposed by $[Co(OH_2)_6]^{2+}$[123] and by certain mononuclear cobalt(III) complexes,[123,151] but it also reacts with some μ-peroxodicobalt(III) complexes. Thus Martin and co-workers[89] observed that the μ-peroxodicobalt(III) complex of L-2,3-diaminopropanoic acid decomposes more rapidly in the presence of hydrogen peroxide; however, Martin also suggested that the H_2O_2 may react with cobalt(II), which is in equilibrium with the μ-peroxo complex, rather than react with the μ-peroxo complex itself.[89]

The binuclear μ-peroxo complex $[(NC)_5CoOOCo(CN)_5]^{6-}$ can be protonated[141,152] ($pK_a \approx 12$) to give the μ-hydroperoxo complex $[(NC)_5Co(O_2H)Co(CN)_5]^{5-}$, which hydrolyzes rapidly even at $0°C$ to yield $[(NC)_5Co(OH_2)]^{2-}$ and H_2O_2 quantitatively. In neutral solution a stable intermediate is observed[152,153] that hydrolyzes on acidification, forming H_2O_2 and $[(NC)_5CoOH_2]^{2-}$; this intermediate is presumed to be $[(NC)_5CoOOH]^{3-}$, as it is identical to a complex prepared[153,154] by the reaction of oxygen with the hydrido complex $[HCo(CN)_5]^{3-}$.

The various reactions that occur in the decomposition of $[(NC)_5CoOOCo(CN)_5]^{6-}$ can be summarized as follows:

$$[(NC)_5CoOOCo(CN)_5]^{6-} + H^+ \rightleftharpoons [(NC)_5Co(O_2H)Co(CN)_5]^{5-} \tag{29}$$

$$[(NC)_5Co(O_2H)Co(CN)_5]^{5-} + H_2O \rightarrow [(NC)_5CoOOH]^{3-} + [(NC)_5Co(OH_2)]^{2-} \tag{30}$$

$$[(NC)_5CoOOH]^{3-} + H_2O + H^+ \rightarrow [(NC)_5Co(OH_2)]^{2-} + H_2O_2 \tag{31}$$

If the decomposition of the unprotonated species

$$[(NC)_5CoOOCo(CN)_5]^{6-} + H_2O \rightarrow [(NC)_5CoOOH]^{3-} + [(NC)_5CoOH]^{2-} \tag{32}$$

occurs at all, then it must be very much slower than the decomposition of the protonated species of reaction (30). It has been suggested[141] that the reaction

$$[(NC)_5CoOOCo(CN)_5]^{6-} + 2[Co(CN)_5]^{3-} + 2H_2O \rightarrow 4[(NC)_5CoOH]^{2-} \tag{33}$$

occurs in the preparation of mononuclear cobalt(III) cyano complexes by the slow passage of oxygen through a solution of $[Co(CN)_5]^{3-}$ so that there is always an excess of $[Co(CN)_5]^{3-}$. The μ-peroxo complex $[(NC)_5CoOOCo(CN)_5]^{6-}$ is obtained only if oxygen is present in excess. If there is a deficiency of oxygen, then the decacyano μ-peroxo complex is reduced by excess $[Co(CN)_5]^{3-}$. Although the individual steps in this presumably multistep process are not yet understood, and the kinetics have not yet been investigated, it has been suggested[141] that the reaction occurs via an outer-sphere activated complex, as there is no evidence for the formation of μ-hydroxo complexes that might be expected if an inner-sphere reaction occurred. Such a reaction would be unusual in that it apparently involves reactants that have large negative charges.

This poses the question of the extent to which the decomposition of the decacyano μ-peroxo complex can be taken as typical of other μ-peroxo complexes. There are at least two major differences in the properties of the cyano complex:

1. The redox potential of the system $[Co(CN)_5]^{3-}/[Co(CN)_5(OH)_2)]^{2-}$ is probably[155,156] about -0.8 V, which is very much lower than that, for example, of $[Co(NH_3)_6]^{2+}/[Co(NH_3)_6]^{3+}$, $+0.1$ V. Thus it is not surprising that $[Co(CN)_5]^{3-}$, which is even able to liberate hydrogen from water,[157] will reduce $[(NC)_5CoOOCo(CN)_5]^{6-}$ to $[(NC)_5Co(OH_2)]^{2-}$.

2. The pH dependence of the rate of decomposition in the cyanide system is quite different from that observed for most other μ-peroxo complexes: the decomposition of $[(NC)_5CoOOCo(CN)_5]^{6-}$ is acid catalyzed and is very slow in alkaline solution.[152] In neutral solution only the first step of the hydrolysis, reaction (30), occurs.

Thus the decomposition of the decacyano complex is not necessarily typical of the decomposition of other μ-peroxodicobalt(III) complexes in alkaline solution, but may be comparable with the decomposition in strong acid referred to earlier (p. 340). Owing to the high negative charge of the decacyano complex, the pK_a is much higher (~ 12) than for amine complexes,[153] so that much less acidic conditions are required for the hydrolysis; indeed, weakly alkaline conditions suffice. It is interesting to note that the other μ-peroxo complexes for which there is definite evidence for mononuclear peroxo (or hydroperoxo) complexes as intermediates are the *trans*-$[X(cyclam-14)CoOOCo(cyclam-14)X]^{2+}$ complexes; the decomposition of these to mononuclear cobalt(III) complexes is also acid catalyzed.[148]

C. Release of oxygen

For a number of systems, especially those with dipeptides as ligands, oxygen is evolved.[123,158] This oxygen may be formed by catalytic decomposition of hydrogen peroxide: although most cobalt(II) complexes [e.g., hexaaquocobalt(II)[123] and bisglycinatocobalt(II)[159]] and a few cobalt(III) complexes[123,151] [e.g., $[Co(gly\text{-}L\text{-}tyr)_2]^-$] catalyze the decomposition of H_2O_2, many mononuclear cobalt(III) complexes [e.g., $[Co(glygly)_2]^-$] do not.[158]

Moreover, it has been observed[97] that for several μ-peroxo complexes containing macrocyclic ligands the decomposition of the peroxo complex is faster than the release of oxygen; this could imply the presence either of a mononuclear peroxo complex or of hydrogen peroxide.

However, in view of the relatively rapid[3] equilibrium between the μ-peroxo dicobalt(III) complex and cobalt(II) and oxygen,

$$[L_5CoOOCoL_5]^{4+} \rightleftharpoons L_5Co_{aq}^{2+} + [L_5CoO_2] \rightleftharpoons 2L_5Co_{aq}^{2+} + O_2 \qquad (34)$$

the oxygen could result simply from the dissociation to a cobalt(II) complex and oxygen. Martin and co-workers suggested[89] that "The simplest interpretation of those systems in which no H_2O_2 can be detected nor oxygen reemitted upon formation of mononuclear Co(III) complexes is that these complexes are not formed from the binuclear peroxocobalt(III) complexes. The peroxo complexes in these cases are unreactive species formed in a side reaction." Several other workers have found that many reactions that apparently involve μ-peroxodicobalt(III) complexes directly are in fact re-actions of their dissociation products. Formation of a μ-hydroxo μ-peroxo complex from the singly bridged μ-peroxo complex, apparently a simple substitution, was shown[128] by $^{18}O_2$ exchange to proceed via dissociation to cobalt(II) and oxygen; kinetic evidence has been known for some years that suggests that cobalt(II) could be an intermediate. A kinetic study indi-cated that the oxidation of $[(H_2O)(cyclam\text{-}14)CoOOCo(cyclam\text{-}14)(H_2O)]^{2+}$ proceeds via dissociation to the cobalt(II) complex rather than by direct oxidation to a μ-superoxo complex[148]; reduction also proceeds via dis-sociation to mononuclear superoxo complex.[148]

It has also been observed[110] that solutions of some μ-peroxo complexes, such as $[(trien)Co(O_2,OH)Co(trien)]^{3+}$, are less stable if there is not a slight excess of the ligand, and from this it might be concluded that the decomposi-tion to mononuclear cobalt(III) is accelerated by Co^{2+}. However, it has also been found[88] that the $[Co^{2+}]$ did not affect the rate of decomposition of $[(en)_2Co(O_2,OH)Co(en)_2]^{3+}$, and that neither $[Co^{2+}]$ nor $[Co(dtda)]$ in-fluenced the rate of decomposition of $[(dtda)CoOOCo(dtda)]$. Whether these results are typical for all μ-peroxo complexes is not yet known, but nevertheless it must be concluded that at present there is no definite evidence for cobalt(II) as an intermediate in the decomposition of μ-peroxodicobalt-(III) complexes to mononuclear cobalt(III).

The mononuclear cobalt(III) complexes could arise from the reaction of the cobalt(II) complex with a molecule of the undissociated μ-peroxo com-plex in a reaction akin to that suggested for the synthesis of cobalt(III) cyano complexes [reaction (33)]. Likewise, "hexol" could be formed by reaction of Co_{aq}^{2+} with the μ-peroxo complex, and $Co(O)(OH)$ from oxidation of the Co_{aq}^{2+} in the alkaline solution. However, this hypothesis is unproved.

Martell and co-workers[145] have also claimed that the proposal by Martin

---

[3] Even for those complexes for which this equilibrium is not rapid in an absolute sense, it is nevertheless rapid compared with the much slower reaction to give mononuclear cobalt(III).

and co-workers[89] is invalid for the complexes with dipeptide ligands; Martell found that the rate of decomposition does not increase as the concentration of unoxygenated cobalt(II) chelate increases.

D. Oxidation of a coordinated ligand

Most μ-peroxodicobalt(III) complexes give a high yield of the mononuclear cobalt(III) complexes, thus excluding any stoichiometric formation of a complex containing an oxidized ligand. In one or two cases, however, satisfactory evidence for the formation of a product containing an oxidized ligand has been obtained. The μ-peroxo complexes of dipeptides have been the most widely studied complexes because they decompose more rapidly than complexes of polyamines. Moreover, the peptides are expected to be more susceptible to oxidation. However, as the μ-peroxo complexes usually cannot be isolated, there is no conclusive evidence that such oxidation as has been observed actually occurred during the decomposition of the μ-peroxo complex, although this may be inferred from the variation in the oxygen uptake with time. In a study on the decomposition of μ-peroxo dicobalt complexes of 18 dipeptides, three types of behavior were observed.[123]

1. Decomposition *with* release of at least part of the oxygen was observed with 5 dipeptides including gly-L-tyr and L-leu-L-tyr.
2. Decomposition accompanied by further, slower oxygen uptake was observed with 4 dipeptides, including gly-L-trp and gly-L-his; oxidation of the dipeptide occurred. Similar reactions had been reported for other ligands by earlier workers.[160,161] A catalytic process is presumably involved, as the total consumption of oxygen is far greater than that required for the formation of a μ-peroxo complex. The only conclusive isolation and identification of an oxidation product was under such catalytic conditions: Spencer[123,162] succeeded in identifying the major product from the oxygenation of a solution containing Co^{2+} and glycyl-L-tryptophan (molar ratio 1:9) as glycyl-L-kynurenine. However, the mechanism of such catalytic processes is totally unknown and does not help solve the problem of the mechanism of the decomposition of isolated μ-peroxo complexes.
3. Decomposition of the μ-peroxo complex *without* release of oxygen: this type of behavior was observed with 9 dipeptides, including gly-gly and gly-L-phe.

This third category of reaction poses a major problem. Although there is oxygen uptake corresponding to the formation of a μ-peroxo complex, there is no appropriate release of oxygen during decomposition.

The complexes of glycylglycine have received the most attention. Despite earlier claims to the contrary,[144] no hydrogen peroxide is formed,[123,145] and neither release of oxygen nor formation of volatile degradation products of the peptide (such as CO_2 or NH_3) has been observed.[123] An earlier suggestion[144] that dehydrogenation occurs, perhaps to give an imine, was resurrected by Martell and co-workers,[145] who assumed that the observed product, $[Co(gly-gly)_2]^-$, was formed only when the oxidized glycylglycine ligand was displaced by the large excess of glycylglycine that they used in their spectrophotometric measurements.[145] In two later publications Harris and Martell described[163,164] mass spectrometric evidence for the presence of traces of hydroxyacetylglycine in the crude mixture prepared by oxygenation of a mixture containing Co^{2+} and glycylglycine. However, hydrolysis of an imine intermediate should give ammonia and glyoxalyl-glycine rather than hydroxyacetylglycine; they were therefore forced to assume that in the reductive cleavage of the complex the imine initially formed was reduced and hydrolyzed.

During extensive column chromatography[138] of the crude product in order to purify the main product, $[Co(gly-gly)_2]^-$, several by-products were isolated, of which only two were consistently present in significant quantities ($> 1\%$). The analytical and spectroscopic results imply that these two products contain coordinated hydroxyacetylglycine (hac-gly) and oxalyl-glycine (ox-gly), respectively, and indicate that they are probably $[Co(gly-gly)(hac-gly)]^{2-}$ and $[Co(gly-gly)(ox-gly)]^{2-}$.

When purified oxygen is bubbled through a solution containing Co^{2+} and gly-gly in a $1:2$ molar ratio, $[Co(gly-gly)_2]^-$ (86–87%, based on cobalt), $[Co(gly-gly)(hac-gly)]^{2-}$ (7%), and $[Co(gly-gly)(hac-gly)]^{2-}$ ($\sim 2\%$) were obtained after chromatography on QAE-Sephadex A-25. A small amount of ammonia (8%) was also detected.[165]

It is clear from these results that a small amount of hydrolysis of glycyl-glycine occurs to give the hydroxyacetylglycine that had been observed by Harris and Martell[163,164] and an approximately equivalent amount of ammonia (which was not detected in earlier work[123]). The small amount of $[Co(gly-gly)(ox-gly)]^{2-}$ formed would account for only about 4% of the "missing" oxygen. As more than 95% of the glycylglycine originally present can be accounted for, it is clear that oxidation of the ligand cannot account for the missing oxygen in this case. It has also been reported[166] that up to 90% of L-ala-gly could be accounted for as the two diastereo-isomers of $[Co(L-ala-gly)_2]^-$ by NMR analysis, although in this case oxygen is released.[123]

The $[Co(gly-gly)(hac-gly)]^{2-}$ observed could originate from a small pro-portion ($< 10\%$) of initial O-terminal coordination of the dipeptide, with the dangling N-terminal group then being hydrolyzed to ammonia and the hydroxyacetylglycine complex; a little of this is then oxidized to the oxalylglycine complex.[165]

Thus, although catalyzed oxidations have been proved,[123] there is to date no evidence for the oxidation or dehydrogenation of a peptide ligand in the decomposition of a μ-peroxodicobalt(III) complex to mononuclear cobalt(III) complexes. The mechanism of the decomposition of the μ-peroxo bis[bis(dipeptidato)cobaltate(III)] complex in alkaline solution has been claimed[123] to involve protonation of the μ-peroxo group to give a red hydroperoxo complex which then decomposes to the bis(dipeptidato)-cobaltate(III) complex. It is therefore probable that the second step of the hydrolytic process to give H_2O_2 [analogous to reaction (31)] is much slower than the first step [cf. reaction (30)], as has already been found for several μ-peroxo complexes.[97,119,141] Thus the formation of hydrogen peroxide and of the oxygen formed by its disproportionation may be occurring over a time too long for them to have been observed by the earlier workers.

E.   Comments on the mechanism

It seems probable that the initial reaction is a hydrolysis to a mononuclear hydroperoxo (or peroxo) complex, either

$$[(L)_5Co(O_2H)Co(L)_5]^{5+} + H_2O \rightarrow [(L)_5CoOOH]^{2+} + [(L)_5Co(OH_2)]^{3+} \qquad (35)$$

or

$$[(L)_5CoOOCo(L)_5]^{4+} + H_2O \rightarrow [(L)_5CoOOH]^{2+} + [(L)_5CoOH]^{2+} \qquad (36)$$

depending on the pH. As the next hydrolytic step

$$[(L)_5CoOOH]^{2+} + H_2O + H^+ \rightarrow [(L)_5Co(OH_2)]^{3+} + H_2O_2 \qquad (37)$$

is much slower, a significant concentration of the intermediate mononuclear hydroperoxo complex, $[(L)_5CoOOH]^{2+}$, may build up; the cobalt(II) complexes that are in equilibrium with the peroxo-bridged complex may, how-ever, reduce this mononuclear hydroperoxo complex[148] before it can be hydrolyzed to hydrogen peroxide. The hydrogen peroxide may be decomposed by either of the two mechanisms.

1.   For stable μ-peroxo complexes, with which there is only an extremely small amount of cobalt(II) in equilibrium, the hydrogen peroxide could react with the μ-peroxo complex, perhaps according to reaction (38):

$$[L_5CoOOCoL_5]^{4+} + H_2O_2 \rightarrow 2[L_5CoOOH]^{2+} + O_2 \qquad (38)$$

Although there are no detailed kinetics on this type of reaction, there are preliminary results[89] that indicate that this reaction is faster than the un-catalyzed decomposition of the μ-peroxo complex; however, it is certainly much slower than reactions (39) and (40).

2. For the less stable μ-peroxo complexes, or at higher temperatures or at extremely low concentration, the proportion of the total cobalt present as cobalt(II) may be increased enough for the reaction between hydrogen peroxide and the cobalt(II) complex to become significant. This reaction is extremely rapid[137] even with ligands that form very stable μ-peroxo complexes and so presumably does not involve μ-peroxo complexes as interme-diates. The mechanism of this complicated reaction has been discussed,[137] but, among other reactions that occur, the processes

$$2L_5Co_{aq}^{2+} + H_2O_2 \rightarrow [L_5Co(II)OOCo(II)L_5]^{2+} + 2H^+ \tag{39}$$

$$[L_5CoOOCoL_5]^{2+} + 2H^+ \rightarrow 2[L_5CoOH]^{2+} \tag{40}$$

may occur and could possibly account for the fact that hydrogen peroxide has not been observed in the various peptide systems. Reaction (39) probably cannot occur in the cyanide system, as $[(NC)_5Co]^{3-}$ is present only if there is a deficiency of oxygen; it has been reported[141] that under these conditions it reacts with $[(NC)_5CoOOCo(CN)_5]^{6-}$.

Either of the two processes, reaction (38) or reactions (39) and (40), could account for the autocatalysis observed[121] in the decomposition of [(dmtr)-Co(O$_2$,OH)Co(dmtr)]$^{3+}$. The kinetics of decomposition of this complex are also concentration dependent, as might be expected if the equilibrium dissociation to cobalt(II) is important in the destruction of the hydrogen peroxide formed in the hydrolytic process.

As the reverse reaction of the hydrolytic process, i.e., the formation of a μ-peroxo complex from mononuclear cobalt(III) and hydrogen peroxide, has been reported[143] (see p. 346), the possibility must be considered that the formation of mononuclear cobalt(III) and hydrogen peroxide may be thermodynamically favorable only if the hydrogen peroxide formed decom-poses to oxygen. The recent measurement of all four rate constants involved in the formation of [(papd)CoOOCo(papd)]$^{4+}$ has allowed the calculation of the equilibrium constant for its formation.[96] Using this result together with the recently determined value for the redox potential for [Co(papd)-(OH$_2$)]$^{2+}$/[Co(papd)(OH$_2$)]$^{3+}$ it has been shown[96] that the reaction

$$[(papd)CoOOCo(papd)]^{4+} + 2H^+ \rightarrow [Co(papd)(OH_2)]^{3+} + H_2O_2$$

is thermodynamically impossible in alkaline solution and is to be expected only in strongly acidic solution. However, these calculations also suggested that the reaction

$$[(papd)CoOOCo(papd)]^{4+} + 2H^+ \rightarrow [Co(papd)(OH_2)]^{3+} + \tfrac{1}{2}O_2$$

should be thermodynamically favorable in both acidic and alkaline solutions.

If, as is to be expected, analogous calculations would give similar results for other μ-peroxo complexes if the corresponding formation constants and redox potentials were known, then it is not surprising that the only examples of the decomposition of μ-peroxo complexes to give hydrogen peroxide are those that occur in acid solution (see p. 349).

VII. FORMATION OF μ-SUPEROXO COMPLEXES

Although there have been several reports[6,141,152,176,178] of the direct synthesis of a μ-superoxo complex by oxygenation of cobalt(II) complexes, these reactions may be considered as occurring in two stages: first, the formation of a μ-peroxo complex, and second, oxidation of this to the superoxo complex either by air[141,152,176] or perhaps by oxidizing anions that are present.[6,178]

A recent report[176] describes the reaction of a binuclear Co(II) cofacial porphyrin complex with oxygen in the presence of a nitrogenous ligand: if the Co(II) complex is first oxidized electrochemically to the mixed valence Co(II)/Co(III) complex, this then reacts very rapidly with oxygen to give the μ-superoxo complex even in the absence of other ligands.

A. By oxidation of μ-peroxo complexes

It is, however, clear from the table of redox potentials of μ-superoxo-dicobalt(III) complexes (Table 10) that the synthesis of most μ-superoxo complexes requires an oxidizing agent much more powerful than air. Many of the most powerful oxidizing agents have been used. The oxidizing agents that have been widely useful in the synthesis of μ-superoxo complexes, such as concentrated nitric acid,[7,170] ozone,[4,171] chlorine,[84,172] cerium-(IV),[34,84] potassium permanganate,[84,169] lead dioxide,[173] and peroxo-disulfate,[8,84,174] all have redox potentials well over 1.16 V, which is the highest potential yet reported for a μ-superoxodicobalt complex[167]

(Table 10). Bromate and nitrate will also rapidly oxidize μ-peroxo to μ-superoxo complexes, but dichromate, though a powerful oxidizing agent ($E° = 1.33$ V), shows no immediate reaction. Weaker oxidizing agents, e.g., hypobromite or even air, have been useful only for the decacyano complex.[141]

The most useful oxidizing agents are those that are effective in strongly acidic solutions because the μ-superoxo complexes are then most stable; concentrated nitric acid, ammonium peroxodisulfate, and ozone have been the most widely used synthetically.

To date there has been little detailed work on the kinetics of this oxidation, and the reactions of hypochlorite[175] and cerium(IV)[85] with $[(en)_2Co-(O_2,NH_2^-)Co(en)_2]^{3+}$ (I) are said to take place instantaneously. The brown μ-peroxo complex I is protonated in strongly acidic solution to the orange μ-hydroperoxo complex II, which then isomerizes to the isomeric red μ-hydroperoxo complex III.[34,85]

$$\tag{41}$$

Salts of the red isomer were first prepared by Werner[7] and the crystal structure of III has been determined.[30] Oxidations of the μ-hydroperoxo complex II are also extremely rapid, but the reaction with the isomeric μ-hydroperoxo complex III is slow, with the same rate as the isomerization of III to II. It may, however, reasonably be presumed that the reaction proceeds via an outer-sphere (i.e., nonbonded) mechanism, as has been consistently found[211] for the reverse process, the reduction of μ-superoxo to μ-peroxo complexes; this is discussed in Section IX,A.

Direct oxygenation of $[Co(CN)_5]^{3-}$ normally gives only a low yield[141] of the μ-superoxo complex $[(NC)_5CoOOCo(CN)_5]^{5-}$, but on prolonged oxygenation the yield increases.[152] The results of a combined ESR and UV absorption study[109,152] of the oxygenation of $[Co(CN)_5]^{3-}$ suggested that in the presence of excess oxygen the concentration of the mononuclear

TABLE 10

Redox potentials for μ-superoxo/μ-peroxo complexes[a]

Complex	$E°$ (V)	Ref.
Singly bridged complexes		
$[(NC)_5CoOOCo(CN)_5]^{5-/6-}$	-0.19	[167]
	$+0.09$	[168]
$[(dien)(gly)CoOOCo(dien)(gly)]^{3+/2+}$	0.69	[167]
$[(H_3N)_5CoOOCo(NH_3)_5]^{5+/4+}$	0.71[b]	[167]
	0.9	[168]
$[(en)(dien)CoOOCo(en)(dien)]^{5+/4+}$	0.90	[167]
$[(H_3N)(trien)CoOOCo(trien)(NH_3)]^{5+/4+}$	0.91	[167]
Hydroxo-bridged complexes		
$[(en)_2Co(O_2,OH)Co(en)_2]^{4+/3+}$	0.93	[167]
$[(trien)Co(O_2,OH)Co(trien)]^{4+/3+}$	1.06	[167]
Amido-bridged complexes		
$[(H_3N)_4Co(O_2,NH_2)Co(NH_3)_4]^{4+/3+}$	0.78	[167]
	0.75	[168]
$[(en)_2Co(O_2,NH_2)Co(en)_2]^{4+/3+}$	0.86[c]	[167]
	<1.0	[169]
$[(tren)Co(O_2,NH_2)Co(tren)]^{4+/3+}$	0.96[c]	[167]
$[(bipy)Co(O_2,NH_2)Co(bipy)]^{4+/3+}$	1.16	[167]
$[(phen)Co(O_2,NH_2)Co(phen)]^{4+/3+}$	1.16	[167]

[a] Relative to the standard hydrogen electrode; 25°C.
[b] pH dependent below pH 1.
[c] pH dependent below pH 3.

superoxo complex $[(NC)_5CoOO]^{3-}$ (**V**) can be quite high.[109] It is presumably formed by dissociation of the μ-peroxo complex **IV** to **V** and $[(NC)_5Co]^{2+}$ (**VI**), which then reacts with oxygen to give more of the mononuclear superoxo complex **V**:

$$[(NC)_5CoOOCo(CN)_5]^{6-} \rightleftharpoons [(NC)_5Co\dot{O}O]^{3-} + [(NC)_5Co]^{3-}$$
$$\quad\text{IV} \qquad\qquad\qquad \text{V} \qquad\qquad \text{VI}$$
(42)

$$[(NC)_5Co]^{3-} + O_2 \rightarrow [(NC)_5Co\dot{O}O]^{3-}$$
$$\quad\text{VI} \qquad\qquad\qquad \text{V}$$
(43)

This could be an intermediate in the formation of the μ-superoxo complex reacting with the μ-peroxo complex **IV** to give the μ-superoxo complex and $[(NC)_5CoOO]^{4-}$.

$$\begin{array}{ccc} [(NC)_5CoOOCo(CN)_5]^{6-} & & [(NC)_5CoOOCo(CN)_5]^{5-} \\ + & \xrightarrow{\qquad\qquad} & + \\ [(NC)_5Co\dot{O}O]^{3-} & & [(NC)_5CoOO]^{4-} \end{array}$$
(44)

The complex $[(NC)_5CoOO]^{4-}$ is the conjugate base[153] ($pK_a \sim 12$) of $[(NC)_5CoOOH]^{3-}$, which is formed by hydrolysis[152] of $[(NC)_5CoOOCo-(CN)_5]^{6-}$ and by oxygenation[153] of $[(NC)_5CoH]^{3-}$. No direct evidence for reaction (44) was obtained, however.[109,152] It is in any case interesting to note that Endicott[148] also postulates a dissociation of the same type as reaction (42) as the first step of the reduction of $[(H_2O)(cyclam)CoOOCo-(cyclam)(OH_2)]^{4+}$.

Werner and Mylius observed that prolonged oxygenation of ammoniacal cobalt(II) nitrate gave[6] the superoxo complex $[(H_3N)_5CoOOCo(NH_3)_5]-(NO_3)_5$, and under other conditions[177] a complex mixture containing a doubly bridged superoxo complex, $[(H_3N)_4Co(O_2,NH_2)Co(NH_3)_4](NO_3)_4$. However, as in these cases, oxidation by the nitrate ion cannot be ruled out: only with cyanide is the formation of the μ-superoxo complex by direct oxygenation of cobalt(II) proved. As the reduction potential[167] (see Table 10) of the μ-superoxo complex to the μ-peroxo complex (-0.19 V) is much lower than for any other μ-superoxo dicobalt complex, this is not surprising.

The synthesis of $[(H_3N)_5CoOOCo(NH_3)_5]^{5+}$ recommended by Brauer[178] involves the oxidation of an ammoniacal solution of cobalt(II) sulfate with a mixture of hydrogen peroxide and ammonium peroxodisulfate. Although the mechanism of this reaction is unknown,[20] it probably involves an initial oxygenation by oxygen formed by the decomposition of the hydrogen peroxide, followed by oxidation by the peroxodisulfate.[8] The whole synthesis is unusual in that it takes place in alkaline solution; it also appears that the synthesis works better if the cobalt(II) sulfate is contaminated with a little nickel(II) salt.

B. By disproportionation of μ-peroxo complexes

As Maquenne[4] claimed that the green complexes, which are now known to be μ-superoxo complexes, were protonated μ-peroxo complexes, Vort-mann[5] concluded that acidification of a μ-peroxo complex should be enough to give the μ-superoxo complex. Vortmann observed that acidification of the μ-peroxo complex $[(H_3N)_5CoOOCo(NH_3)_5]^{4+}$ with dilute nitric acid or with dilute sulfuric acid gave the μ-peroxo complex $[(H_3N)_5-CoOOCo(NH_3)_5]^{5+}$; with sulfuric acid the yield is very small, and with nitric acid oxidation by the nitric acid may occur instead of disproportion-ation.

However, a more recent reinvestigation[179] of the decomposition of $[(H_3N)_5CoOOCo(NH_3)_5](SO_4)_2 \cdot 3H_2O$ in 1 M H_2SO_4 confirmed that although Co^{2+} accounts for 54% of the total cobalt, and mononuclear cobalt(III) complexes account for a further 32%, 6% of the μ-superoxo complex $[(H_3N)_5CoOOCo(NH_3)_5]^{5+}$ is formed. As the rate of decomposition of $[(H_3N)_5CoOOCo(NH_3)_5]^{4+}$ in 0.02–0.2 M perchloric acid is independent of the proton concentration, it has been suggested[110] that the first step of the reaction could be the decomposition discussed in Section IV.

$$[(H_3N)_5CoOOCo(NH_3)_5]^{4+} + H_2O \rightarrow [(H_3N)_5Co(OH_2)]^{2+} + [(H_3N)_5CoO_2]^{2+} \quad (45)$$
$$\text{VII} \qquad\qquad \text{VIII}$$

The pentamminecobalt(II) complex **VII** decomposes immediately in the acid solution, whereas the mononuclear superoxo complex **VIII** can decompose to cobalt(II) and oxygen or could possibly react with the μ-peroxo complex, perhaps as the protonated species:

$$[(H_3N)_5Co(O_2H)Co(NH_3)_5]^{5+} \qquad [(H_3N)_5CoOOCo(NH_3)_5]^{5+}$$
$$+ \qquad\qquad \longrightarrow \qquad\qquad + \qquad\qquad (46)$$
$$[(H_3N)_5CoO\dot{O}]^{2+} \qquad\qquad [(H_3N)_5CoOOH]^{2+}$$

This reaction is analogous to that occurring with the decacyano μ-peroxo complex [reaction (44), Section VII,A], apart from the protonation caused by the much lower pH.

Werner[7] reported that the red protonated μ-peroxo complex $[(en)_2Co(O_2H,NH_2)Co(en)_2]^{4+}$ reacted with bromide to give the green μ-superoxo complex $[(en)_2Co(O_2,NH_2)Co(en)_2]^{4+}$. Thompson and Wilmarth[169] found the stoichiometry to be as follows:

$$3[(en)_2Co(O_2,NH_2)Co(en)_2]^{3+} + 3H^+ \rightarrow$$
$$2[(en)_2Co(O_2,NH_2)Co(en)_2]^{4+} + [(en)_2Co(OH,NH_2)Co(en)_2]^{4+} + H_2O \quad (47)$$

Davies and Sykes[175] found that chloride likewise catalyzes this disproportionation, and that the μ-hydroxo complex is in equilibrium:

$$[(en)_2Co(OH,NH_2)Co(en)_2]^{4+} \overset{HX}{\rightleftharpoons} [(en)_2XCo(NH_2)Co(OH_2)(en)_2]^{4+} \quad (48)$$

In the absence of halide ions no μ-superoxo complex is formed, and the only reaction is the slow decomposition of the μ-peroxo complex. Davies and Sykes also studied[175] the pH dependence of the rate with both bromide and chloride. As the reaction is extremely slow at pH 4 and as the rate increases considerably as the $[H^+]$ increases, they suggested that only the protonated complex (see p. 356) reacts, and that of the two isomeric μ-hydroperoxo complexes **II** and **III**, the red isomer **III** is probably more reactive than **II**. The same relative reactivity has been observed for the reactions of other reducing agents with the μ-hydroperoxo complexes **II** and **III** (see Section IX,B, p. 369).

It was suggested many years ago that the hypohalous acids HOCl and HOBr are intermediates in the oxidation of halides by hydrogen peroxide[181] and by peroxo acids.[182] It was therefore proposed[175] that the reaction of halide with the protonated μ-peroxo complex in acidic solution also gives hypohalous acids,

$$[(en)_2Co(O_2H,NH_2)Co(en)_2]^{4+} + Cl^- \rightarrow [(en)_2Co(OH,NH_2)Co(en)_2]^{4+} + HOCl \quad (49)$$

which then oxidize the μ-peroxo complex to the μ-superoxo complex:

$$2[(en)_2Co(O_2,NH_2)Co(en)_2]^{3+} + HOCl + H^+ \rightarrow$$

$$2[(en)_2Co(O_2,NH_2)Co(en)_2]^{4+} + H_2O + Cl^- \quad (50)$$

Although hypochlorous acid has not yet been detected as an intermediate, it does oxidize the μ-peroxo complex to the μ-superoxo complex.

Disproportionation of the analogous μ-peroxo μ-hydroxo complexes has not yet been observed, but the singly bridged complex [(dien)(gly)CoOOCo-(dien)(gly)]$^{2+}$ disproportionates in $> 10^{-2}$ M perchloric acid to give the μ-superoxo complex in 50% yield[110]; the detailed mechanism of this reaction is still unclear, as some μ-superoxo complex is formed even in the absence of chloride.

It is notable that the three systems for which disproportionation has been observed all have relatively low redox potentials for the μ-superoxo complexes (Table 10). Unfortunately, no redox potentials are yet known for half-reactions such as

$$[(en)_2Co(O_2H,NH_2)Co(en)_2]^{4+} + H^+ + 2e^- \rightarrow$$

$$[(en)_2Co(OH,NH_2)Co(en)_2]^{4+} + OH^- \quad (51)$$

or

$$[(dien)(gly)CoOOCo(dien)(gly)]^{2+} + 4H^+ + 2e^- \rightarrow 2[Co(dien)(gly)(OH_2)]^{2+} \quad (52)$$

It is thus not yet possible to predict those systems for which disproportionation should occur.

C. By ligand replacement in μ-superoxo complexes

Owing to the instability of μ-superoxo complexes (except for the various polycyano complexes) in alkaline solution, ligand replacement is of limited use in their synthesis. However, the reaction of cyanide with [(H_3N)_5-CoOOCo(NH_3)_5]$^{5+}$ has been used[84,183] to prepare K_5[(NC)_5CoOOCo-(CN)_5]·H_2O. Likewise, the reaction of an excess of cyanide with [(H_3N)_4-Co(O_2,NH_2)Co(NH_3)_4](NO_3)_4 gives[183] K_4[(NC)_4Co(O_2,NH_2)Co(CN)_4]; incomplete replacement also occurs[183] to give the diamminehexacyano and ammineheptacyano complexes, which can be separated by chromatography on alumina.

The reaction[84,169,184] of ethylenediamine with $[(H_3N)_4Co(O_2,NH_2)Co(NH_3)_4](NO_3)_4$ at 60°C for 2 h, followed by acidification with concentrated nitric acid, might seem to be a simple substitution; however, it is clear from Werner's detailed description[184] of the color changes that occur that this reaction proceeds via initial reduction of the green μ-superoxo complex to $[(H_3N)_4Co(O_2,NH_2)Co(NH_3)_4]^{3+}$, followed by the formation of the brown μ-peroxo complex $[(en)_2Co(O_2,NH_2)Co(en)_2]^{3+}$, which can be isolated[184] as the red, protonated μ-peroxo complex $[(en)_2Co(O_2H,NH_2)Co(en)_2](NO_3)_4$. This is then reoxidized[185] by the nitric acid to the green μ-superoxo complex $[(en)_2Co(O_2,NH_2)Co(en)_2]^{4+}$. Similar reactions of the μ-superoxo octaammine complex with 1,2-diaminopropane,[170] bipyridyl,[186,187] and phenanthroline[186,187] have also been reported. These also involve successive reduction, substitution, and reoxidation with concentrated nitric acid.

Addition of liquid ammonia to halo μ-superoxo complexes such as $[(H_3N)_3BrCo(O_2,NH_2)Co(NH_3)_3Br]Br_2$ is reported to give[188] the octaammine μ-superoxo complex $[(H_3N)_4Co(O_2,NH_2)Co(NH_3)_4]Br_4$, apparently without the need of any reoxidizing agent. Werner also reported[7] several examples of the apparent substitution of μ-superoxo complexes by halide ions in acidic solutions, including[189]

$$[(H_3N)_3Co(O_2,OH,NH_2)Co(NH_3)_3]X_3 + HX \rightarrow$$

$$[(H_3N)_3XCo(O_2,NH_2)Co(NH_3)_3X]X_2 \qquad X = Cl \text{ or } Br \quad (53)$$

The structures both of the starting material and of the product are uncertain; further work on this reaction is obviously required.

Thus the only proved reactions involving ligand replacement in μ-superoxo complexes are those involving cyanide and perhaps the reactions in liquid ammonia. There is therefore no evidence as to whether true substitution occurs in μ-superoxo complexes or whether such reactions proceed via redox reactions such as those found for the μ-peroxo complexes (see Section V, p. 342).

VIII. PROPERTIES OF μ-SUPEROXO COMPLEXES

A. Spectra

With the exception of the cyano complexes, which are red, all μ-superoxo complexes are dark green and have a characteristic absorption band at ~ 700 nm.[20]

Complex	λ_{max} (nm)	ε_{max} (M^{-1} cm^{-1})	Ref.
Green			
$[(H_3N)_5CoOOCo(NH_3)_5]^{5+}$	670	890	[84]
$[(tetren)CoOOCo(tetren)]^{5+}$	704	1330	[172]
$[(H_3N)_4Co(O_2,NH_2)Co(NH_3)_4]^{4+}$	700	306	[84]
$[(en)_2Co(O_2,NH_2)Co(en)_2]^{4+}$	687	485	[84]
$[(H_3N)_3Co(O_2,OH,NH_2)Co(NH_3)_3]^{3+}$	730	206	[190]
Red			
$[(NC)_5CoOOCo(CN)_5]^{5-}$	485	1130	[152]
	485	745	[191]
$[(NC)_4Co(O_2,NH_2)Co(CN)_4]^{4-}$	519	200	[192]

The brown compound described by Werner[193] as an imido-bridged μ-superoxo complex, $[(H_3N)_4Co(O_2,NH)Co(NH_3)_4]I_3$, does not have[194] this characteristic absorption band and therefore is probably an amido-bridged μ-peroxo complex; its reported[193] reactions with acid would also seem to confirm this.[195]

Werner[196] succeeded in resolving $[(en)_2Co(O_2,NH_2)Co(en)_2]^{4+}$, and several other μ-superoxo complexes have been resolved more recently.[170,186,187] The circular dichroism (CD) spectra of several μ-superoxo complexes have also been measured[170,187,194,195]: the absorption band at ~ 700 nm has only a very low CD absorption. The absolute configuration of $\Delta\Delta$-($-$)-$[(en)_2Co(O_2,NH_2)Co(en)_2]^{4+}$ has been deduced[195,197] by degradation via $\Delta\Delta$-($-$)-$[(en)_2Co(OH,NH_2)Co(en)_2]^{4+}$ to Δ-($-$)-$[Co(en)_2$-$(NH_3)_2]^{3+}$; this has been confirmed by comparison[170] of the CD spectrum of the tetrakis(ethylenediamine) μ-superoxo complex with that of the analogous propylenediamine complex, $\Delta\Delta$-($-$)-$[\{(-)-pn\}_2Co(O_2,NH_2)Co\{(-)-pn\}_2]^{4+}$, formed stereospecifically from the resolved ligands.

All μ-superoxodicobalt(III) complexes are paramagnetic[18,20] and have a magnetic moment corresponding to one unpaired electron (1.7 BM). A 15-line ESR spectrum is characteristic of the μ-superoxo complexes,[172,183,200] and detailed studies[198] using ^{17}O-enriched $[(H_3N)_4Co(O_2,NH_2)Co(NH_3)_4]^{4+}$ indicate a high residence time of the unpaired electron on the oxygen bridge, consistent with the superoxide formulation. The O–O stretching frequencies observed by Raman spectroscopy and the O–O distances determined crystallographically are also similar to those in potassium superoxide (see Table 2).

B. Thermal decomposition in acidic solution

Many μ-superoxo complexes are unstable even in the solid state, decomposing slowly at least to the μ-peroxo complexes or even to the mononuclear cobalt(III) complexes; the rate of decomposition may depend on the anion in an as yet unexplained manner.[199]

In marked contrast to the μ-peroxo complexes, which are usually most stable in alkaline solution and which decompose rapidly in acidic solutions, the μ-superoxo complexes are usually very stable in strongly acidic solutions but decompose rapidly in neutral or alkaline solutions;[4] e.g., solutions of $[(H_3N)_5CoOOCo(NH_3)_5]^{5+}$ in concentrated sulfuric or phosphoric acid are stable for several weeks at room temperature,[200,202] but aqueous solutions decompose rapidly at pH >3. UV absorption[85] and ESR[85,200] studies provide no evidence for any protonation of the μ-superoxo group. This is in contrast to the protonation and isomerism of a protonated μ-peroxo group.[85] In hot, acidic solution, aquation yields equimolar amounts of cobalt(III), cobalt(II), and oxygen[171,180]:

$$[(H_3N)_5CoOOCo(NH_3)_5]^{5+} + 5H^+ + H_2O \rightarrow$$
$$[Co(NH_3)_5(OH_2)]^{3+} + Co^{2+} + 5NH_4^+ + O_2 \quad (54)$$

The kinetics were studied[180] in aqueous perchloric acid (0.02–2.0 M); the rate is first order with respect to the complex and is independent of the hydrogen ion concentration in these strongly acidic solutions. The first step of the reaction is believed to be

$$[(H_3N)_5CoOOCo(NH_3)_5]^{5+} + H_2O \rightarrow [(H_3N)_5Co(OH_2)]^{3+} + [(H_3N)_5CoO_2]^{2+} \quad (55)$$

The $[(H_3N)_5CoO_2]^{2+}$ then decomposes to cobalt(II) and oxygen, as is normally observed in the decomposition of μ-peroxo complexes in acid [reaction (16), Section IV,A]. However, between pH 3 and 7 the rate of decomposition of $[(H_3N)_5CoOOCo(NH_3)_5]^{5+}$ is inversely proportional to the proton concentration, and a range of analogous polyamine complexes of the type $[(Z)CoOOCo(Z)]^{5+}$ [Z = (dien)(en), dien(NH_3)_2, and (trien)-(NH_3)] behave similarly.[110] The mechanism of this reaction is, however,

---

[4] However, the negatively charged cyano complexes $[(NC)_5CoOOCo(CN)_5]^{5-}$ and $[(NC)_4Co(O_2,NH_2)Co(CN)_4]^{4-}$ are also stable in alkaline solution; indeed, the visible absorption spectra of $[(NC)_5CoOOCo(CN)_5]^{5-}$ in 10 M KOH, in water, and in concentrated acids are identical.[183]

TABLE 11

Thermal decomposition of μ-superoxo complexes in acidic solution[a]

Complex	k (min^{-1})	Ref.	Products/mole complex
$[(H_3N)_5CoOOCo(NH_3)_5]^{5+}$	1.4×10^{-5}	[171]	$Co^{2+} + [Co(NH_3)_5(OH_2)]^{3+} + O_2$
$[(H_3N)_4Co(O_2,NH_2)Co(NH_3)_4]^{4+}$	3×10^{-7}	[171]	$Co^{2+} + O_2 + 0.5[Co(NH_3)_5(OH_2)]^{3+}$
			$+ 0.5cis/trans-[Co(NH_3)_4(OH_2)_2]^{3+}$
$[(NC)_5CoOOCo(CN)_5]^{5-}$	1.2×10^{-4}	[171]	$2[Co(CN)_5(OH_2)]^{2-}$
$[(NC)_4Co(O_2,NH_2)Co(CN)_4]^{4-}$	—	[171]	"Almost indefinitely stable"

[a] 25°C, 0.1–2.0 M.

not yet known, but is probably identical to that of the even faster decomposition that occurs in alkaline solution (Section VIII,D, p. 365). Also unexplained is the effect of chloride ion on the decomposition of $[(dien)(NH_3)_2$-$CoOOCo(dien)(NH_3)_2]^{5+}$, for which the rate is 1000-fold slower in the presence of 1 M KCl.

The doubly bridged complex $[(H_3N)_4Co(O_2,NH_2)Co(NH_3)_5](SO_4)_2$ is much more stable but nevertheless decomposes in hot (85°C) 2 M perchloric acid and also gives equimolar amounts of cobalt(II), cobalt(III), and oxygen [180]

$$2[(H_3N)_4Co(O_2,NH_2)Co(NH_3)_4]^{4+} + 3H_2O + 9H^+ \rightarrow$$
$$2Co^{2+} + 7NH_4^+ + 2O_2 + [Co(NH_3)_5(OH_2)]^{3+} + cis/trans-[Co(NH_3)_4(OH_2)_2]^{3+} \quad (56)$$

In concentrated sulfuric acid less oxygen is formed, and an appropriately larger amount of cobalt(III) complexes and then decomposes further to cobalt(II) sulfate and nitrogen on further heating. Werner used this reaction as an analytical method to determine the oxidation states present in μ-superoxo complexes.[201]

The rates of thermal decomposition of some typical μ-superoxo complexes in acidic solution, and details of the products formed, are compared in Table 11.

C. Photochemical decomposition in acidic solution

The stoichiometry of the photochemical decomposition of $[(H_3N)_5$-$CoOOCo(NH_3)_5]^{5+}$ (IX) in dilute hydrochloric[203] or perchloric[171] acid has been found to be the same as that of the thermal decomposition [reaction (54)]. In 0.1 M HClO$_4$ the quantum yield varies[171] from 10^{-3} at 700 nm to 0.3 at 320 nm; in 0.1 M hydrochloric acid the quantum yield is higher,[203] 0.65 at 365 nm and 0.83 at 313 nm. Photolysis of IX in concentrated sulfuric acid and in 85% phosphoric acid gave $[Co(NH_3)_5(SO_4H)]^{2+}$ and $[Co(NH_3)_5-(PO_4H_2)]^{2+}$, respectively[171]; NO_3^-, Cl^-, CF_3COO, and CCl_3COO could also be incorporated as $[Co(NH_3)_5X]^{2+}$ in the photolysis of **IX** or of the octaammine complex $[(H_3N)_4Co(O_2,NH_2)Co(NH_3)_4]^{4+}$ (**X**). This suggests that the penta-coordinated $[Co(NH_3)_5]^{3+}$ may be formed as an intermediate. Moreover, the ratio of the amount of chloride ion or water incorporated, as indicated by the ratio $[Co(NH_3)_5Cl]^{2+} : [Co(NH_3)_5(OH_2)]^{3+}$, is the same as in other reactions[204] involving $[Co(NH_3)_5]^{3+}$.

It is interesting that although the thermal decomposition of the octa-ammine complex **X** yields an equilibrium mixture of *cis*- and *trans*-[Co-$(NH_3)_4(OH_2)_2]^{3+}$, among other products [reaction (56)], the photochemical decomposition yields no detectable $[Co(NH_3)_4(OH_2)_2]^{3+}$.

$$[(H_3N)_4Co(O_2,NH_2)Co(NH_3)_4]^{4+} + 5H^+ + H_2O \xrightarrow{h\nu}$$

$$[Co(NH_3)_5(OH_2)]^{3+} + Co^{2+} + O_2 + 4NH_4^+ \quad (57)$$

Photolysis of the decacyano complex yields hydrogen peroxide,[171]

$$2[(NC)_5CoOOCo(CN)_5]^{5-} + 2H^+ + 4H_2O \xrightarrow{h\nu} 4[(NC)_5Co(OH_2)]^{2-} + H_2O_2 + O_2 \quad (58)$$

which could not be detected in the thermal studies.

The amido-bridged octacyano complex $[(NC)_4Co(O_2,NH_2)Co(CN)_4]^{4-}$ is extremely stable, and neither thermal nor photochemical decomposition was observed.[171]

D. Decomposition in alkaline solution

All μ-superoxodicobalt(III) complexes, other than the cyano complexes, decompose rapidly to the corresponding μ-peroxo complexes in alkaline solution, even in the absence of conventional reducing agents. Indeed, the reaction of aqueous ammonia with the μ-superoxo complex $[(H_3N)_4Co-(O_2,NH_2)Co(NH_3)_4]Cl_4$ provides a good synthesis of the μ-peroxo complex $NH_4[(H_3N)_4Co(O_2,NH_2)Co(NH_3)_4](NO_3)_4$ in 65% yield.[34] The analogous reaction with $[(en)_2Co(O_2,NH_2)Co(en)_2]^{4+}$ has been investigated by polar-ography[205] and by analysis of the products.[195] The stoichiometry seems to be approximately

$$3[(en)_2Co(O_2,NH_2)Co(en)_2]^{4+} + 2OH^- \rightarrow$$

$$2[(en)_2Co(O_2,NH_2)Co(en)_2]^{3+} + [(en)_2(HO)Co(NH_2)Co(en)_2(OH)]^{4+} + O_2 \quad (59)$$

although some oxidation of the ligand may also occur.[20] The stoichiometry in alkaline solution seems to be different from that found in acidic solution [reactions (54) and (55)].

The decomposition of a number of singly bridged μ-superoxo complexes of polyamine ligands is apparently proportional to the hydroxide ion concentration over a wide pH range.[110] This is consistent with an earlier suggestion[20] that this decomposition might involve the formation of the conjugate base, which could be enhanced by the high positive charge of these complexes. It is noteworthy that the only μ-superoxo complexes that are stable in alkaline solution are the negatively charged cyanide complexes. However, the detailed mechanism of this complicated reaction is still unknown, and it clearly merits further investigation.

The formation of the μ-superoxo complex $[(H_3N)_5CoOOCo(NH_3)_5]^{5+}$ in acidic solution and its reduction by hydroxide ion in alkaline solution to give a mixture containing the μ-peroxo complex $[(H_3N)_5CoOOCo-(NH_3)_5]^{4+}$ led to the original[4] erroneous interpretation of this reaction as a protonation.[5]

IX. REDUCTION OF μ-SUPEROXO AND μ-PEROXO COMPLEXES

A. One-electron reduction of μ-superoxo complexes to μ-peroxo complexes

The reduction of μ-superoxo complexes, especially of $[(en)_2Co(O_2,NH_2)-Co(en)_2]^{4+}$, has been studied in great detail, both from preparative[195] and, more recently, from kinetic and mechanistic[211] points of view.

The separation and identification of the reaction products are of considerable importance in deciding whether the reaction proceeds by an inner-sphere or an outer-sphere mechanism. The reduction products of some μ-superoxo complexes are listed in Table 12. Although various products are formed, detailed studies seem to indicate that the initial product is always the μ-peroxo complex, although this may decompoose or undergo further reduction. This implies that the reduction of μ-superoxo complexes proceeds by an outer-sphere mechanism.

As the μ-superoxo complexes apparently cannot be protonated (see p. 363), the rate of reduction is pH dependent only if the reductant undergoes protonation (e.g., with SO_3^{2-} or NO_2^{-}) The pH may, however, affect the degree of ion pairing, and this may influence the rate.

For example, the reduction of $[(H_3N)_5CoOOCo(NH_3)_5]^{5+}$ with a wide range of reducing agents gives cobalt(II) and oxygen according to reaction (60).

$$[(H_3N)_5CoOOCo(NH_3)_5]^{5+} + e^- + 10H^+ \rightarrow 2Co^{2+} + 10NH_4^+ + O_2 \qquad (60)$$

TABLE 12

Reduction of μ-superoxodicobalt(III) complexes

Complex	Reductant	Product	Ref.
$[(H_3N)_5CoOOCo(NH_3)_5]^{5+}$	Fe^{2+}	μ-peroxo[a]	[199, 206]
	VO^{2+}, Sn^{2+}	μ-peroxo	[199]
	I^-, NO_2^-, $S_2O_3^{2-}$	μ-peroxo	[199]
	Cr^{2+}, V^{2+}, Eu^{2+}	μ-peroxo	[207]
	I^-	μ-peroxo	[208]
	AsO_2^-/OsO_4	μ-peroxo	[8][b]
	SO_3^{2-}	$[Co(NH_3)_5SO_4]^+$ $+ Co^{2+}$	[199, 209][c]
	NO_2^-	$[Co(NH_3)_5NO_3]^{2+}$ $+ Co^{2+}$	[20, 199][c]
$[(H_3N)_4Co(O_2,NH_2)Co(NH_3)_4]^{4+}$	I^-	μ-peroxo[d]	[208][c]
	Fe^{2+}	μ-peroxo	[206, 210][c]
	OH^-	μ-peroxo	[34][f]
	NO_2^-	$(\mu\text{-}NO_2,\mu\text{-}NH_2)$[e]	[211][f]
	SO_2	$(\mu\text{-}SO_4,\mu\text{-}NH_2)$[e]	[7, 84, 212][f]
$[(H_3N)_3Co(O_2,OH,NH_2)Co(NH_3)_3]^{3+}$	I^-	μ-peroxo[d]	[208][c]
	Fe^{2+}	μ-peroxo	[206][c]
	NO_2^-	$(\mu\text{-}NO_2,\mu\text{-}NH_2)$	[7][f]
	SO_2	$(\mu\text{-}NH_2)$	[7][f]
$[(en)_2Co(O_2,NH_2)Co(en)_2]^{4+}$	I^-	μ-peroxo[d]	[208][c]
	Br^-	μ-peroxo	[175]
	Fe^{2+}	μ-peroxo	[206][c]
	Fe^{2+}, Sn^{2+}, I^-	μ-peroxo	[195]
	Cr^{2+}	μ-peroxo[d]	[213, 214][c]
	V^{2+}	μ-peroxo	[214][c]
	NO	μ-peroxo	[212]
	NO_2^-	μ-peroxo	[215]
	SO_3^{2-}	μ-peroxo[d]	[215]
	SO_2	$(\mu\text{-}SO_4,\mu\text{-}NH_2)$	[195][c]
	SeO_2	$(\mu\text{-}SeO_4,\mu\text{-}NH_2)$	[195][c]
	AsO_2^-	No reduction[g]	[215]
$[(bipy)_2Co(O_2,NH_2)Co(bipy)_2]^{4+}$	Fe^{2+}	μ-peroxo	[216]
$[(phen)_2Co(O_2,NH_2)Co(phen)_2]^{4+}$	Fe^{2+}	μ-peroxo	[216]
$[(NC)_5CoO_2Co(CN)_5]^{5-}$	I^-	$2[Co(CN)_5(OH_2)]^{2-}$	[141][c]
	$[Co(CN)_5]^{3-}$	μ-peroxo[c]	[152][c]
	NH_2OH	μ-peroxo	[152][c]
	$N_2H_5^+$	$[Co(CN)_5]_n^{2n-}$	[152]
	$S_2O_4^{2-}$	Products unknown	[152]

[a] Which then decomposes to Co^{2+} and oxygen.
[b] The reduction is extremely slow without the addition of OsO_4. The oxidation of arsenite by several common oxidizing agents is also catalyzed by OsO_4.
[c] Product identified spectroscopically.
[d] Which is then reduced further (see Table 13).
[e] The μ-peroxo complex may be formed initially and then reduced further.
[f] Product was isolated.
[g] The corresponding μ-peroxo complex is, however, reduced (see also footnote b).

The μ-peroxo complex formed initially is very unstable. The relative rates of reduction found[199,207] with Fe^{2+}, V^{2+}, Cr^{2+}, and Eu^{2+} are the same as have been found for other reductions that are thought to involve an outer-sphere mechanism.[21]

The products of the reactions with sulfite[199,209] and with nitrite[20,199] seem on initial examination to be anomalous.

$$[(H_3N)_5CoOOCo(NH_3)_5]^{5+} + 2SO_3^{2-} \rightarrow$$

$$[Co(NH_3)_5(SO_4)]^+ + Co^{2+} + SO_4^{2-} + 5NH_3 \quad (61)$$

$$[(H_3N)_5CoOOCo(NH_3)_5]^{5+} + 2NO_2^- \rightarrow$$

$$[Co(NH_3)_5(NO_3)]^{2+} + Co^{2+} + NO_3^- + 5NH_3 \quad (62)$$

Experiments [209] with $^{18}O_2$-labeled $[(H_3N)_5CoOOCo(NH_3)_5]^{5+}$ showed that the labeled oxygen was retained both by the free sulfate and by $[Co(NH_3)_5(SO_4)]^+$. However, it has been suggested[211] that these sulfate and nitrato pentaammine cobalt(III) complexes arise from further reduction of the μ-peroxo complex initially formed.

Reduction of $[(H_3N)_4Co(O_2,NH_2)Co(NH_3)_4]^{4+}$ has been studied less extensively, but the reactions with I^- and with Fe^{2+} appear to be outer sphere.[21] The reactions with hydroxide,[34] nitrite,[211] and sulfur dioxide[84] provide useful syntheses of the μ-peroxo, μ-nitrito, and μ-sulfato complexes, respectively.

Reduction of the more stable $[(en)_2Co(O_2,NH_2)Co(en)_2]^{4+}$ with many reducing agents has been studied kinetically[21] and by using circular dichroism spectra on the resolved complex.[195] The reaction is believed to occur via an outer-sphere activated complex, as the initial product is normally the μ-peroxo complex and no product containing chromium bound to the peroxo group is observed in the reaction with Cr^{2+}. Again, the reaction with sulfite is apparently an exception,[215] as both the μ-peroxo complex $[(en)_2Co(O_2,NH_2)Co(en)_2]^{3+}$ and the μ-sulfato complex $[(en)_2Co(SO_4,NH_2)Co(en)_2]^{3+}$ are formed; however, the μ-sulfato complex is probably formed by further reduction[215] of the μ-peroxo complex (see p. 371).

There have been a number of reports of polarographic reduction of μ-superoxo complexes.[205,217]

Although reduction of the μ-superoxo complexes to the μ-peroxo complexes is usually followed rapidly either by decomposition of the μ-peroxo complex or by its further reduction, the one-electron reduction can be studied in isolation, even in the presence of an excess of the reducing agent, by following the decrease in absorption of the 700-nm band, as the products usually do not absorb in this region. The effect of pH on the rate of reduction has been discussed in detail in a review by Sykes and Weil.[20] The most interesting feature seen in the kinetics of reduction of $[(H_3N)_5CoOOCo-$

$(NH_3)_5]^{5+}$ is the strong catalysis by anions,[218] which decreases 20,000-fold in the series $F^- > Cl^- > SO_4^{2-} > Br^- \gg NO_3^-$.

As the electron exchange[219] between the μ-peroxo complex $[(en)_2Co-(O_2,NH_2)Co(en)_2]^{3+}$ and the five μ-superoxo complexes is independent of pH, protonated forms of the peroxo complex do not participate and proton bridging between the μ-peroxo and the μ-superoxo ligands is apparently insignificant.

B. Reduction of μ-peroxo complexes

The reduction of μ-peroxo complexes has received less attention than that of the μ-superoxo complexes. Some examples of typical reducing agents and of the reduced cobalt complexes that are formed are given in Table 13.

Reduction of μ-peroxo complexes is usually pH dependent, in contrast to the reduction of μ-superoxo complexes. This is a result of the protonation of the μ-peroxo group, which has been studied in detail for $[(en)_2Co(O_2,NH_2)-Co(en)_2]^{3+}$; in this case it has been suggested[34,85] that the protonated μ-peroxo complex isomerizes [see reaction (41), Section VII,A]. It has also been suggested[221] that reducing agents react faster with the red isomer of the μ-hydroperoxo complex **III** than with either the normal protonated form **II** or the unprotonated μ-peroxo complex **I**.

Reductions of the μ-peroxo complexes normally proceed by an inner-sphere mechanism.[211] This can be clearly seen in the reduction of $[(en)_2-Co(O_2,NH_2)Co(en)_2]^{3+}$ by Cr^{2+}, in which the absorption spectra of the intermediates give ample evidence for an inner-sphere mechanism. The first step has second-order kinetics[214] and is thought to be the reaction

$$[(en)_2Co(O_2,NH_2)Co(en)_2]^{3+} + Cr_{aq}^{2+} + 5H^+ \rightarrow$$

$$[(en)_2(H_3N)CoOOCr(OH_2)_5]^{4+} + Co^{2+} + 2H_2en^{2+} \quad (63)$$

The mechanism of the remaining steps that are observed kinetically is somewhat uncertain, but probably involves reduction of the μ-peroxo group to a μ-hydroxo group, perhaps via protonation and isomerization of the μ-peroxo group, as has been suggested for the dicobalt(III) complexes **I** $\rightleftharpoons$ **II** $\rightleftharpoons$ **III** discussed in Section VII,A [reaction (41)].

$$[(en)_2(H_3N)CoOOCr(OH_2)_5]^{4+} \xrightarrow{H^+} [(en)_2(H_3N)Co-O-Cr(OH_2)_5]^{4+}$$
$$\overset{|}{OH}$$

$$\xrightarrow{Cr^{2+}} [(en)_2(H_3N)Co(OH)Cr(OH_2)_5]^{5+} \quad (64)$$

TABLE 13

Reduction of μ-peroxodicobalt(III) complexes

Complex	Reductant	Product	Ref.
$[(H_3N)_5CoOOCo(NH_3)_5]^{4+}$	Fe^{2+}	—[a]	[199]
	I^+	No reduction in 7 M NH_3	[20]
$[(en)_2Co(O_2,OH)Co(en)_2]^{3+}$	Fe^{2+}	—[a]	[220]
	SO_2	$(\mu\text{-}SO_4,\mu\text{-}NH_2)$	[212]
	I^-	$[Co(en)_2(OH_2)_2]^{3+}$	[122]
$[(trien)Co(O_2,OH)Co(trien)]^{3+}$	Fe^{2+}	—[a]	[220]
$[(tren)Co(O_2,OH)Co(tren)]^{3+}$	SO_2	$(\mu\text{-}SO_4,\mu\text{-}OH)$	[212][b]
$[(H_3N)_4Co(O_2,NH_2)Co(NH_3)_4]^{3+}$	I^-	$(\mu\text{-}OH,\mu\text{-}NH_2)$	[208][b]
$[(en)_2Co(O_2,NH_2)Co(en)_2]^{4+}$	I^-	$(\mu\text{-}OH,\mu\text{-}NH_2)$	[221]
	Fe^{2+}, Sn^{2+}	$[(en)_2(H_2O)Co(NH_2)Co(en)_2(OH_2)]^{5+}$	[195]
	Cr^{2+}	$[(en)_2(H_3N)CoOOCr(OH_2)_5]^{4+}$	[214]
		$+ [(en)_2(H_3N)Co(OH)Cr(OH_2)_5]^{5+}$	
	NO_2^-	$(\mu\text{-}NO_2,\mu\text{-}NH_2)$	[215][b]
	SO_2	$(\mu\text{-}SO_4,\mu\text{-}NH_2)$	[195][b]
	SeO_2	$(\mu\text{-}SeO_4,\mu\text{-}NH_2)$	[195][c]
	AsO_2^-	$(\mu\text{-}OH,\mu\text{-}NH_2)$	[215]
$[(en)_2Co(O_2,H,NH_2)Co(en)_2]^{5+}$	I^-	$(\mu\text{-}OH,\mu\text{-}NH_2)$	[7,[c] 221]
$[(tren)Co(O_2,NH_2)Co(tren)]^{3+}$	SO_2	$(\mu\text{-}SO_4,\mu\text{-}NH_2)$	[212][b]
$[(H_3N)_3Co(O_2,OH,NH_2)Co(NH_3)_3]^{2+}$	I^-	Probably $(\mu\text{-}OH,\mu\text{-}OH,\mu\text{-}NH_2)$	[208]

[a] No reduction is observed as the proton-catalyzed decomposition is faster.

[b] Product identified spectroscopically.

[c] Product was isolated.

The hydroxo-bridged cobalt(III) chromium(III) complex is relatively stable and is only slowly reduced to cobalt(II) by an outer-sphere mechanism:

$$[(en)_2(H_3N)Co(OH)Cr(OH_2)_5]^{5+} + Cr^{2+} \xrightarrow{H^+/H_2O}$$

$$Co^{2+} + 2Cr^{3+} + 2H_2en^{2+} + NH_4^+ \quad (65)$$

It was suggested many years ago[195] that, as the reduction of $[(en)_2Co(O_2,NH_2)Co(en)_2]^{3+}$ by SO_2 or SeO_2 occurs with retention of the configurations of both cobalt atoms, the reaction can be considered as a direct addition to the peroxo bridge; whether the SO_2 or SeO_2 (or also NO) reacts by inserting into the O–O bond, as has recently been suggested,[311] is as yet unproved.

It has been observed[122] that the rates of reduction of the meso and racemic isomers of $[(en)_2Co(O_2,OH)Co(en)_2]^{3+}$ by iodide are the same as the rates of decomposition in acidic solution (the latter are shown in Fig. 12); this also excludes an outer-sphere mechanism.

The different mechanisms of reduction suggested for μ-peroxo and μ-superoxo complexes can be explained[211] by the different donor properties of the peroxo and superoxo bridges: the peroxo bridge is readily protonated ($pK_a = 0.8$ for $[(en)_2Co(O_2,NH_2)Co(en)_2]^{3+}$),[85] whereas there is no evidence for the protonation of a superoxo bridge (see Section VIII,B, p. 361).

X. CONCLUSIONS AND UNANSWERED QUESTIONS

A new dimension in the study of the kinetics and mechanism of the reactions of μ-peroxodicobalt(III) complexes has been entered with the realization that, even for those complexes that are thermodynamically stable, the equilibrium with cobalt(II) and dioxygen is faster than most, if not all, other reactions of singly bridged μ-peroxo complexes. This equilibrium therefore influences all reactions of the singly bridged complexes, and also needs to be considered for those doubly bridged complexes that also have a bridging hydroxo group, even though these are more inert. To what extent this is also true for the even less reactive amido-bridged complexes is still unknown.

The mechanism of the decomposition of μ-peroxo complexes to mononuclear cobalt(III) complexes in alkaline solution is clearly very uncertain and requires much more work, especially detailed kinetic work. Perhaps the most interesting development in this area is the suggestion that, at least for certain μ-peroxo complexes, the decomposition to mononuclear cobalt(III) and hydrogen peroxide is thermodynamically unfavorable and that it can occur only if the reverse reaction is prevented by the decomposition of the hydrogen peroxide.

Although the reduction of the μ-superoxo complexes to the μ-peroxo complexes is fairly well understood, and equilibria involving cobalt(II) probably have little effect on these fast reactions, such equilibria may well be involved in the redox reactions of the μ-peroxo complexes, and this aspect clearly requires further study. Certain other redox processes, such as the disproportionation of μ-peroxo complexes to μ-superoxo complexes in acid solution and the decomposition of μ-superoxo complexes to μ-peroxo complexes in alkaline solution, also need further clarification.

It is clear that future work will also involve the use of ligands individually tailored to modify the properties of the dioxygen complexes in order to increase the resemblance to biological dioxygen adducts, which is at present rather tenuous.

Acknowledgments

This work was supported by the Swiss National Science Foundation (Project 2.432-0.79) and by Ciba–Geigy, Basel, Switzerland. We would like to express our thanks to those individuals who supplied results prior to publication. Thanks are also due to co-workers in the Institut für Anorganische Chemie, whose efforts have enabled our own work to contribute to this interesting and stimulating area.

References

[1] Tassaert, B. M. *Ann. Chim. (Paris)* **1798**, *28*, 92–107.
[2] Thenard, *Ann. Chim. (Paris)* **1802**, *42*, 210–219.
[3] Frémy, E. *Ann. Chem. Pharm.* **1852**, *83*, 227–249 and 289–317.
[4] Maquenne, L. *C. R. Hebd. Seances Acad. Sci.* **1883**, *96*, 344–345.
[5] Vortmann, G. *Monatsh. Chem.* **1885**, *6*, 404–445.
[6] Werner, A.; Mylius, A. *Z. Anorg. Chem.* **1898**, *16*, 245–267.
[7] Werner, A. *Justus Liebigs Ann. Chem.* **1910**, *375*, 1–144.
[8] Gleu, K.; Rehm, K. *Z. Anorg. Allg. Chem.* **1938**, *237*, 79–88.
[9] Dunitz, J. D.; Orgel, L. E. *J. Chem. Soc.* **1953**, 2594–2596.
[10] Summerville, D. A.; Jones, R. D.; Hoffman, B. M.; Basolo, F. *J. Chem. Educ.* **1979**, *56*, 157–162.
[11] Drago, R. S.; Corden, B. B.; Zombeck, A. *Comments Inorg. Chem.* **1981**, *1*, 53–70.
[12] Goodman, G. L.; Hecht, H. G.; Weil, J. A. *Adv. Chem. Ser.* **1962**, *36*, 90–97.
[13] Erskine, R. W.; Field, B. O. *Struct. Bonding (Berlin)* **1976**, *28*, 1–50.
[14] Vaska, L. *Acc. Chem. Res.* **1976**, *9*, 175–183.
[15] Jones, R. D.; Summerville, D. A.; Basolo, F. *Chem. Rev.* **1979**, *79*, 139–179.
[16] Sykes, A. G. *Adv. Inorg. Bioinorg. Mech.* **1982**, *1*, 121–178.
[17] McLendon, G.; Martell, A. E. *Coord. Chem. Rev.* **1976**, *19*, 1–39.

[18] Connor, J. A.; Ebsworth, E. A. V. *Adv. Inorg. Chem. Radiochem.* **1964**, *6*, 279–381, especially pp. 327–341.
[19] Garbett, K.; Gillard, R. D. *J. Chem. Soc. A* **1968**, 1725–1735.
[20] Sykes, A. G.; Weil, J. A. *Prog. Inorg. Chem.* **1970**, *13*, 1–106.
[21] Sykes, A. G. *Chem. Br.* **1974**, *10*, 170–175.
[22] Stewart, R. F.; Estep, P. A.; Sebastian, J. J. S. *Inf. Circ.—U.S. Bur. Mines.* **1959**, 7906.
[23] Braun-Steinle, D.; Mäcke, H.; Fallab, S. *Helv. Chim. Acta* **1976**, *59*, 2032–2038.
[24] Martell, A. E. *Acc. Chem. Res.* **1982**, *15*, 155–162.
[25] House, D. A. *Coord. Chem. Rev.* **1977**, *23*, 223–322.
[26] Gainsford, A. R.; House, D. A. *Inorg. Nucl. Chem. Lett.* **1968**, *4*, 621–623.
[27] Gainsford, A. R.; House, D. A. *Inorg. Chim. Acta* **1969**, *3*, 33–40 and 367–372.
[28] Mitchell, P. R. Cobalt. *In* "Methodicum Chemicum", Vol. 8; Niedenzu, K.; Zimmer, H., Eds.; Academic Press: New York, 1976, pp. 307–344.
[29] Halpern, J. *Inorg. Chim. Acta* **1982**, *62*, 31–37.
[30] Thewalt, U.; Marsh, R. *J. Am. Chem. Soc.* **1967**, *89*, 6364–6365.
[31] Gubelmann, M. H.; Williams, A. F. *Struct. Bonding (Berlin)* **1983**, *55*, 1–65.
[32] Simplicio, J.; Wilkins, R. G. *J. Am. Chem. Soc.* **1969**, *91*, 1325–1329.
[33] Michaelis, L. *Arch. Biochem.* **1948**, *17*, 201–203.
[34] Mori, M.; Weil, J. A.; Ishiguro, M. *J. Am. Chem. Soc.* **1968**, *90*, 615–621.
[35] Zehnder, M. Ph.D. Thesis, University of Basel, **1973**.
[36] Bekaroglu, O.; Fallab, S. *Helv. Chim. Acta* **1963**, *46*, 2120–2125.
[37] Caraco, R.; Braun-Steinle, D.; Fallab, S. *Coord. Chem. Rev.* **1975**, *16*, 147–152.
[38] Bedell, S. A.; Timmons, J. H.; Martell, A. E.; Murase, I. *Inorg. Chem.* **1982**, *21*, 874–878.
[39] Fallab, S. *Chimia* **1969**, *23*, 177–179.
[40] Donatsch, P.; Gerber, K. H.; Zuberbühler, A.; Fallab, S. *Helv. Chim. Acta* **1970**, *53*, 262–268.
[41] Timmons, J. H.; Martell, A. E.; Harris, W. R.; Murase, I. *Inorg. Chem.* **1982**, *21*, 1525–1529.
[42] Powell, H. K. J.; Nancollas, G. H. *J. Am. Chem. Soc.* **1972**, *94*, 2664–2668.
[43] McLendon, G.; Martell, A. E. *J. Chem. Soc., Chem. Commun.* **1975**, 223–225.
[44] Goodwin, H. A.; Lions, F. *J. Am. Chem. Soc.* **1960**, *82*, 5013–5023.
[45] Bosnich, B.; Gillard, R. D.; McKenzie, E. D.; Webb, G. A. *J. Chem. Soc. A* **1966**, 1331–1339.
[46] Zehnder, M.; Thewalt, U. *Z. Anorg. Allg. Chem.* **1980**, *461*, 53–60.
[47] Kee, Teng Sew; Powell, H. K. J. *J. Chem. Soc., Dalton Trans.* **1975**, 2023–2027.
[48] Gold, M.; Powell, H. K. J. *J. Chem. Soc., Dalton Trans.* **1976**, 1418–1421.
[49] Tsumaki, T. *Bull. Chem. Soc. Jpn.* **1938**, *13*, 252–260; *Chem. Abstr.* **1938**, *32*, 3719.
[50] Martell, A. E.; Calvin, M. "Chemistry of the Metal Chelate Compounds"; Prentice-Hall; Englewood Cliffs, New Jersey, 1952, pp. 337–352.
[51] Floriani, C.; Calderazzo, F. *J. Chem. Soc. A* **1969**, 946–953.
[52] Crumbliss, A. L.; Basolo, F. *Science* **1969**, *164*, 1168–1170.
[53] Walker, F. A. *J. Am. Chem. Soc.* **1970**, *92*, 4235–4244.
[54] Brown, L. D.; Raymond, K. N. *Inorg. Chem.* **1975**, *14*, 2595–2601.
[55] Zehnder, M. Unpublished results.
[56] Machida, R.; Kimura, E.; Kodama, M. *Inorg. Chem.* **1983**, *22*, 2055–2061.
[57] Pickens, S. R.; Martell, A. E.; McLendon, G.; Lever, A. B. P.; Gray, H. B. *Inorg. Chem.* **1978**, *17*, 2190–2192.
[58] Ochiai, E. *J. Inorg. Nucl. Chem.* **1973**, *35*, 3375–3389.
[59] Basolo, F.; Hoffman, B. M.; Ibers, J. A. *Acc. Chem. Res.* **1975**, *8*, 384–392.
[60] Thöny, D. Ph.D. Thesis, University of Basel, **1974**.

[61] Ochiai, E. *J. Inorg. Nucl. Chem.* **1973,** *35*, 1727–1739.

[62] Schaefer, W. P. *Inorg. Chem.* **1968,** *7*, 725–731.

[63] Shibahara, T.; Mori, M. *Bull. Chem. Soc. Jpn.* **1978,** *51*, 1374–1379.

[64] Marsh, R. E.; Schaefer, W. P. *Acta Crystallogr., Sect. B* **1968,** *B24*, 246–251.

[65] Fronczek, F. R.; Schaefer, W. P. *Inorg. Chim. Acta* **1974,** *9*, 143–151.

[66] Fronczek, F. R.; Schaefer, W. P.; Marsh, R. E. *Inorg. Chem.* **1975,** *14*, 611–617.

[67] Wells, A. F. "Structural Inorganic Chemistry"; Oxford Univ. Press: London and New York, 1962; pp. 406–409.

[68] Nakamoto, K. "Infrared Spectra of Inorganic and Coordination Compounds"; 2nd ed., 1970, p. 78.

[69] Creighton, J. A.; Lippincott, E. R. *J. Chem. Phys.* **1964,** *40*, 1779–1980.

[70] Busing, W. R.; Levy, H. A. *J. Chem. Phys.* **1965,** *42*, 3054–3059.

[71] Miller, R. L.; Hornig, D. F. *J. Chem. Phys.* **1961,** *34*, 265–272.

[72] Evans, J. C. *Chem. Commun.* **1969,** 682–683.

[73] Fantucci, P.; Valenti, V. *J. Am. Chem. Soc.* **1976,** *98*, 3832–3838.

[74] Hoffmann, R.; Chen, M. M.-L.; Thorn, D. L. *Inorg. Chem.* **1977,** *16*, 503–511.

[75] Daul, C.; Schläpfer, C. W.; von Zelewsky, A. *Struct. Bonding (Berlin)* **1979,** *36*, 129–171.

[76] Boca, R. *Coord. Chem. Rev.* **1983,** *50*, 1–72.

[77] Smith, T. D.; Pilbrow, J. R. *Coord. Chem. Rev.* **1981,** *39*, 295–383.

[78] Drago, R. S. *Coord. Chem. Rev.* **1980,** *32*, 97–110.

[79] Drago, R. S.; Corden, B. B. *Acc. Chem. Res.* **1980,** *13*, 353–360.

[80] Zehnder, M.; Fallab, S. *Helv. Chim. Acta* **1972,** *55*, 1691–1696.

[81] Sasaki, Y.; Fujita, J.; Saito, K. *Bull. Chem. Soc. Jpn.* **1971,** *44*, 3373–3378.

[82] Thewalt, U.; Struckmeier, G. *Z. Anorg. Allg. Chem.* **1976,** *419*, 163–170.

[82a] Cabani, S.; Ceccanti, N.; Conti, G. *J. Chem. Soc., Dalton Trans.* **1983,** 1247–1251.

[83] Thewalt, U.; Marsh, R. E. *Inorg. Chem.* **1972,** *11*, 351–356.

[84] Davies, R.; Mori, M.; Sykes, A. G.; Weil, J. A. *Inorg. Synth.* **1970,** *12*, 197–214.

[85] Mori, M.; Weil, J. A. *J. Am. Chem. Soc.* **1967,** *89*, 3732–3744.

[86] Zehnder, M.; Thewalt, U.; Fallab, S. *Helv. Chim. Acta* **1979,** *62*, 2099–2108.

[87] Zehnder, M.; Thewalt, U.; Fallab, S. *Helv. Chim. Acta* **1976,** *59*, 2290–2294.

[88] Mitchell, P. R. Unpublished observations.

[89] Stadtherr, L. G.; Prados, R.; Martin, R. B. *Inorg. Chem.* **1973,** *12*, 1814–1818.

[90] Lever, A. B. P.; Gray, H. B. *Acc. Chem. Res.* **1978,** *11*, 348–355.

[91] Miskowski, V. M.; Robbins, J. L.; Treitel, I. M.; Gray, H. B. *Inorg. Chem.* **1975,** *14*, 2318–2321.

[92] Fallab, S.; Zehnder, M. *Helv. Chim. Acta* **1984,** *67*, 392–398.

[93] Kuroya, H.; Tsuchida, R. *Bull. Chem. Soc. Jpn.* **1940,** *15*, 427–439.

[94] Kubokura, K.; Okawa, H.; Kida, S. *Bull. Chem. Soc. Jpn.* **1978,** *51*, 2036–2040.

[95] Miller, F.; Simplicio, J.; Wilkins, R. G. *J. Am. Chem. Soc.* **1969,** *91*, 1962–1967.

[96] Maeder, M. Unpublished observation.

[97] Wong, C. L.; Switzer, J. A.; Balakrishnan, K. P.; Endicott, J. F. *J. Am. Chem. Soc.* **1980,** *102*, 5511–5518.

[98] Miller, F.; Wilkins, R. G. *J. Am. Chem. Soc.* **1970,** *92*, 2687–2691.

[99] Wilkins, R. G. *Adv. Chem. Ser.* **1971,** *100*, 111–134 and 5511–5518.

[100] Simplicio, J.; Wilkins, R. G. *J. Am. Chem. Soc.* **1967,** *89*, 6092–6095.

[101] Mäcke, H. Unpublished observations.

[102] Mäcke, H. *Helv. Chim. Acta* **1981,** *64*, 1579–1598.

[103] Merbach, A. *Pure Appl. Chem.* **1982,** *54*, 1479–1493.

[104] Endicott, J. F.; Lilie, J.; Kuszaj, J. M.; Ramaswamy, B. S.; Schmousees, W. G.; Simic, M. G.; Glick, M. D.; Rilema, D. P. *J. Am. Chem. Soc.* **1977,** *99*, 429–439.

[105] Mäcke, H.; Fallab, S. *Chimia* **1977**, *31*, 10–11.
[106] Paoletti, P.; Ciampolini, M.; Sacconi, L. *J. Chem. Soc.* **1963**, 3589–3593.
[107] Exnar, I.; Mäcke, H. *Helv. Chim. Acta* **1977**, *60*, 2504–2513.
[108] Yang, N.-L.; Oster, G. *J. Am. Chem. Soc.* **1970**, *92*, 5265–5266.
[109] Bayston, J. H.; Looney, F. D.; Winfield, M. D. *Aust. J. Chem.* **1963**, *16*, 557–564.
[110] Fallab, S. Unpublished observations.
[111] Zehnder, M.; Mäcke, H.; Fallab, S. *Helv. Chim. Acta* **1975**, *58*, 2306–2312.
[112] Mäcke, H.; Zehnder, M.; Thewalt, U.; Fallab, S. *Helv. Chim. Acta* **1979**, *62*, 1804–1815.
[113] Fallab, S.; Thewalt, U.; Zehnder, M. To be published.
[114] Ferrer, M.; Hand, T. D.; Sykes, A. G. *J. Chem. Soc., Dalton Trans.* **1980**, 14–18.
[115] Thewalt, U.; Zehnder, M.; Fallab, S. *Helv. Chim. Acta* **1977**, *60*, 867–873.
[116] Fallab, S. *Chimia* **1970**, *24*, 76–77.
[117] Maeder, M.; Fallab, S. *Chimia*, submitted.
[118] Fallab, S. *Chimia* **1983**, *37*, 128–134.
[119] Sasaki, Y.; Susuki, K. Z.; Matsumoto, A.; Saito, K. *Inorg. Chem.* **1982**, *21*, 1825–1828.
[120] Fritch, J. R.; Christoph, G. G.; Schaefer, W. P. *Inorg. Chem.* **1973**, *12*, 2170–2175.
[121] Maeder, M.; Mäcke, H. *Nat. Meet., Am. Chem. Soc.* **1982**, Abstr. No. 60.
[122] Brüstlein-Banks, P.; Fallab, S. *Helv. Chim. Acta* **1977**, *60*, 1601–1606.
[123] Gillard, R. D.; Spencer, A. *J. Chem. Soc. A* **1969**, 2718–2725.
[124] Fallab, S. *Angew. Chem., Int. Ed. Engl.* **1967**, *6*, 496–507.
[125] Heiniger, R., Ph.D. Thesis, University of Basel, **1983.**
[126] Scott, K. L.; Taylor, R. S.; Wharton, R. K.; Sykes, A. G. *J. Chem. Soc., Dalton Trans.* **1975**, 2119–2123.
[127] Fallab, S.; Zehnder, M.; Thewalt, U. *Helv. Chim. Acta* **1980**, *63*, 1491–1498.
[128] Fallab, S.; Hunold, H.-P.; Maeder, M.; Mitchell, P. R. *J. Chem. Soc., Chem. Commun.* **1981**, 469–471.
[129] Sasaki, Y.; Tachibana, M.; Saito, K. *Bull. Chem. Soc. Jpn.* **1982**, *55*, 3651–3652.
[130] Wilkins, R. G. "The Study of Kinetics and Mechanism of Reactions of Transition Metal Complexes"; Allyn & Bacon: Boston, 1974, p. 188.
[131] Stranks, D. R.; Yandell, J. K. *Inorg. Chem.* **1970**, *9*, 751–757.
[132] Pearson, R. G.; Boston, C. R.; Basolo, F. *J. Am. Chem. Soc.* **1953**, *75*, 3089–3092.
[133] Michailidis, M. S.; Martin, R. B. *J. Am. Chem. Soc.* **1969**, *91*, 4683–4689.
[134] Mäcke, H.; Fallab, S. *Chimia* **1972**, *26*, 422–424.
[135] Mäcke, H. Ph.D. Thesis, University of Basel, **1971.**
[136] Mellor, D. P.; Stephenson, N. C. *Inorg. Nucl. Chem. Lett.* **1967**, *3*, 431–435.
[137] Yalman, R. G. *J. Phys. Chem.* **1961**, *65*, 556–560.
[138] Boas, L. V.; Evans, C. A.; Gillard, R. D.; Mitchell, P. R.; Phipps, D. A. *J. Chem. Soc., Dalton Trans.* **1979**, 582–595.
[139] Zehnder, M.; Fallab, S. *Helv. Chim. Acta* **1974**, *57*, 1493–1498.
[140] Werner, A.; Berl, E.; Zinggeler, E.; Jantsch, G. *Ber. Dtsch. Chem. Ges.* **1907**, *40*, 2103–2125.
[141] Haim, A.; Wilmarth, W. K. *J. Am. Chem. Soc.* **1961**, *83*, 509–516.
[142] Beck, M. T. *Naturwissenschaften* **1958**, *45*, 162.
[143] Zehnder, M.; Fallab, S. *Helv. Chim. Acta* **1975**, *58*, 2312–2317.
[144] Caglioti, V.; Silvestroni, P.; Furlani, C. *J. Inorg. Nucl. Chem.* **1960**, *13*, 95–100.
[145] Harris, W. R.; Bess, R. C.; Martell, A. E.; Ridgeway, T. H. *J. Am. Chem. Soc.* **1977**, *99*, 2958–2963.
[146] Shibahara, T.; Kuroya, H.; Mori, M. *Bull. Chem. Soc. Jpn.* **1980**, *53*, 2834–2838.
[147] Bosnich, B.; Poon, C. K.; Tobe, M. L. *Inorg. Chem.* **1966**, *5*, 1514–1517.
[148] Wong, C.-L.; Endicott, J. F. *Inorg. Chem.* **1981**, *20*, 2233–2239.

[149] Munakata, M. *Bull. Chem. Soc. Jpn.* **1971**, *44*, 1791–1796.
[150] Gillard, R. D.; Spencer, A. *J. Chem. Soc. A* **1970**, 1761–1763.
[151] Sigel, H.; Prijs, B.; Rapp, P.; Dinglinger, F. *J. Inorg. Nucl. Chem.* **1977**, *39*, 179–184.
[152] Bayston, J. H.; Beale, R. N.; Kelso King, N.; Winfield, M. E. *Aust. J. Chem.* **1963**, *16*, 954–968.
[153] Bayston, J. H.; Winfield, M. E. *J. Catal.* **1964**, *3*, 123–128.
[154] Roberts, H. L.; Symes, W. R. *J. Chem. Soc. A* **1968**, 1450–1453.
[155] Sharpe, A. G. "The Chemistry of Cyano Complexes of the Transition Metals"; Academic Press: London, 1976, p. 172.
[156] Grube, G. *Z. Elektrochem.* **1926**, *32*, 561–566.
[157] Manchot, W.; Herzog, J. *Ber. Dtsch. Chem. Ges.* **1900**, *33*, 1742–1748.
[158] Gillard, R. D.; Spencer, A. *Discuss. Faraday Soc.* **1968**, *46*, 213–221.
[159] Yalman, R. G.; Warga, M. B. *J. Am. Chem. Soc.* **1958**, *80*, 1011.
[160] Gilbert, J. B.; Otey, M. C.; Price, V. E. *J. Biol. Chem.* **1951**, *190*, 377–389.
[161] Hearon, J. Z.; Burk, D.; Schade, A. L. *J. Natl. Cancer Inst. (U.S.)* **1949**, *9*, 337–377; *Chem. Abstr.* **1949**, *43*, 7856c.
[162] Spencer, A., Ph.D. Thesis, University of Kent, Canterbury, **1970**.
[163] Martell, A. E.; Harris, W. R. *J. Mol. Catal.* **1980**, *7*, 99–105.
[164] Harris, W. R.; Martell, A. E. *J. Coord. Chem.* **1980**, *10*, 107–113.
[165] Mitchell, P. R. *Proc. Int. Conf. Coord. Chem., 21st* **1980**, 438.
[166] Stadtherr, L. G.; Martin, R. B. *Inorg. Chem.* **1973**, *12*, 1810–1814.
[167] Richens, D. T.; Sykes, A. G. *J. Chem. Soc., Dalton Trans.* **1982**, 1621–1624.
[168] McLendon, G.; Mooney, W. F. *Inorg. Chem.* **1980**, *19*, 12–15.
[169] Thompson, L. R.; Wilmarth, W. K. *J. Phys. Chem.* **1952**, *56*, 5–9.
[170] Sasaki, Y.; Fujita, J.; Saito, K. *Bull. Chem. Soc. Jpn.* **1967**, *40*, 2206; **1969**, *42*, 146–152.
[171] Valentine, J. S.; Valentine, D. *J. Am. Chem. Soc.* **1971**, *93*, 1111–1117.
[172] Duffy, D. L.; House, D. A.; Weil, J. A. *J. Inorg. Nucl. Chem.* **1969**, *31*, 2053–2058.
[173] Emmenegger, F. P. Ph.D. Thesis, University of Zürich, **1963**.
[174] Linhard, M.; Weigel, M. *Z. Anorg. Allg. Chem.* **1961**, *308*, 254–262.
[175] Davies, R.; Sykes, A. G. *J. Chem. Soc. A* **1968**, 2840–2847.
[176] Le Mest, Y.; L'Her, M.; Courtot-Coupez, J.; Collman, J. P.; Evitt, E. R.; Bencosme, C. S. *Chem. Commun.* **1983**, 1286–1287.
[177] Werner, A. *Ber. Dtsch. Chem. Ges.* **1907**, *40*, 4605–4615.
[178] Brauer, G. "Handbuch der Präparativen Anorganischen Chemie"; 2nd ed., Vol. 2, Enke Verlag: Stuttgart, 1962, pp. 1340–1341; "Handbook of Preparative Inorganic Chemistry"; 2nd ed., Vol. 2, 1965, Academic Press: London, 1965, p. 1540.
[179] Charles, R. G.; Barnartt, S. *J. Inorg. Nucl. Chem.* **1961**, *22*, 69–76.
[180] Mast, R. D.; Sykes, A. G. *J. Chem. Soc. A* **1968**, 1031–1034.
[181] Mohammad, A.; Liebhafsky, H. A. *J. Am. Chem. Soc.* **1934**, *56*, 1680–1685.
[182] Fortnum, D. H.; Battaglia, C. J.; Cohen, S. R.; Edwards, J. O. *J. Am. Chem. Soc.* **1960**, *82*, 778–782.
[183] Mori, M.; Weil, J. A.; Kinnaird, J. K. *J. Phys. Chem.* **1967**, *71*, 103–108.
[184] Werner, A. *Justus Liebigs Ann. Chem.* **1910**, *375*, 70–74.
[185] Stevenson, M. B.; Sykes, A. G. *J. Chem. Soc. A* **1969**, 2293–2295.
[186] Sasaki, Y.; Fujita, J. *Bull. Chem. Soc. Jpn.* **1969**, *42*, 2089.
[187] Sasaki, Y.; Fujita, J.; Saito, K. *Bull. Chem. Soc. Jpn.* **1970**, *43*, 3462–3467.
[188] Werner, A. *Justus Liebigs Ann. Chem.* **1910**, *375*, 109 and 113.
[189] Werner, A. *Justus Liebigs Ann. Chem.* **1910**, *375*, 107–108.
[190] Sykes, A. G.; Mast, R. D. *J. Chem. Soc. A* **1967**, 784–789.

[191] Miskowski, V. M.; Robbins, J. L.; Treitel, I. M.; Gray, H. B. *Inorg. Chem.* **1975**, *14*, 2318–2321.

[192] Li, T.; Weil, J. A. Unpublished work cited by reference [20].

[193] Werner, A. *Justus Liebigs Ann. Chem.* **1910**, *375*, 74–75.

[194] Matthieu, J.-P. *Bull. Chim. Soc. Fr.* **1938**, *5*, 105–113.

[195] Garbett, K.; Gillard, R. D. *J. Chem. Soc. A* **1968**, 1725–1735.

[196] Werner, A. *Ber. Dtsch. Chem. Ges.* **1914**, *47*, 1961–1979.

[197] Garbett, K.; Gillard, R. D. *Chem. Commun.* **1966**, 99–100.

[198] Weil, J. A.; Kinnaird, J. K. *J. Phys. Chem.* **1967**, *71*, 3341–3343.

[199] Sykes, A. G. *Trans. Faraday Soc.* **1963**, *59*, 1325–1333.

[200] Ebsworth, E. A. V.; Weil, J. A. *J. Phys. Chem.* **1959**, *63*, 1890–1900.

[201] Werner, A. *Justus Liebigs Ann. Chem.* **1910**, *375*, 64–66, 134–135.

[202] Bernal, I.; Ebsworth, E. A. V.; Weil, J. A. *Proc. Chem. Soc., London* **1959**, 57–58.

[203] Barnes, J. E.; Barrett, J.; Brett, R. W.; Brown, J. *J. Inorg. Nucl. Chem.* **1968**, *30*, 2207–2210.

[204] Haim, A.; Taube, H. *Inorg. Chem.* **1963**, *2*, 1199–1203.

[205] Graff, H.; Wilmarth, W. K. *Proc. Symp. Coord. Chem., 1964* **1965**, 255–264; *Chem. Abstr.* **1966**, *64*, 18968h.

[206] Davies, R.; Sykes, A. G. *J. Chem. Soc. A* **1968**, 2831–2836.

[207] Hoffman, A. B.; Taube, H. *Inorg. Chem.* **1968**, *7*, 1971–1976.

[208] Davies, R.; Sykes, A. G. *J. Chem. Soc. A* **1968**, 2237–2241.

[209] Davies, R.; Hagopian, A. K. E.; Sykes, A. G. *J. Chem. Soc. A* **1969**, 623–629.

[210] Sykes, A. G. *Trans. Faraday Soc.* **1962**, *58*, 543–550.

[211] Werner, A. *Justus Liebigs Ann. Chem.* **1910**, *375*, 54.

[212] Yang, C.-H.; Keeton, D. P.; Sykes, A. G. *J. Chem. Soc., Dalton Trans.* **1974**, 1089–1093.

[213] Hyde, M. R.; Sykes, A. G. *Chem. Commun.* **1972**, 1340–1341.

[214] Hyde, M. R.; Sykes, A. G. *J. Chem. Soc., Dalton Trans.* **1974**, 1550–1561.

[215] Edwards, J. D.; Yang, C.-H.; Sykes, A. G. *J. Chem. Soc., Dalton Trans.* **1974**, 1561–1568.

[216] Davies, K. M.; Sykes, A. G. *J. Chem. Soc. A* **1971**, 1414–1417.

[217] Vlcek, A. A. *Collect. Czech. Chem. Commun.* **1960**, *25*, 3036–3054.

[218] Sykes, A. G. *Trans. Faraday Soc.* **1963**, *59*, 1334–1342.

[219] Davies, K. M.; Sykes, A. G. *J. Chem. Soc. A* **1971**, 1418–1423.

[220] Al-Shatti, N.; Ferrer, M.; Sykes, A. G. *J. Chem. Soc., Dalton Trans.* **1980**, 2533–2535.

[221] Davies, R.; Stevenson, M. B.; Sykes, A. G. *J. Chem. Soc. A* **1970**, 1261–1266.

Index

A

Abstraction, from coordinated ligands, 302–304

Activation enthalpy, 25

Activation entropy, 25

Adatom site, 108, 109

Addition, to coordinated ligands, 302–304

Algal peroxidase, 182

Alkylation, of metalloporphyrins, 165–168

Alkylnitroso ligands, structural chemistry, 219–220, 223

Arsenato complexes, formation and reactivity of, 299–300

Arylnitroso ligands, structural chemistry, 219–220, 223

Ascorbate oxidase, redox properties, 161–162

Association, 2

Association constant, 25

    for formation of cytochrome c reaction complexes, 34–37

Azurin, redox properties, 159–160

B

Bacterial peroxidase, 180–181

Binuclear dioxygen complexes of cobalt, *see* Cobalt, μ-peroxo complexes; Cobalt, μ-superoxo complexes

Bromoperoxidase, 182

Building block model, 108–109

C

Carbamato complexes, formation and reactivity of, 298–299

Carbonato complexes, reactivity, 277–281

Carbon dioxide uptake reactions, kinetics, 278–279

Carboxylato complexes, formation and reactivity of, 294–296

Catalases

    catalytic mechanisms, 190–204

    Compound I intermediates,

        formation, 191–192, 199

        structures, 193–195

    Compound II intermediates, 195–196

    Compound III intermediates, 195–196

    donor substrate oxidation, 196–198

    model systems, 202–203

    occurrence, characterization, and function, 177–186

    role in catalysis, 198–202

    X-ray structure determination, 187–189

Chloride, effect on cytochrome c oxidation, 41

Chloroperoxidase, 181–182, 199

Chromate complexes, formation and reactivity of, 300–302

Chromium-containing reagents, structure and properties, 20–21

Cobalt, μ-peroxo complexes

    decomposition, to cobalt(II) and dioxygen, 332–341

        hydrogen peroxide formation and, 347–349

        to mononuclear cobalt(III) species, 344–355

        stoichiometry, 346–347

    disproportionation, 358–360

    doubly bridged

        decomposition, 338–341

        formation, 320–322

    electronic spectra, 322–324

    kinetics of formation, 324–332